Long Memory in Economics

Gilles Teyssière · Alan P. Kirman
Editors

Long Memory in Economics

With 116 Figures and 50 Tables

Springer

Dr. Gilles Teyssière
Université Paris 1
Laboratoire de Statistique Appliquée
et de Modélisation Stochastique (SAMOS)
Panthéon Sorbonne
Centre Pierre Mendès
90 rue de Tolbiac
75634 Paris Cedex 13
France

Professor Alan P. Kirman
Université Aix-Marseille III
L'Ecole des Hautes Etudes
en Sciences Sociales (EHESS)
2 rue de la Charité
13002 Marseille
France

ISBN-10 3-540-22694-X Springer Berlin Heidelberg New York
ISBN-13 978-3-540-22694-9 Springer Berlin Heidelberg New York

Cataloging-in-Publication Data
Library of Congress Control Number: 2005930460

Springer is a part of Springer Science+Business Media

springeronline.com

Printed in Germany

Hardcover-Design: Erich Kirchner, Heidelberg

SPIN 11308669 43/3100-5 4 3 2 1 0 – Printed on acid-free paper

Preface

Long–range dependent, or long–memory, time series are stationary time series displaying a statistically significant dependence between very distant observations. We formalize this dependence by assuming that the autocorrelation function of these stationary series decays very slowly, hyperbolically, as a function of the time lag.

Many economic series display these empirical features: volatility of asset prices returns, future interest rates, etc. There is a huge statistical literature on long–memory processes, some of this research is highly technical, so that it is cited, but often misused in the applied econometrics and empirical economics literature. The first purpose of this book is to present in a formal and pedagogical way some statistical methods for studying long–range dependent processes.

Furthermore, the occurrence of long–memory in economic time series might be a statistical artefact as the hyperbolic decay of the sample autocorrelation function does not necessarily derive from long–range dependent processes. Indeed, the realizations of non-homogeneous processes, e.g., switching regime and change–point processes, display the same empirical features. We thus also present in this book recent statistical methods able to discriminate between the long–memory and change–point alternatives.

Going beyond the purely statistical analysis of economic series, it is of interest to determine which economic mechanisms are generating the strong dependence properties of economic series, whether they are genuine, or spurious. The regularities of the long–memory and change–point properties across economic time series, e.g., common degree of long–range dependence and/or common change–points, suggest the existence of a common economic cause. A promising approach is the use of the class of micro–based models in which the set of economic agents is represented as a (self)–organizing and interacting society whose composition evolves over time, i.e., something resembling the realization of a non–homogeneous stochastic process. Some of these models, inspired by the works of entomologists, are able to mimic some empirical properties of financial and non–financial markets. This implicitly suggests that what is termed as "long–memory" in economics is more complex than a standard (nonlinear) long–range dependent process, and mastering a wide range of statistical tools is a great asset for studying economic time series.

This volume starts with the chapter by Liudas Giraitis, Remis Leipus and Donatas Surgailis, who have been awarded this year the Lithuanian National Prize for Science for their work on "Long–memory: models, limit distributions, statistical inference". Donatas Surgailis and his numerous former students have made, and are still making, essential contributions to this topic. Their (encyclopedic) survey chapter reviews some recent theoretical findings on ARCH type volatility models. They focus mainly on covariance stationary models which display empirically observed properties which have come to be recognized as "stylized facts". One of the major issues to determine is whether the corresponding model for squares r_k^2 of ARCH sequences has long–memory or short memory. It is pointed out that for several ARCH-type models the behavior of $\mathrm{Cov}(r_k^2, r_0^2)$ alone is sufficient to derive the limit distribution of $\sum_k (r_k^2 - Er_k^2)$ and statistical inferences, without imposing any additional (e.g. mixing) assumptions on the dependence structure. This first chapter also discusses ARCH(∞) processes and their modifications such as linear ARCH, bilinear models and stochastic volatility, regime switching stochastic volatility models, random coefficient ARCH and aggregation. They give an overview of the theoretical results on the existence of a stationary solution, dependence structure, limit behavior of partial sums, leverage effect and long–memory property of these models. Statistical estimation of ARCH parameters and testing for change-points are also discussed.

Bhansali, Holland and Kokoszka consider a new and an entirely different approach to modeling phenomena exhibiting long–memory, intermittency and heavy-tailed marginal distributions, namely through the use of chaotic intermittency maps. This class of maps has witnessed considerable development in recent years and it represents an important emerging branch of the subject area of Dynamical Systems Theory. Three principal properties of these maps are relevant and these properties qualify them as a plausible class of models for financial returns. First, unlike some of the standard chaotic maps, the intermittency maps display long–memory and have correlations decaying at a sub-exponential rate, meaning at a polynomial rate or even slower. Secondly, the invariant density of these maps can display 'Pareto' tails and thus go to zero at a polynomial rate. Thirdly, as their name implies, these maps display intermittency and generate time series, called the orbit of the map, which display intermittent chaos, meaning the orbit of the map alternates between laminar and chaotic regions.

Brousseau analyzes the time series of the euro–dollar exchange rate as the realization of a continuous–time physical process, which implies the use of different degrees of time resolution. The analysis takes into account various statistical indicators, but puts special emphasis on the spectrum of the process. Brousseau finds that this spectrum has an identifiable pattern, which is a core characteristic of the process. Then he simulates a process having the same spectrum, and the behavior of the actual process and of the simulated process are compared using various statistical indicators. It appears that the simulated process provides a good, but not perfect, replication of the behavior of the actual euro–dollar exchange rate.

The next two chapters deal with the issue of change–point detection. Račkauskas and Suquet present the invariance principle by Donsker and Prokhorov, which can be used for analyzing structural changes. They focus on invariance principles with respect to Hölder topologies, as Hölder spaces bring out well variations properties of processes. They present some applications of the Hölderian invariance principles to the problem of testing the stability of a sample against epidemic change–points alternatives.

Lavielle and Teyssière consider the multiple change–point problem for time series, including strongly dependent processes, with an unknown number of change–points. They propose an adaptive method for finding the sequence of change–points $\boldsymbol{\tau}$ with the optimal level of resolution. This optimal segmentation is obtained by minimizing a standard penalized contrast function $J(\boldsymbol{\tau}, \boldsymbol{y}) + \beta \text{pen}(\boldsymbol{\tau})$. The adaptive procedure is such that the optimal segmentation does not strongly depend on the penalization parameter β. This algorithm is applied to the problem of detection of change–points in the volatility of financial time series, and compared with the binary segmentation procedure by Vostrikova.

The chapter by Henry is on the issue of bandwidth selection for semiparametric estimators of the long–memory parameter. The spectral based estimators are derived from the shape of the spectral density at low frequencies, where all but the lowest harmonics of the periodogram are discarded. This allows one to ignore the specification of the short range dynamic structure of the time series, and avoid the bias incurred when the latter is misspecified. Such a procedure entails an order of magnitude loss of efficiency with respect to parametric estimation, but may be warranted when long series (earth scientific or financial) can be obtained. This chapter presents strategies proposed for the choice of bandwidth, i.e., the number of periodogram harmonics used in estimation, with the aim of minimizing this loss of efficiency.

Teyssière and Abry present and study the performance of the semiparametric wavelet estimator for the long–memory parameter devised by Veitch and Abry, and compare this estimator with two semiparametric estimators in the spectral domain: the local Whittle and the "log–periodogram" estimators. The wavelet estimator performs well for a wide range of nonlinear long–memory processes in the conditional mean and the conditional variance, and is reliable for discriminating between change–points and long–range dependence in volatility. The authors also address the issue of the selection of the range of octaves used as regressors by the weighted least squares estimator. It appears that using the feasible optimal bandwidths for either the spectral estimators, surveyed by Henry in the previous chapter, is a useful rule of thumb for selecting the lowest octave. The wavelet estimator is applied to volatility and volume financial time series.

Kateb, Seghier and Teyssière study a fast version of the Levinson–Durbin algorithm, derived from the asymptotic behavior of the first column of the inverse of $T_N(f)$, the $(N+1) \times (N+1)$ Toeplitz matrix with typical element f, the spectral density of a long–memory process. The inversion of $T_N(f)$ with Yule–Walker type equations requires $O(N^3)$ operations, while the Levinson–Durbin algorithm requires $O(N^2)$ elementary operations. In this chapter, an asymptotic behavior of $(T_N(f))$, for large values of N is given so that the computations of the inverse elements are performed in $O(N)$ operations. The numerical results are compared with those given by the Levinson–Durbin algorithm, with particular emphasis on problems of predicting stationary stochastic long–range dependent processes.

Gaunersdorfer and Hommes study a simple nonlinear structural model of endogenous belief heterogeneity. News about fundamentals is an independent and identically distributed random process, but nevertheless volatility clustering occurs as an endogenous phenomenon caused by the interaction between different types of traders, fundamentalists and technical analysts. The belief types are driven by adaptive, evolutionary dynamics according to the success of the prediction strategies as measured by accumulated realized profits, conditioned upon price deviations from the rational expectations fundamental price. Asset prices switch irregularly between two different regimes – periods of small price fluctuations and periods of large price changes triggered by random news and reinforced by technical trading – thus, creating time varying volatility similar to that observed in real financial data.

Cont attempts to model the volatility clustering of asset prices: large changes in prices tend to cluster together, resulting in persistence of the amplitudes of their changes. After recalling various methods for quantifying and modeling this phenomenon, this chapter discusses several economic mechanisms which have been proposed to explain the origin of this volatility clustering in terms of behavior of market participants and the news arrival process. A common feature of these models seems to be a switching between low and high activity regimes with heavy-tailed durations of regimes. Finally, a simple agent-based model is presented, which links such variations in market activity to threshold behavior of market participants and suggests a link between volatility clustering and investor inertia.

Kirmans chapter is devoted to an account of the sort of micro-economic founded models that can give rise to long memory and other stylised facts about financial and economic series. The basic idea is that the market is populated by individuals who have different views about future prices. As time unfolds they may change their ways of forecasting the future. As they do so they change their demands and thus prices. In many of these models the typical rules are "chartist" or extrapolative rules and those based on the idea that prices will revert to some "fundamental" values. In fact, any finite number of rules can be considered. The switching between rules may result in "herding" on some particular rule for a period of time and give rise to long memory and volatility clustering. The first models were used to generate data which was then tested to see if the stylised facts were generated. More recently theoretical results have been obtained which characterise the long run distribution of the price process.

Alfarano and Lux present a very simple model of a financial market with heterogeneous interacting agents capable of reproducing empirical statistical properties of returns. In this model, the traders are divided into two groups, fundamentalists and chartists, and their interactions are based on herding mechanism. The statistical analysis of the simulated data shows long-term dependence in the auto-correlations of squared and absolute returns and hyperbolic decay in the tail of the distribution of the raw returns, both with estimated decay parameters in the same range like empirical data. Theoretical analysis, however, excludes the possibility of "true" scaling behavior because of the Markovian nature of the underlying process and the finite set of possible realized returns.

The purpose of the chapter by de Peretti is to determine whether hysteretic series can be confused with long–memory series. The hysteretic effect is a persistence in the series like the long–memory effect, although hysteretic series are not mean reverting whereas long–memory series are. Hysteresis models have in fact a short memory, since dominant shocks erase the memory of the series, and the persistence is due to permanent and non-reverting state changes at a microstructure level. In order to check whether hysteretic models display spurious long–range dependence, a model possessing the hysteresis property is used for simulating hysteretic data to which statistical tests for short–memory against long–memory alternatives are applied.

We wish to thank all contributors to this book, all the referees for their careful and fast reports which made the selection of the chapters easier, the Springer Verlag editor, Dr Martina Bihn, for efficient and patient management, and Mrs Ruth Milewski for correcting the first series of manuscripts and the final version. Teyssière managed chapters 1 to 8 and 10, while Kirman managed chapters 9, 11, 12 and 13.

Gilles Teyssière thanks Paul Doukhan for his help to return to the academic system on serious grounds, and the SAMOS (Statistique Appliquée et MOdélisation Stochastique) University Paris 1. He wants to express his gratitude to Rosa Maria to whom he owes bunches of flowers as the completion of this volume spilt over into their spare time. Alan Kirman would like to thank the other people with whom he has investigated this sort of problem, and in particular, Hans Foellmer, Ulrich Horst, Richard Topol and Romain Ricciotti.

Paris,
June 2006

Gilles Teyssière
Alan Kirman

Contents

Part II Economic Models

Part I

Statistical Methods

Recent Advances in ARCH Modelling

Liudas Giraitis[1] *, Remigijus Leipus[2], and Donatas Surgailis[3]

[1] University of York, Department of Mathematics and Economics. lg8@york.ac.uk
[2] Vilnius University, Institute of Mathematics and Informatics. Remigijus.Leipus@maf.vu.lt
[3] Vilnius Institute of Mathematics and Informatics. sdonatas@ktl.mii.lt

1 Introduction

Econometric modelling of financial data received a broad interest in the last 20 years and the literature on ARCH and related models is vast. Starting with the path breaking works by Engle (1982) and Bollerslev (1986), one of the most popular models became the Generalized AutoRegressive Conditionally Heteroscedastic (GARCH) process. The classical GARCH(p,q) model is given by equations

$$r_t = \sigma_t \varepsilon_t, \quad \sigma_t^2 = \alpha_0 + \sum_{i=1}^{p} \beta_i \sigma_{t-i}^2 + \sum_{j=1}^{q} \alpha_j r_{t-j}^2, \tag{1}$$

where $\alpha_0 > 0, \alpha_j \geq 0, \beta_i \geq 0,\ p \geq 0, q \geq 0$ are model parameters and $\{\varepsilon_j, j \in \mathbf{Z}\}$ are independent identically distributed (i.i.d.) zero mean random variables. The variables r_t, σ_t, ε_t in (65) are usually interpreted as financial (log)returns (r_t), their volatilities or conditional standard deviations (σ_t), and so-called innovations or shocks (ε_t), respectively; in (65) the innovations are supposed to follow a certain fixed distribution (e.g., standard normal). Later, a number of modifications of (65) were proposed, which account for asymmetry, leverage effect, heavy tails and other "stylized facts". For statistical and econometric aspects of ARCH modelling, see the surveys of Bollerslev *et al.* (1992), Shephard (1996), Bera and Higgins (1993), Bollerslev *et al.* (1994); for specific features of modelling the financial data, including ARCH, see Pagan (1996), Rydberg (2000), Mikosch (2003). Berkes *et al.* (2002b) review some recent results. One should mention here, besides the classical reference to Taylor (1986), the related monographs by Gouriéroux (1997), Fan and Yao (2002), Tsay (2002). Let us note that the GARCH model for returns is also related to the Autoregressive Conditional Duration (ACD) model proposed by Engle and Russell (1998) for modelling of durations between events.

* Supported by the ESRC grant R000239538.

Under some additional conditions, similarly as in the case of ARMA models, the GARCH model can be written as ARCH(∞) model (see (3) below), i.e., σ_t^2 can be represented as a moving average of the past squared returns $r_s^2, s < t$, with exponentially decaying coefficients (see Bollerslev, 1988) and absolutely summable exponentially decaying autocovariance function.

However, empirical studies of financial data show that sample autocorrelations of power series and volatilities (such as absolute values $|r_t|$ or squares r_t^2) remain non-zero for very large lags; see, e.g., Dacorogna *et al.* (1993), Ding *et al.* (1993), Baillie *et al.* (1996a), Ding and Granger (1996), Breidt *et al.* (1998), Mikosch and Stărică (2003), Andersen *et al.* (2001). These studies provide a clearcut evidence in favor of models with autocovariances decaying slowly with the lag as $k^{-\gamma}$, for some $0 < \gamma < 1$.

A number of such models (FIGARCH, LM-ARCH, FIEGARCH) were suggested in the ARCH literature. The long memory property was rigorously established for some of these models including the Gaussian subordinated stochastic volatility model (Robinson, 2001), with general form of nonlinearity, the FIEGARCH and related exponential volatility models (Harvey, 1998; Surgailis and Viano, 2002), the LARCH model (Giraitis *et al.*, 2000c), the stochastic volatility model of Robinson and Zaffaroni (1997, 1998). The long memory property (and even the existence of stationary regime) of some other models (FIGARCH, LM-ARCH) has not been theoretically established; see Giraitis *et al.* (2000a) Mikosch and Stărică (2000, 2003), Kazakevičius *et al.* (2004). Covariance long memory was also proved for some regime switching SV models (Liu, 2000; Leipus *et al.*, 2005). One should also mention that some authors (Mikosch and Stărică, 1999, 2004) argue that the observed long memory in sample autocorrelations can be explained by short memory GARCH models with structural breaks and/or slowly changing trends.

The present paper reviews some recent theoretical findings on ARCH type models. We focus mainly on covariance stationary models which display empirically observed properties known as "stylized facts". One of the major issues to determine is whether the corresponding model r_t^2 for squares has *long memory* or *short memory*, i.e. whether $\sum_{k=0}^{\infty} |\mathrm{Cov}(r_k^2, r_0^2)| = \infty$ or $\sum_{k=0}^{\infty} |\mathrm{Cov}(r_k^2, r_0^2)| < \infty$ holds. It is pointed out that for several ARCH-type models the behavior of $\mathrm{Cov}(r_k^2, r_0^2)$ alone is sufficient to derive the limit distribution of $\sum_{j=1}^{N}(r_j^2 - Er_j^2)$ and statistical inferences, without imposing any additional (e.g. mixing) assumptions on the dependence structure.

The review discusses ARCH(∞) processes and their modifications such as linear ARCH (LARCH), bilinear models, long memory EGARCH and stochastic volatility, regime switching SV models, random coefficient ARCH and aggregation. We give an overview of the theoretical results on the existence of a stationary solution, dependence structure, limit behavior of partial sums, leverage effect and long memory property of these models. Statistical estimation of ARCH parameters and testing for change-points are also discussed.

2 ARCH(∞)

A random sequence $\{r_t,\ t \in \mathbf{Z}\}$ is said to satisfy ARCH(∞) equations if there exists a sequence of i.i.d. zero mean random variables $\{\varepsilon_t,\ t \in \mathbf{Z}\}$ and a deterministic sequence $b_j \geq 0$, $j = 0, 1, \ldots$ such that for any t

$$r_t = \sigma_t \varepsilon_t,\ \sigma_t^2 = b_0 + \sum_{j=1}^{\infty} b_j r_{t-j}^2. \tag{2}$$

Clearly, if $E(\varepsilon_t | r_s, s < t) = 0$, $E(\varepsilon_t^2 | r_s, s < t) = 1$ then r_t has conditional mean zero and a random conditional variance σ_t^2, i.e.

$$E(r_t | r_s, s < t) = 0, \quad \operatorname{Var}(r_t | r_s, s < t) = \sigma_t^2.$$

The general framework leading to the model (2) was introduced by Robinson (1991) in the context of testing for strong serial correlation and has been subsequently studied by Kokoszka and Leipus (2000) in the change-point problem context. The class of ARCH(∞) models include the finite order ARCH and GARCH models of Engle (1982) and Bollerslev (1986). For instance, the GARCH(p, q) process $\{r_t, t \in \mathbf{Z}\}$ of (65) can be written as $r_t = \sigma_t \varepsilon_t$,

$$\sigma_t^2 = (1 - \beta(1))^{-1} \alpha_0 + (1 - \beta(L))^{-1} \alpha(L) r_t^2, \tag{3}$$

where $\beta(L) = \beta_1 L + ... + \beta_p L^p$ and L stands for the back-shift operator, $L^j X_t = X_{t-j}$. This leads to ARCH(∞) representation (2) for GARCH(p, q) model with $b_0 = (1 - \beta(1))^{-1} \alpha_0$ and with positive *exponentially* decaying weights $b_j, j \geq 1$ defined by the generating function $\alpha(z)/(1 - \beta(z)) = \sum_{i=1}^{\infty} b_i z^i$. It is interesting to note that the non-negativity of the regression coefficients α_j, β_j in (65) is not necessary for non-negativity of b_j in the corresponding ARCH(∞) representation, see Nelson and Cao (1992).

2.1 Existence of Second and Fourth Order Stationary Solutions

One of the first questions which usually arise in the study of recursion equations of the type (2) is to find conditions for the existence of a stationary solution. We first discuss conditions on the coefficients b_j and the random variables ε_t which guarantee the existence of a stationary solution to equations (2) with finite second or fourth moments.

Formally, recursion relations (2) give the following Volterra series expansion of r_t^2:

$$\begin{aligned} r_t^2 \equiv \varepsilon_t^2 \sigma_t^2 &= \varepsilon_t^2 b_0 \Big(1 + \sum_{k=1}^{\infty} \sum_{j_1, \ldots, j_k = 1}^{\infty} b_{j_1} \ldots b_{j_k} \varepsilon_{t-j_1}^2 \ldots \varepsilon_{t-j_1-\cdots-j_k}^2\Big) \\ &= \varepsilon_t^2 b_0 \Big(1 + \sum_{k=1}^{\infty} \sum_{-\infty < s_k < \cdots < s_1 < t}^{\infty} b_{t-s_1} b_{s_1-s_2} \ldots b_{s_{k-1}-s_k} \varepsilon_{s_1}^2 \ldots \varepsilon_{s_k}^2\Big). \end{aligned} \tag{4}$$

By taking the expectation on both sides and using the independence of ε_t's, one obtains

$$Er_t^2 = (E\varepsilon_t^2)b_0\Big\{1+\sum_{k=1}^{\infty}\sum_{-\infty<s_k<\cdots<s_1<t}^{\infty} b_{t-s_1}b_{s_1-s_2}\dots b_{s_{k-1}-s_k}E\varepsilon_{s_1}^2\dots E\varepsilon_{s_k}^2\Big\}$$
$$= (E\varepsilon_t^2)b_0\Big\{1+\sum_{k=1}^{\infty}\Big(E\varepsilon_0^2\sum_{j=1}^{\infty}b_j\Big)^k\Big\} = \frac{b_0E\varepsilon_0^2}{1-E\varepsilon_0^2\sum_{j=1}^{\infty}b_j}.$$

Hence it easily follows that

$$E\varepsilon_0^2\sum_{j=1}^{\infty}b_j<1 \tag{5}$$

is sufficient for the existence of stationary solution (4) with $Er_t^2<\infty$. The uniqueness and the necessity of (5) for the existence of such a solution also follow easily, see Kokoszka and Leipus (2000), Giraitis *et al.* (2000a).

It is also easy to obtain a sufficient condition for the existence of a stationary solution with finite fourth moment. To that end, apply to (4) the norm (Minkowski) inequality: $(E\big(\sum_i\xi_i\big)^2)^{1/2}\le\sum_i(E\xi_i^2)^{1/2}$. Similarly as above, this yields

$$(Er_t^4)^{1/2}\le(E\varepsilon_t^4)^{1/2}$$
$$\times b_0\Big\{1+\sum_{k=1}^{\infty}\sum_{-\infty<s_k<\cdots<s_1<t}^{\infty} b_{t-s_1}b_{s_1-s_2}\dots b_{s_{k-1}-s_k}(E\varepsilon_{s_1}^4)^{1/2}\dots(E\varepsilon_{s_k}^4)^{1/2}\Big\}$$
$$=\frac{b_0(E\varepsilon_0^4)^{1/2}}{1-(E\varepsilon_0^4)^{1/2}\sum_{j=1}^{\infty}b_j}.$$

Hence if condition

$$(E\varepsilon_0^4)^{1/2}\sum_{j=1}^{\infty}b_j<1 \tag{6}$$

is satisfied, then r_t of (4) is a fourth order stationary solution to (2), see Giraitis *et al.* (2000a). A similar norm inequality works in the case of $E(r_t^2)^p$ and arbitrary $p\ge1$, yielding a sufficient condition $(E|\varepsilon_0|^{2p})^{1/p}\sum_{j=1}^{\infty}b_j<1$.

Condition (6) is not necessary for the existence of fourth order stationary solution. For example, in the case of GARCH(1,1) $r_t=\varepsilon_t\sigma_t$, $\sigma_t^2=\alpha_0+\alpha r_{t-1}^2+\beta\sigma_{t-1}^2$, (6) translates to $\alpha\lambda_2^{1/2}+\beta<1$, $\lambda_i=E\varepsilon^{2i}, i=1,2$, while a fourth order stationary solution is known to exist under the weaker conditions

$$\alpha\lambda_1+\beta<1,\qquad \alpha^2\lambda_2+\beta^2<1, \tag{7}$$

see Karanasos (1999), He and Teräsvirta (1999). To obtain a sufficient and necessary condition in the general case, one needs to study *orthogonal* Volterra representation of r_t^2.

The orthogonal Volterra representation (9) of r_t^2 is obtained by centering the innovations in the (nonorthogonal) representation (4), i.e. by replacing the ε_j^2's by $\kappa\zeta_j + \lambda_1 = \varepsilon_j^2$, where the standardized $\zeta_j = (\varepsilon_j^2 - E\varepsilon_j^2)/\kappa$, $\kappa^2 = \mathrm{Var}(\varepsilon_0^2)$ have zero mean and unit variance.

The resulting expression appears rather complicated, but nevertheless it can be identified and studied (Giraitis and Surgailis, 2002). In order to describe it, denote g_j the coefficients of the generating function

$$\sum_{j=0}^{\infty} g_j z^j = \left(1 - \lambda_1 \sum_{i=1}^{\infty} b_i z^i\right)^{-1}.$$

More explicitly,

$$g_j = \sum_{k=1}^{j} \lambda_1^k \sum_{0<i_1<\ldots<i_{k-1}<j} b_{i_1} b_{i_2-i_1} \ldots b_{i_{k-2}-i_{k-1}} b_{j-i_{k-1}} \quad (j \geq 1), \qquad g_0 = 1. \tag{8}$$

Also introduce $h_j = (\kappa/\lambda_1) g_j$, $j \geq 1$,

$$B = \sum_{j=1}^{\infty} b_j, \qquad H^2 = \sum_{j=1}^{\infty} h_j^2.$$

Then

$$r_t^2 = \mu + (\kappa/\lambda_1)\mu \sum_{k=1}^{\infty} \sum_{s_k<\cdots<s_2<s_1\leq t} g_{t-s_1} h_{s_1-s_2} \ldots h_{s_{k-1}-s_k} \zeta_{s_1} \ldots \zeta_{s_k}, \tag{9}$$

where $\mu = Er_t^2 = \lambda_1 b_0/(1-\lambda_1 B)$. The series (9) converges in mean square if and only if

$$\lambda_1 B < 1, \qquad H < 1 \tag{10}$$

hold, and define a stationary solution of (2).

In fact, conditions (10) are sufficient and necessary for the existence of fourth order stationary solution of (2) (Giraitis and Surgailis, 2002). By orthogonality, it easily follows that

$$\begin{aligned}
\mathrm{Cov}(r_t^2, r_0^2) &= (\kappa/\lambda_1)^2 \mu^2 \sum_{k=1}^{\infty} \sum_{s_k<\cdots<s_1\leq 0} g_{-s_1} g_{t-s_1} h_{s_1-s_2}^2 \ldots h_{s_{k-1}-s_k}^2 \\
&= (\kappa/\lambda_1)^2 \mu^2 \sum_{s\leq 0} g_s g_{t-s} \sum_{k=1}^{\infty} H^{2(k-1)} \\
&= \frac{(\kappa/\lambda_1)^2 \mu^2}{1-H^2} \sum_{s=0}^{\infty} g_s g_{s+t}.
\end{aligned} \tag{11}$$

For the GARCH(1,1) model

$$r_t = \varepsilon_t \sigma_t, \quad \sigma_t^2 = \alpha_0 + \alpha r_{t-1}^2 + \beta \sigma_{t-1}^2$$

the above formulas are more explicit. The model itself can be rewritten in the ARCH(∞) form

$$\sigma_t^2 = \alpha_0 (1-\beta)^{-1} + \alpha \sum_{j=1}^{\infty} \beta^{j-1} r_{t-j}^2.$$

In this case, $g_j = \alpha\lambda_1(\lambda_1\alpha + \beta)^{j-1}$, conditions (7) and (10) coincide, and (9) becomes

$$\begin{aligned} r_t^2 &= \mu + \mu\kappa\lambda_1^{-1}\zeta_t\Big(1 + \sum_{k=1}^{\infty}(\alpha\kappa/\gamma)^k \sum_{s_k<\cdots<s_1<t} \gamma^{t-s_k}\zeta_{s_1}\ldots\zeta_{s_k}\Big) \\ &\quad + \mu\sum_{k=1}^{\infty}(\alpha\kappa/\gamma)^k \sum_{s_k<\cdots<s_1<t} \gamma^{t-s_k}\zeta_{s_1}\ldots\zeta_{s_k}, \end{aligned} \tag{12}$$

where $\gamma = \lambda_1\alpha + \beta$, $\mu = \alpha_0\lambda_1/(1-\gamma)$. From (12) or (11) one can explicitly find the variance and covariance function of the GARCH(1,1) model in terms of the coefficients α_0, α, β and the moments λ_1, λ_2:

$$\begin{aligned} \operatorname{Var} r_0^2 &= \frac{\alpha_0^2\kappa^2(1-\gamma^2+\gamma\alpha\lambda_1)}{(1-\gamma)^2(1-\gamma^2-\alpha^2\kappa^2)}, \\ \operatorname{Cov}(r_k^2, r_0^2) &= \frac{\alpha_0^2\alpha\lambda_1\kappa^2(1-\gamma^2+\gamma\alpha\lambda_1)}{(1-\gamma)^2(1-\gamma^2-\alpha^2\kappa^2)}\,\gamma^{k-1}, \quad k \ge 1, \end{aligned}$$

which were also obtained in Teräsvirta (1996). We also note an alternative approach in Kazakevičius *et al.* (2004) to the problem of the existence of fourth order stationary solution of ARCH(∞), which leads to equivalent necessary and sufficient conditions as (10).

For necessary and sufficient conditions of the existence of high order moments for the family of GARCH processes see Ling and McAleer (2002a, 2002b). Ling and McAleer (2003a) studied theoretical properties of the multivariate ARMA-GARCH model.

2.2 Dependence Structure, Association and Limit Theorems

The equation (11) for the covariance of ARCH(∞) squares r_t^2 allows to directly study its summability and decay properties. From (8) and the summability of b_j's it follows the summability of g_j's which in turn implies by (11) the summability of the autocovariances of r_t^2:

$$\sum_{k=-\infty}^{\infty} \operatorname{Cov}(r_k^2, r_0^2) < \infty. \tag{13}$$

(Note that $\operatorname{Cov}(r_k^2, r_0^2) \ge 0$ for all k, which follows from (11) and also from the associativity property of r_t^2, see below.) Equation (13) indicates that the *squares r_t^2 of a fourth order stationary solution of ARCH(∞) have short memory.* The above mentioned papers Giraitis *et al.* (2000a), Giraitis and Surgailis (2002) also prove that a hyperbolic decay $b_j \sim Cj^{-\gamma}$ with $\gamma > 1$ implies

$$\mathrm{Cov}(r_k^2, r_0^2) \asymp k^{-\gamma}.$$

(Here and below, $x_k \sim y_k$ means $x_k/y_k \to 1$ while $x_k \asymp y_k$ means that there are positive constants C_1 and C_2 such that $C_1 y_k < x_k < C_2 y_k$ for all k.) Thus, even though condition (6) implies absolute summability of the covariances, it allows for a very slow rate of decay of the autocorrelation function when $\gamma > 1$ is close to 1. The last property may be characterized as *moderate* memory. Near epoch dependence and moderate memory property of the so-called HYGARCH model were studied by Davidson (2004).

The above discussion basically concerns second-order properties of r_t^2 only. Some further insight about these properties can be obtained from the *moving average representation*

$$r_t^2 = Er_t^2 + \sum_{j=0}^{\infty} g_j \nu_{t-j}, \tag{14}$$

where g_j (8) and $\nu_t \equiv \sigma_t^2(\varepsilon_t^2 - E\varepsilon_t^2)$ are *martingale differences.* The above representation is a direct consequence of (9), from which the ν_t's can be also expressed as a Volterra series in the standardized variables $\zeta_s, s < t$. Of course, (14) yields the same covariance formula as (11). On the other hand, the ν_t's are *not* independent, meaning that "higher order" dependence and distributional properties of (14) may be very different from the usual moving average in i.i.d. random variables.

ARCH(∞) sequences have important property of associativity. A random sequence $\{X_t\}$ is said to be *associated* (or *positively correlated*) if the inequality

$$\mathrm{Cov}(f(X_{t_1}, \ldots, X_{t_n}), g(X_{t_1}, \ldots, X_{t_n})) \geq 0,$$

holds for any coordinate nondecreasing functions $f, g : \mathbf{R}^n \to \mathbf{R}$ and any $t_1, \ldots, t_n, n = 1, 2, \ldots$. In particular, the covariance function (if it exists) of associated sequence is nonnegative: $\mathrm{Cov}(X_s, X_t) \geq 0$ for any s, t. Associated sequences are widely encountered in applications, see e.g. Barlow and Proschan (1981), Newman (1984), Cox and Grimmett (1984). Association is a very strong property, under which uncorrelatedness implies independence similarly as in Gaussian case. A number of limit theorems have been proved for associated sequences under covariance restrictions only. One of the most celebrated results, due to Newman and Wright (1981), says that if $\{X_t, t \in \mathbf{Z}\}$ is strictly stationary and associated and $\sigma^2 = \sum_{t \in \mathbf{Z}} \mathrm{Cov}(X_0, X_t) < \infty$ then the partial sums' process

$$\left\{ N^{-1/2} \sum_{t=1}^{[N\tau]} (X_t - EX_t), \tau \in [0,1] \right\} \to_{D[0,1]} \{\sigma W(\tau), \tau \in [0,1]\}, \tag{15}$$

in the Skorokhod space $D[0,1]$, where $\{W(\tau)\}$ is a standard Brownian motion.

It is well known that independent random variables are associated, and that this property is preserved by coordinate-nondecreasing (nonlinear) transformations. In particular, the ARCH(∞) process of (4) is a coordinate-nondecreasing transformation of the i.i.d. sequence $\{\varepsilon_t^2\}$. It is clear from non-negativity $b_j \geq 0$, $j \geq 0$ that r_t^2 can only increase if any of $\varepsilon_s^2, s \leq t$ on the r.h.s. of (4) is replaced by some larger quantity. Therefore the *ARCH(∞) process* (4) *is associated.*

An immediate consequence of (15), (13) and the association property of is the functional central limit theorem for squares r_t^2 of ARCH(∞):

$$\left\{N^{-1/2}\sum_{t=1}^{[N\tau]}(r_t^2 - Er_t^2),\ \tau \in [0,1]\right\} \to_{D[0,1]} \{\sigma W(\tau), \tau \in [0,1]\}, \tag{16}$$

where σ^2 equals to the sum in (13). This result is quite surprising given a rather complicated nonlinear structure of ARCH(∞), since it holds for any stationary solution r_t such that $Er_t^4 < \infty$. Giraitis *et al.* (2000a) obtained a similar result by using finite memory approximation to ARCH(∞).

It seems that the implications of association property to the study of ARCH models have been not yet fully explored. This remark applies e.g. to the covariance structure and dependence properties of general nonlinear transformations of ARCH(∞), Rosenthal inequalities, rate of convergence, empirical processes and many other questions. See the dissertation of Louichi (1998) for references.

2.3 Stationary Solution of ARCH(∞) Without Moment Assumptions

A rather unusual feature of ARCH equations is the fact that they may admit a stationary solution which does not have any moments, even if the i.i.d. "shocks" ε_t's are $N(0,1)$. In such case, the Volterra series (4) converge in probability but not in any moment sense, and the properties of the infinite series are much more difficult to study. Nelson (1990) showed, using the theory of products of random matrices, that a necessary and sufficient condition for the existence of a strictly stationary GARCH(1,1) process is

$$E\log(\alpha\varepsilon_0^2 + \beta) < 0. \tag{17}$$

This condition is of course much weaker than any of conditions (5), (6), (7) given above (which imply in particular the existence of finite moment $Er_t^2 < \infty$ or $Er_t^4 < \infty$). Nelson's result was extended to the GARCH(p,q) case by Bougerol and Picard (1992), who showed that, under condition $E\varepsilon_0^2 = 1$, a stationary solution to (65) exists if and only if the top Lyapunov exponent γ is strictly negative. Moreover, in such case there exists only one stationary GARCH(p,q) process. The top Lyapunov exponent is defined by

$$\gamma = \lim_{n \to \infty} n^{-1} \log \|A_1 \cdots A_n\|, \tag{18}$$

where $\{A_k\}$ are i.i.d. $(p+q-1) \times (p+q-1)$ random matrices (depending only on ε_k^2) such that the $(p+q-1)$-valued process $X_t = (\sigma_{t+1}^2, \ldots, \sigma_{t-p+2}^2, r_t^2, \ldots, r_{t-q+2}^2)'$ satisfies the random coefficient matrix AR(1) equation

$$X_t = A_t X_{t-1} + B,$$

with $B = (\alpha_0, 0, \ldots, 0)'$; see Bougerol and Picard (1992).

Further progress in this direction was made by Kazakevičius and Leipus (2002), who discussed the general case of ARCH(∞). They observed that the volatility can be written as

$$\sigma_t^2 = b_0 \left(1 + \sum_{n=1}^{\infty} \sigma_{t,n}^2 \right), \tag{19}$$

the convergence of the series being equivalent to the existence of a stationary solution $r_t = \varepsilon_t \sigma_t$, where

$$\sigma_{t,n}^2 = \sum_{k=1}^{n} \sum_{\substack{i_1, \ldots, i_k \geq 1 \\ i_1 + \cdots + i_k = n}} b_{i_1} b_{i_2} \cdots b_{i_k} \varepsilon_{t-i_1}^2 \varepsilon_{t-i_1-i_2}^2 \cdots \varepsilon_{t-i_1-\cdots-i_k}^2 \quad (n \geq 1),$$

$\sigma_{t,n}^2 = 0$ $(n \leq 0)$, satisfy the recurrent equation

$$\sigma_{t,n}^2 = \varepsilon_{t-n}^2 \sum_{i=1}^{n} b_i \sigma_{t,n-i}^2, \quad n \geq 1. \tag{20}$$

Equation (20) is a "stochastic" version of the corresponding equation satisfied by $g_n = E\sigma_{t,n}^2$ of (8) in the case $\lambda_1 = E\varepsilon_0^2 < \infty$. For a fixed t (say, $t = 0$), equation (20) can be written in the matrix form:

$$\Sigma_n = B_n \Sigma_{n-1},$$

where $\Sigma_n = (\sigma_{0,n}^2, \sigma_{0,n-1}^2, \ldots, \sigma_{0,1}^2, 0, \ldots)'$ and where $\{B_n\}$ are random i.i.d. (infinite) matrices:

$$B_n = \begin{pmatrix} b_1 \varepsilon_{-n}^2 & b_2 \varepsilon_{-n}^2 & b_3 \varepsilon_{-n}^2 & \cdots \\ 1 & 0 & 0 & \cdots \\ 0 & 1 & 0 & \cdots \\ \cdots & \cdots & \cdots & \cdots \end{pmatrix}. \tag{21}$$

Kazakevičius and Leipus (2002) showed that in the GARCH(p, q) case, the top Lyapunov exponent satisfies

$$\gamma = -\log R, \tag{22}$$

where $R = \left(\limsup \sqrt[n]{\sigma_{0,n}^2}\right)^{-1} \in [0, \infty]$ is the convergence radius of the random power series

$$\sigma_t^2(z) = b_0 \left(1 + \sum_{n=1}^{\infty} \sigma_{t,n}^2 z^n\right), \tag{23}$$

see equation (19). It turns out that in general ARCH(∞) case, R is *nonrandom* and therefore γ defined by (22) can be considered an the analog of of the top Lyapunov exponent. Similarly as in the GARCH case, $\gamma < 0$ implies the existence of a stationary solution $r_t = \varepsilon_t \sigma_t$ of ARCH(∞), with σ_t^2 given by (19). The last condition is of course not necessary and a stationary solution may as well exist if $\gamma = 0$. In fact, using the argument of Kazakevičius and Leipus (2003) one can show that a stationary solution of ARCH(∞) with finite mean and coefficients $b_j \asymp j^{-q}, q > 1$ satisfies $\gamma = 0$. This shows that vanishing of the top Lyapunov exponent is typical for stationary solutions of ARCH(∞) with a power-like decay of coefficients.

Kazakevičius and Leipus (2002) also proved the uniqueness of the above stationary solution under some additional condition on the coefficients b_j (which is satisfied, for example, if these coefficients monotonically decay starting with j large enough).

2.4 Existence of Integrated ARCH(∞) Process

To invoke the widely discussed and notable analogy between ARCH and ARMA, one has to rearrange the GARCH(p, q) equation (65) so that r_t^2 satisfies ARMA($p + q, q$) equation with martingale difference innovations $\nu_t = r_t^2 - \sigma_t^2$:

$$(1 - \alpha(L) - \beta(L))r_t^2 = \alpha_0 + (1 - \beta(L))\nu_t \tag{24}$$

(we assume $E\varepsilon_0^2 = 1$ for simplicity). Then

$$\sum_{i=1}^{q} \alpha_i + \sum_{j=1}^{p} \beta_j = 1 \tag{25}$$

is the unit root condition. The corresponding GARCH model, called the *Integrated GARCH(p, q)*, was introduced by Engle and Bollerslev (1986) in analogy with integrated ARMA model in order to explain the observed IGARCH effect in financial data when the estimated parameters $\alpha_1, \ldots, \alpha_q$ and $\beta_1, \ldots, \beta_p$ of the (65) model sum up to a value which is close to one. A similar rearrangement of (2) leads to the notion *Integrated ARCH(∞)*, or *IARCH(∞)*, which is defined as a solution to (2) with

$$E\varepsilon_0^2 \sum_{j=1}^{\infty} b_j = 1. \tag{26}$$

Integrated processes constitute an important class of ARCH processes where the similarities and the differences between ARCH and ARMA most distinctly appear. It is easily seen (see also (5)) that stationarity and (26) imply $Er_t^2 = \infty$ and therefore a stationary IARCH(∞) process r_t necessarily has infinite variance. Moreover, in this case even the interpretation of ν_t as martingale innovations becomes peculiar as the ν_t's have infinite unconditional absolute mean. Nevertheless, differently from AR(∞) case, the IGARCH(∞) equation *may admit a stationary solution.*

The most famous example of IARCH processes is the *FIGARCH process* defined by

$$r_t = \sigma_t \varepsilon_t, \; \sigma_t^2 = b_0 + (1 - (1 - L)^d) r_t^2, \tag{27}$$

where $b_0 > 0$ and $(1 - L)^d, 0 < d < 1$ is the fractional differencing operator. This model, introduced by Baillie *et al.* (1996a) in order to capture long memory effect in volatility, allows a hyperbolic decay of the coefficients b_j which are positive, summable, and satisfy the unit root condition (26). However, the proof of existence of stationary solution to (27) given in Baillie *et al.* (1996a) does not seem to be correct. The question of the existence a stationary solution to FIGARCH equation (the "FIGARCH problem") is open at the present and seems very hard. See Giraitis *et al.* (2000a), Mikosch and Stărică (2000, 2003), for discussion and controversies surrounding the FIGARCH case.

Bougerol and Picard (1992) obtained the existence of a stationary Integrated GARCH(p, q) process as a corollary to their more general result discussed above. Essentially, their conditions (which ensure that the top Lyapunov exponent $\gamma < 0$) require that all coefficients α_i, β_i are strictly positive. Kazakevičius and Leipus (2003) discussed a similar problem in the general IARCH(∞) case, by using the definition of γ (22) via the convergence radius R. They replaced the positivity condition of the coefficients of Bougerol and Picard (1992) by the requirement that the weights b_j decay exponentially, more precisely, that

$$\sum_{j=1}^{\infty} b_j q^j < \infty \quad \text{for some} \quad q > 1.$$

If γ is negative then $r_t = \varepsilon_t \sigma_t$ with σ_t (19) is a stationary IARCH(∞) solution. They also showed (assuming $E \log^- \varepsilon_0^2 < \infty$) that if this decay condition of b_j is not satisfied (as in the FIGARCH case), then $\gamma = 0$. The last result can be considered as a further confirmation of the difficulty of the FIGARCH problem, and indicates the limitations of the random matrices approach to the existence problem of stationary solution of ARCH equations.

Our final remark concerns the question of long memory. The result (13) can be interpreted as the fact that the squares r_t^2 of covariance stationary ARCH(∞) always have short memory. However, the last fact does not rule out the possibility that absolute values $|r_t|$ or some (fractional) powers $|r_t|^\delta, \delta > 0$ may have non-summable autocorrelations, under the same conditions which guarantee the existence of covariance stationary solution and especially in the case of IARCH when r_t^2 has infinite variance. Such possibility seems unlikely, because of the summability of the coefficients b_j in ARCH(∞) and the associativity property discussed above. Nevertheless, several empirical studies indicate that sample autocovariances of absolute powers of asset returns exhibit "maximal memory" for $\delta = 1$, see Ding *et al.* (1993). While absolute powers $|r_t|^\delta$ of ARCH(∞) are mathematically hard to handle (unless δ is an even integer), this was one of the reasons for introducing stochastic volatility models, which allow modelling of long memory; see below.

3 Other ARCH and Related Models

In a wider sense, the term "ARCH process" refers to the class of processes r_t with zero conditional mean $E(r_t|r_s, s < t) = 0$ and the conditional variance $\sigma_t^2 = \mathrm{Var}(r_t|r_s, s < t)$ being a general function of the past information set $r_s, s < t$ (which may also include some additional "exogenous" variables). The possibilities of choices of functional forms of σ_t^2 are very numerous, leading to a vast family of ARCH models and modifications, only part of which is mentioned in our reference list. This review focuses on the probabilistic aspects of ARCH modelling, and we do not attempt to cover the whole econometric literature on ARCH. Of course, the topics and models discussed below are motivated in some sense by the interests of the authors of the review.

3.1 The LARCH Model

The *Linear ARCH (LARCH)* model, introduced by Robinson (1991), is defined by

$$r_t = \sigma_t \varepsilon_t, \quad \sigma_t = \alpha + \sum_{j=1}^{\infty} \beta_j r_{t-j}, \tag{28}$$

where $\{\varepsilon_t, \ t \in \mathbf{Z}\}$ is an i.i.d. sequence with zero mean and finite variance, and the coefficients β_j satisfy

$$\beta_j \sim c j^{d-1}, \tag{29}$$

for some $0 < d < 1/2, c > 0$. The particular case

$$r_t = \sigma_t \varepsilon_t, \quad \sigma_t = (1 - L)^{-d} r_t$$

corresponds to the LARCH equation with FARIMA$(0, d, 0)$ coefficients.

The main advantage of LARCH is that it allows modelling of long memory as well as some characteristic asymmetries (the "leverage effect"). Both these properties cannot be modeled by the classical ARCH(∞) with finite fourth moment. The condition (29) implies only $\sum_j \beta_j^2 < \infty$ which is weaker than assumption $\sum_j b_j < \infty$ for the ARCH(∞) model (2). Neither α nor the β_j are assumed positive and, unlike in (2), σ_t (not σ_t^2), is a linear combination of the past values of r_t, rather than their squares.

A not so pleasant feature of the LARCH model is that σ_t may be negative or vanish, being a linear combination of martingale differences r_t with zero mean, and so it lacks some of the usual volatility interpretation. Recently, Koulikov (2003) considered a particular case of LARCH which he claims has the property that $\sigma_t > 0$ a.s. If Koulikov's claim is correct, this would certainly increase the interest to the LARCH model and statistical inferences.

By the definition (28), the conditional variance is $\mathrm{Var}(r_t|r_s, s < t) = \sigma_t^2$, where

$$\sigma_t^2 = \Big(\alpha + \sum_{j=1}^{\infty} \beta_j r_{t-j}\Big)^2$$

is the squared linear combination of $r_s, s < t$ and so the LARCH model formally appears a particular case of Sentana's *Quadratic ARCH (QARCH)* models (Sentana, 1995). In the QARCH case, the conditional variance is an arbitrary nonnegative second order polynomial of the past information set $r_s, s < t$, which may contain both linear and quadratic terms. While the discussion in Sentana (1995) seems limited to finite memory Markov models, his model allows considerable flexibility and asymmetry. It is not clear if the method of Volterra expansions can be applied to the study of the QARCH model, as these expansions might appear intractable.

Long memory properties of the LARCH model were studied in Giraitis *et al.* (2000c), Giraitis *et al.* (2004). Similarly as the ARCH(∞) case, it is easy to show that a covariance stationary solution r_t to (28) exists if and only if

$$b = \Big\{\sum_{j=1}^{\infty} \beta_j^2\Big\}^{1/2} < 1 \tag{30}$$

(we assume $E\varepsilon^2 = 1$), in which case it can be represented by the convergent orthogonal Volterra series

$$r_t = \sigma_t \varepsilon_t, \quad \sigma_t = \alpha\Big(1 + \sum_{k=1}^{\infty} \sum_{j_1,\ldots,j_k=1}^{\infty} \beta_{j_1} \ldots \beta_{j_k} \varepsilon_{t-j_1} \ldots \varepsilon_{t-j_1-\cdots-j_k}\Big). \tag{31}$$

The above fact holds independently of (29) and requires (30) only. Whence or directly from the LARCH equations (28), the representation

$$\mathrm{Cov}(\sigma_t, \sigma_0) = b^{-2} \sum_{j=1}^{\infty} \beta_j \beta_{t+j}$$

immediately follows, which coincides with the covariance of the linear filter with coefficients $b^{-1}\beta_j$. Clearly, under condition (29), σ_t displays the covariance long memory similar to FARIMA$(0, d, 0)$ models.

The long memory property of the "observable" process r_t is more difficult to establish, as it requires the study of dependence properties of *nonlinear* functions of r_t. One of the simplest such functions is r_t^2, however, even in such case, the expression for $\mathrm{Cov}(r_k^2, r_0^2)$ is quite complicated and involves a combinatorial formalism of *diagrams* (Giraitis *et al.*, 2000c). As shown in this paper, under some additional boundedness condition on b (30), the squares r_t^2 of the LARCH model (28), (29) exhibit long memory, in the sense that

$$\mathrm{Cov}(r_k^2, r_0^2) \sim Ck^{2d-1}, \tag{32}$$

as $k \to \infty$, where the constant $C > 0$ is explicitly written in terms of parameters α, b, c, d. The above paper also obtains the convergence of the partial sums' process of r_t^2 to a *fractional Brownian motion* $W_d(\tau)$, $EW_d^2(\tau) = \tau^{2d+1}$:

$$\left\{ \frac{1}{N^{1/2+d}} \sum_{j=1}^{[N\tau]} (r_j^2 - Er_j^2) \right\} \to_{D[0,1]} \{c_d W_d(\tau)\}, \tag{33}$$

where c_d is a positive constant. Similar results under increasingly stringent bounds on b (30) were proved for arbitrary powers $r_t^k, \sigma_t^k, k = 2, 3, \ldots$.

These result indicate that long memory properties of LARCH may be somewhat similar to the properties of linear moving average process with coefficients (29). In fact, the first term of the expansion (31) (corresponding to $k = 1$) is exactly the linear process $\sum_{j=1}^{\infty} \beta_j \varepsilon_{t-j}$, up to the constant α. The nonlinearity of the LARCH comes out when analyzing the behavior of higher-order multiple sums in (31). It turns out that every term $\sum_{j_1,\ldots,j_k} \beta_{j_1} \ldots \beta_{j_k} \varepsilon_{t-j_1} \ldots \varepsilon_{t-j_k}$ behaves similarly as the first (linear) term and contributes to the limiting constants C and c_d in (32) and (33) respectively, although these "contributions decay geometrically" with k. Further important results on long memory LARCH processes can be found in Berkes and Horváth (2003b).

Short memory versions LARCH(p), GLARCH(p, q) of (28) can be introduced, similarly as in ARCH(∞) case. In fact, LARCH(1) turns out to be a particular case of the asymmetric ARCH model of Engle (1990), and GLARCH(p, q) a particular case of Sentana's QARCH model. As mentioned above, the main reasons for introducing these models was the desire to model asymmetric behavior, in particular, the *leverage effect.* This effect, first described by Black (1976), is the empirically observed property for volatility of asset returns to move in the opposite direction to returns, after a delay, as happens when the conditional variance is negatively correlated with past returns. Engle (1990) and Sentana (1995) discussed the (potential) leverage property in their models. For the LARCH model, the leverage property was rigorously proved in Giraitis *et al.* (2004). They showed that if the coefficient α and $\beta_1, \ldots, \beta_k$ have *opposite signs*, i.e.

$$\alpha\beta_j < 0, \quad j = 1, \ldots, k,$$

and some additional conditions are satisfied, then σ_t^2 and $r_{t-1}, \ldots, r_{t-k}$ are negatively correlated:

$$\mathrm{Cov}(\sigma_t^2, r_{t-j}) < 0, \quad j = 1, \ldots, k,$$

where $k = 1, 2, \dots$ can be arbitrary.

A heuristic explanation of this phenomenon from the LARCH equations is the following. Consider LARCH(1) $\sigma_t = \alpha + \beta_1 r_{t-1}$ and $\alpha > 0, \beta_1 < 0$. Clearly an increase of r_{t-1} results in a decrease of σ_t because of $\beta_1 < 0$. On the other hand, because of $\alpha > 0$, σ_t spends more time on the positive side, or $\sigma_t > 0$ is more likely to happen than $\sigma_t < 0$. This means that a decrease of σ_t is more likely to cause a decrease of σ_t^2, too, or that r_{t-1} and σ_t^2 should be negatively correlated.

Giraitis *et al.* (2003) developed R/S-type tests for long memory in LARCH processes against their short memory counterparts. Giraitis *et al.* (2000b) discussed the estimation of the long memory parameter in the LARCH model.

3.2 Bilinear Process

Formally, the classes AR, ARCH, LARCH (at least, their finite memory counterparts ARMA, GARCH, GLARCH) all belong to the general class of *bilinear models* introduced by Granger and Andersen (1978). The existing literature on bilinear time series models is quite large, see the monographs by Subba Rao and Gabr (1984) and Terdik (1999), however, it does not have much in common with the ARCH literature, probably because it focuses on homoscedastic case. Giraitis and Surgailis (2002) studied the heteroscedastic bilinear equation

$$X_t = \zeta_t \left(a_0 + \sum_{j=1}^{\infty} a_j X_{t-j} \right) + c_0 + \sum_{j=1}^{\infty} c_j X_{t-j}, \tag{34}$$

where $\{\zeta_t, t \in \mathbf{Z}\}$ are i.i.d. random variables, with zero mean and variance 1, and $a_j, c_j, j \geq 0$ are real (not necessary nonnegative) coefficients. Equation (34) appears naturally when studying the class of processes with the property that the conditional mean $\mu_t = E(X_t | X_s, s < t)$ is a linear combination of $X_s, s < t$, and the conditional variance $\sigma_t^2 = \mathrm{Var}(X_t | X_s, s < t)$ is the *square of a linear combinations of* $X_s, s < t$, as it is in the case of (34):

$$\sigma_t^2 = \left(a_0 + \sum_{j=1}^{\infty} a_j X_{t-j} \right)^2, \qquad \mu_t = c_0 + \sum_{j=1}^{\infty} c_j X_{t-j}.$$

Clearly, the case $a_j \equiv 0$, $j \geq 1$ gives the linear AR(∞) equation, while $c_j \equiv 0$ $(j \geq 0)$ results in the LARCH equation (28).

It is less obvious that the ARCH(∞) equation (2) is also a special case of the bilinear equation (34). To see this, put $X_t = r_t^2$, $\zeta_t := (\varepsilon_t^2 - \lambda_1)/\kappa$, $\lambda_1 = E\varepsilon_0^2$, $\kappa^2 = \mathrm{Var}(\varepsilon_0^2)$. Then (2) can be rewritten as

$$X_t = \zeta_t \left(\kappa b_0 + \kappa \sum_{j=1}^{\infty} b_j X_{t-j} \right) + \lambda_1 b_0 + \lambda_1 \sum_{j=1}^{\infty} b_j X_{t-j},$$

which is a particular case of (34).

Heteroscedastic models with non-zero conditional mean (combinations of the type ARMA-ARCH) have been studied in the literature; see e.g. Baillie *et al.* (1996b), Ling and Li (1998), Teyssière (2000), Li *et al.* (2002). The paper Giraitis and Surgailis (2002) attempts a systematic study of bilinear models (34) with long memory. For (34), it defines *long memory in conditional mean* and *long memory in conditional variance*, and describes a class of fractional bilinear models which exhibit both types of long memory, with arbitrary (fractional) parameters $0 < d_1, d_2 < 1/2$.

The bilinear model (34) in the cases $c_0 \neq 0$ and $c_0 = 0$ has different properties. The first case (to which the ARCH(∞) reduces) does not essentially allow for covariance stationary long memory in (34). Let $c_0 = 0$ below. Introduce the generating functions

$$G(z) := (1 - C(z))^{-1} = \sum_{j=0}^{\infty} g_j z^j, \qquad H(z) := A(z)G(z) = \sum_{j=1}^{\infty} h_j z^j,$$

where $A(z) := \sum_{j=1}^{\infty} a_j z^j$, $C(z) := \sum_{j=1}^{\infty} c_j z^j$. Under some natural conditions on coefficients g_j, h_j, the most important of which is

$$H^2 = \sum_{j=1}^{\infty} h_j^2 \; < \; 1,$$

it was proved that equation (34) admits a unique stationary and ergodic solution $X_t = \zeta_t \sigma_t + \mu_t$ given by the convergent orthogonal Volterra series

$$\sigma_t = a_0 \Big(1 + \sum_{k=1}^{\infty} \sum_{s_k < \cdots < s_1 < t} h_{t-s_1} h_{s_1 - s_2} \ldots h_{s_{k-1} - s_k} \zeta_{s_1} \ldots \zeta_{s_k} \Big), \tag{35}$$

$$\mu_t = a_0 \sum_{k=1}^{\infty} \sum_{s_k < \cdots < s_1 < t} g_{t-s_1} h_{s_1 - s_2} \ldots h_{s_{k-1} - s_k} \zeta_{s_1} \ldots \zeta_{s_k}. \tag{36}$$

Note that here $\sigma_t \neq \sqrt{\sigma_t^2}$, similarly as in the LARCH case, as σ_t may assume negative values. From (35), (36) it follows that the processes X_t, μ_t, σ_t admit the moving average representations:

$$\sigma_t = a_0 + \sum_{j=1}^{\infty} h_j \nu_{t-j}, \qquad \mu_t = \sum_{j=1}^{\infty} g_j \nu_{t-j}, \qquad X_t = \sum_{j=0}^{\infty} g_j \nu_{t-j} \tag{37}$$

w.r.t. martingale differences $\nu_s = \zeta_s \sigma_s$, implying in particular

$$EX_t = 0, \;\; \mathrm{Cov}(X_k, X_0) = a_0^2 (1 - H^2)^{-1} \sum_{j=0}^{\infty} g_j g_{j+k}.$$

If $|g_j|$ and $|h_j|$ are summable, then X_t, μ_t and σ_t have short memory (absolutely summable autocovariances). For example, this holds when

$$(\sum_{j=1}^{\infty} |a_j|^2)^{1/2} + \sum_{j=1}^{\infty} |b_j|^2 < 1.$$

The long memory in conditional mean (respectively, in conditional variance) is defined in terms of the coefficients of the linear filters (37), namely

$$y_j \sim C_1 j^{d_1 - 1} \qquad \text{(respectively,} \quad h_j \sim C_2 j^{d_2 - 1}), \tag{38}$$

where $C_i \neq 0$, $0 < d_i < 1/2$, $i = 1, 2$. Giraitis and Surgailis (2002) presented concrete examples of generating functions of the form $C(z) = 1 - P_1(z)(1-z)^{d_1}$, $A(z) = P_2(z)(1-z)^{d_1 - d_2}$, where $P_i(z)$, $i = 1, 2$ satisfy some root conditions, for which the corresponding $G(z), H(z)$ satisfy (38). Consequently, the bilinear equation (34) may exhibit *double long memory* (i.e. long memory both in conditional mean and in conditional variance). A natural question arises which of these two long memories plays a dominating role. The asymptotic behavior of the covariance of X_t depends on d_1 only ($\mathrm{Cov}(X_k, X_0)$ decays as $k^{2d_1 - 1}$), so the above question concerns nonlinear functionals, in particular the squared process X_t^2. The answer to this question depends on which of the two quantities $2(1 - 2d_1), 1 - 2d_2$ is larger: if $2(1 - 2d_1) < 1 - 2d_2$, then the long memory in conditional mean dominates, meaning that the covariance $\mathrm{Cov}(X_k^2, X_0^2)$ decays as $k^{2(2d_1 - 1)}$ and that a suitably normalized sum $\sum_{t=1}^{N} (X_t^2 - EX_t^2)$ converges to a non Gaussian distribution, similarly as if X_t were a linear FARIMA process. On the other hand, if $2(1 - 2d_1) > 1 - 2d_2$ holds, then the long memory in conditional variance dominates, in the sense that the covariance $\mathrm{Cov}(X_k^2, X_0^2)$ decays as $k^{2d_2 - 1}$ and $\sum_{t=1}^{N} (X_t^2 - EX_t^2)$ converges after normalization to a Gaussian distribution, similarly as in the LARCH model. An econometric discussion of double long memory and other models can be found in Teyssière (2000).

3.3 EGARCH and Stochastic Volatility Models

By *stochastic volatility (SV)* one usually means a model of the form

$$r_t = \varepsilon_t \sigma_t, \qquad \varepsilon_t \ \text{ i.i.d.}, \quad E\varepsilon_t = 0, \quad \mathrm{Var}\,\varepsilon_t = 1, \tag{39}$$

where $\sigma_t > 0$ is a measurable function of the past "information set" F_{t-1} which contains all information $r_s, \varepsilon_s \leq t - 1$ up to time $t - 1$, and may contain some other "unobservable information" as well, with the property, that $\varepsilon_s, s \geq t$ are independent of F_{t-1}. This implies of course $E(r_t | F_{t-1}) = 0$ and $\sigma_t^2 = \mathrm{Var}(r_t | F_{t-1})$. It is often assumed that the volatility is of the form

$$\sigma_t = f(\eta_t), \tag{40}$$

where f is a (nonlinear) function, and η_t is a stationary process of some familiar type, e.g. Gaussian or ARMA.

The choice of σ_t as a function of Gaussian process imposes distributional assumptions on the volatility which one would like to avoid, of course. On the other hand, it has several important advantages, as explained in Robinson (2001). Firstly, it allows modelling of long memory volatility, by taking η_t a long memory Gaussian process. Secondly, it allows a very general form of nonlinearity f in (40) by using the techniques of expansions of nonlinear functions of Gaussian random variables in Hermite polynomials. It is clear that a nonlinear function of σ_t, say $|\sigma_t|^\delta$, where $\delta > 0$ is arbitrary, is again of the form (40) and so this choice of volatility form allows the study of covariance behavior and other properties of general nonlinear transformations of r_t and σ_t. An appropriate choice of Gaussian nonlinearity leads to long memory SV models (39) which have *infinite fourth moment* and for which the autocovariances of r_t^2 are not well-defined. Robinson (2001) discusses a very general class of SV models with Gaussian nonlinearities, where $\varepsilon_t = f_1(\eta_{1t})$ in (39) is also a nonlinear function of another Gaussian process η_{1t} which is uncorrelated with $\eta_s, s \le t$, and the processes η_{1t}, η_t may be vector-valued. He obtains the asymptotic formulas for autocovariances of general nonlinear functions of $f_1(\eta_{1t})f(\eta_t)$ and exhibits various forms of long memory behavior.

Another popular choice of (40) is $f(\eta_t) = e^{\eta_t}$, where η_t is a moving average process of ARMA or FARIMA type. The corresponding SV model known as the *Exponential Generalized ARCH (EGARCH)* model, was proposed by Nelson (1991). More precisely, the EGARCH model is given by equations

$$r_t = \sigma_t \varepsilon_t, \ \sigma_t = \exp\left\{a + \sum_{j=1}^{\infty} b_j g(\varepsilon_{t-j})\right\}, \tag{41}$$

where

$$g(z) = \theta z + \gamma(|z| - E|\varepsilon_0|) \tag{42}$$

and where θ, γ are parameters which account for certain asymmetries observed in financial data. The particular case EGARCH(p,q) corresponds to $\eta_t = \log \sigma_t$ being an ARMA(p,q) process, i.e. a stationary solution of $\phi(L)\eta_t = \psi(L)g(\varepsilon_t)$, where $\phi(z)$, $\psi(z)$ are polynomials of order p and q, respectively. The case of FARIMA(p,d,q) η_t solving $\phi(L)(1-L)^d\eta_t = \psi(L)g(\varepsilon_t)$ corresponds to the *Fractional Integrated Exponential GARCH (FIEGARCH)* model of Bollerslev and Mikkelsen (1996).

A related class of long memory SV models was introduced in Breidt *et al.* (1998) and Harvey (1998):

$$r_t = \sigma_t \varepsilon_t, \quad \sigma_t = e^{\eta_t}, \quad \eta_t = a + \sum_{j=1}^{\infty} b_j \xi_{t-j}, \tag{43}$$

where $\xi_t, t \in \mathbf{Z}$ is a sequence of standard i.i.d. random variables, independent of the sequence $\varepsilon_t, t \in \mathbf{Z}$, and where b_j are as in FARIMA case.

Long memory properties of powers $|r_t|^\delta, \delta > 0$ of the SV model of (43) were first obtained in Harvey (1998). In the case when ξ_t are normal i.i.d., he obtained the autocorrelation function

$$\mathrm{Cov}(|r_t|^\delta, |r_0|^\delta) = (E|r_0|^\delta)^2(e^{\delta^2 \mathrm{Cov}(\eta_t,\eta_0)} - 1) \sim (\delta E|r_0|^\delta)^2 \mathrm{Cov}(\eta_t, \eta_0), \quad (44)$$

which shows that the decay of autocovariances of $|r_t|^\delta$ is proportional to the decay of autocovariances of (FARIMA) process η_t.

Surgailis and Viano (2002) obtained similar results without imposing distributional assumptions. They considered a generalization of (43), where ε_t and ξ_t are not necessary independent; it is only assumed that the bivariate sequence (ε_t, ξ_t), $t \in \mathbf{Z}$ is i.i.d., with zero means $E\varepsilon_t = E\xi_t = 0$ and unit variances. Their model also includes the FIEGARCH model as the particular case $\xi_t = g(\varepsilon_t)$. By assuming

$$b_j \sim c_0 j^{d-1}, \quad (c_0 \neq 0, 0 < d < 1/2) \quad (45)$$

and appropriate moment conditions on ε_0, ξ_0, they proved for any real $\delta > 0$

$$\mathrm{Cov}(|r_t|^\delta, |r_0|^\delta) \sim (\delta E|r_0|^\delta c_1)^2 t^{2d-1}, \quad (46)$$

similarly as in (44), where c_1 depends only on c_0, d. They also proved the convergence to a fractional Brownian motion:

$$\left\{ N^{-d-1/2} \sum_{s=1}^{[N\tau]} (|r_s|^\delta - E|r_s|^\delta) \right\} \rightarrow_{D[0,1]} \{c_2 E|r_0|^u W_d(\tau)\}, \quad (47)$$

where $c_2 = c_1/(d(2d+1))$.

The last paper also discussed the case where the process η_t in (43) is *short memory*, in the sense that the coefficients b_j are summable: $\sum_{j=1}^\infty |b_j| < \infty$. Under similar moment conditions, it proved that in such case the autocovariances of powers $|r_t|^\delta$ are also summable, that partial sums of $|r_t|^\delta$ converge to a standard Brownian motion, under normalization growing as $N^{1/2}$. The proof of the last result uses cumulants and some combinatorial formulas for cumulants of exponents of linear combinations in independent random variables which seem to be new even in the Gaussian case.

As mentioned above, the specific form $\xi_s = g(\zeta_s)$ of (42) allows to model certain asymmetries observed in financial data (leverage effect). Nelson (1991) obtained the formula for the covariance between $\log \sigma_t$ and ζ_{t-k}. This leverage effect was also discussed by Bollerslev and Mikkelsen (1996). Surgailis and Viano (2002) showed that if the distribution of ε_0 in the EGARCH model (41) is symmetric, then the covariance $\mathrm{Cov}(\sigma_t^\delta, r_{t-k})$ has always the sign of the product θb_k, in particular $\mathrm{Cov}(\sigma_t^\delta, r_{t-k}) < 0$ if $\theta b_k < 0$.

3.4 Regime Switching SV and Related Models

Mikosch and Stărică (1999, 2004) argue that the observed long memory in financial data is spurious and can be explained by structural breaks in GARCH models. Dahlhaus and Subba Rao (2003) investigated properties and estimation of nonstationary ARCH models with time-varying coefficients. A popular approach to modelling of structural breaks is Markov switching (Hamilton, 1989). Markov switching ARCH models are discussed in Hamilton and Susmel (1994), Cai (1994), Dueker (1997), Francq *et al.* (2001).

An alternative to Markov regime switching (which generally leads to short memory processes) is renewal regime switching, with independent and heavy tailed consecutive regime durations. Empirical evidence of heavy tailed regime durations is discussed in Jensen and Liu (2001) (lengths of the US business cycle's), Chow and Liu (1999) (dividend series from the CRSP data), Liu (2000) (daily S&P composite price index). Jensen and Liu (2001), Gourieroux and Jasiak (2001) argue that regime switching with heavy tails may lead to a new forecasting methodology, as an alternative to FARIMA forecasting. Various regime switching models leading to the long memory property and related econometrical issues were discussed in Parke (1999), Granger and Hyung (1999), Diebold and Inoue (2001), Liu (2000), Jensen and Liu (2001), Gourieroux and Jasiak (2001), Leipus and Viano (2003).

Liu (2000), Leipus *et al.* (2005) discussed regime switching SV model $r_t = \sigma_t \varepsilon_t$, where volatility

$$\sigma_t = \zeta_j, \qquad S_{j-1} < t \le S_j \tag{48}$$

assumes i.i.d. values $\zeta_j > 0$ on random intervals $(S_{j-1}, S_j]$ of a stationary renewal process with i.i.d. durations $U_j = S_j - S_{j-1}$. If $E\zeta^2 < \infty$ and the tail distribution $P(U > u)$ decays as

$$P(U > u) \sim c_1 u^{-\beta} \qquad (u \to \infty) \tag{49}$$

with some $c_1 > 0, \beta > 1$, then the autocovariance of σ_t decays as $t^{1-\beta}$ so that for $1 < \beta < 2$, the SV model in (48) has covariance long memory (Liu, 2000). Moreover, for any $\delta > 0$

$$|r_t|^\delta - E|r_t|^\delta = \sigma_t^\delta \nu_t + (\sigma_t^\delta - E\sigma_t^\delta) E|\varepsilon|^\delta, \tag{50}$$

where $\nu_t = |\varepsilon_t|^\delta - E|\varepsilon_t|^\delta$ are zero mean i.i.d., and σ_t^δ has the same form (48) with ζ_j replaced by ζ_j^δ. Let $E|\varepsilon|^{2\delta} < \infty$, $E\zeta^{2\delta} < \infty$. Then $\sum_{t=1}^N \sigma_t^\delta \nu_t = O_p(N^{1/2})$ and the limit distribution of the partial sums process $\sum_{t=1}^{[N\tau]}(|r_t|^\delta - E|r_t|^\delta)$ is determined by the second term on the r.h.s. of (50) which is asymptotically β–stable; more precisely,

$$\left\{ N^{-1/\beta} \sum_{s=1}^{[N\tau]} (|r_s|^\delta - E|r_s|^\delta) \right\} \to_{\text{f.d.d.}} \{Z_\beta(\tau)\} \tag{51}$$

in the sense of convergence of finite dimensional distributions, where $Z_\beta(\tau)$ is a β–stable Lévy process with independent increments, see Taqqu and Levy (1986), Pipiras *et al.* (2004).

The result (51) is typical for "renewal type long memory" and is in deep contrast with the fBM asymptotics of the corresponding partial sums processes in (47) and (33) for the EGARCH and LARCH models. The limit process $Z_\beta(\tau)$ has infinite variance while $|r_t|^\delta$ has finite variance, which means an increase of variability in the distributional limit (51). On the other hand, the limit process in (51) has independent increments while the summands have (covarance) long memory, meaning that this long memory does not persist in the distributional limit. A similar lack of persistence of long memory seems characteristic to some other econometric models, in particular to Parke's (1999) model (see Davidson and Sibbertsen, 2002). Similar properties were proved in Mikosch *et al.* (2002) for some models arising in telecommunications.

A general renewal regime switching scheme leading to a similar "increase of variability" and stable limit distribution of partial sums is discussed in Leipus *et al.* (2005), in particular, the linear model

$$X_t = \mu_t + a_t X_{t-1} + \sigma_t \varepsilon_t$$

with renewal switching in levels (μ_t), slope (a_t) and/or volatility (σ_t). The case of AR(1) equation with a_t perfoming a heavy tailed regime switching in the interval $[0,1]$ including the unit root $a_t = 1$ and its neighborhood is discussed in Leipus and Surgailis (2003). Let us note that the existence of covariance long memory in AR(1) model with a_t switching between two values 0 and 1 was first observed by Pourahmadi (1988).

As noted above, similar results can be expected for finite memory ARCH models with heavy tailed switching coefficients. The simplest case is the GARCH(1,1) equation

$$r_t^2 = \varepsilon_t^2 \left(\alpha_{0t} + \alpha r_{t-1}^2 + \beta \sigma_{t-1}^2\right), \tag{52}$$

where α, β are nonrandom, and $\{\alpha_{0t}\}$ is a (stationary) process, independent of $\{\varepsilon_t\}$. In particular, α_{0t} may assume only two values 0 and 1 on consecutive intervals of a stationary renewal process with a heavy-tailed inter-renewal distribution U, similarly as in Pourahmadi (1988). Equation (52) can be rewritten as

$$r_t^2 = \alpha_{0t} r_{0t}^2,$$

where $\{r_{0t}^2\}$ is a GARCH(1,1) process with (nonrandom) coefficients $\alpha_0 = 1$, α, β, which is independent of $\{\alpha_{0t}\}$. Clearly,

$$\mathrm{Cov}(r_t^2, r_0^2) = (Er_{00}^2)^2 \mathrm{Cov}(\alpha_{0t}, \alpha_{00}) + E[\alpha_{0t}\alpha_{00}] \mathrm{Cov}(r_{0t}^2, r_{00}^2),$$

see equation (53) where the second term on the r.h.s. vanishes exponentially, but the first term may decay very slowly, e.g. as $t^{1-\beta}$ in the case of renewal switching α_{0t} with inter-renewal distribution (49). Moreover, in the above example (52) one can show a similar covariance decay for arbitrary powers $|r_t|^\delta$, provided a_t assumes values 0, 1 only. While this example might be too simple and not characteristic, it also demonstrates the possibility of modeling long memory with the help of classical GARCH models with time-varying random coefficients.

4 Random Coefficient ARCH and Aggregation

A natural generalization of GARCH(1,1) (more generally, of ARCH(∞)) is to assume the coefficients α_0, α, β *random* and/or *time-varying*. There exists a considerable literature on random coefficient AR models (AR(1) in particular), and the interest in such generalizations recently has increased in connection with the studies of models which involve regime switching and structural breaks. The corresponding ARCH models with *constant* random coefficients lead to non-ergodic processes whose parameters cannot be consistently estimated. The motivation for such studies follows the important Granger's idea of aggregation.

4.1 Aggregation

The basic scheme of *contemporaneous aggregation* usually starts with N "elementary" individual processes $\{X_t^{(1)}\}, \ldots, \{X_t^{(N)}\}$, which evolve according to a random parametric short memory dynamics, and the aggregated process is the limit of the normalized averages $N^{-1}\sum_{i=1}^N X_t^{(i)}$, as $N \to \infty$. Granger (1980) found that aggregation of random coefficient autoregressive AR(1) process can lead to Gaussian long memory aggregated process. Since Granger's pioneering work, this question has attracted considerable attention in the econometric and time series literature, see Gonçalves and Gouriéroux (1988), Lippi and Zaffaroni (1999), and Oppenheim and Viano (1999) among others.

A related problem of generating long memory by aggregation of short memory models in ARCH set-up was analyzed in Ding and Granger (1996). They studied the so-called N-component GARCH(1,1) aggregation scheme

$$r_{N,t} = \sigma_t \varepsilon_t, \quad \sigma_t^2 = N^{-1}\sum_{i=1}^N \sigma_{i,t}^2$$

based on averaging of GARCH(1,1) *volatilities*

$$\sigma_{i,t}^2 = \sigma^2(1-\alpha_i)(1-\beta_i) + \alpha_i(1-\beta_i) r_{N,t-1}^2 + \beta_i \sigma_{i,t-1}^2,$$

where ε_t are i.i.d. $(0,1)$ random variables and α_i, β_i are i.i.d. random coefficients. Then $\sigma_{i,t}^2$ and σ_t^2 can be written in ARCH(∞) form

$$\sigma_{i,t}^2 = a_0^{(i)} + \sum_{k=1}^\infty a_k^{(i)} r_{N,t-k}^2, \quad \sigma_t^2 = \bar{a}_0 + \sum_{k=1}^\infty \bar{a}_k r_{N,t-k}^2,$$

where $a_0^{(i)} = \sigma^2(1-\alpha_i)$, $a_k^{(i)} = \alpha_i(1-\beta_i)\beta_i^{k-1}$, and $\bar{a}_k = N^{-1}\sum_{i=1}^N a_k^{(i)}$, $k \geq 0$. By ergodicity, $\bar{a}_k \to Ea_k^{(i)}$. Ding and Granger (1996) conjectured that, similarly as in AR(1) case, the limiting aggregated model

$$r_t^2 = \varepsilon_t^2\sigma_t^2 = \varepsilon_t^2\left(b_0 + \sum_{k-1}^{\infty} b_k r_{t-k}^2\right),$$

which is an ARCH(∞) model with deterministic coefficients $b_k = Ea_k^{(i)}$, can exhibit long memory, if α_i and β_i are properly chosen. Further investigation of this model was conducted by Kazakevičius *et al.* (2004), who assumed that $\sigma^2 > 0$ is a constant, α_k are i.i.d. random variables on $[0,1]$ with the mean $E\alpha_k = \mu > 0$ and β_k are i.i.d. Beta(p,q) random variables independent of α_k. For $p+q=1$ and $0<q<1/2$, they showed that

$$\bar{a}_0 \to Ea_0^{(i)} = \sigma^2(1-\mu), \quad \bar{a}_k \to Ea_k^{(i)} = \mu\tilde{b}_k, \quad k \geq 1, \quad \text{a.s.},$$

where $\tilde{b}_k = E(1-\beta_i)\beta_i^{k-1} \sim ck^{-q-1}$ $(k \to \infty)$ are coefficients of the expansion $\sum_{k=1}^{\infty}\tilde{b}_k\, z^k = 1-(1-z)^q$, $\sum_{k=1}^{\infty}\tilde{b}_k = 1$. If $\mu\sqrt{\lambda_2} < 1$, then[4] $r_{N,t}^2 \xrightarrow{L_2} r_t^2 = \sigma_t^2\varepsilon_t^2$, see Kazakevičius *et al.* (2004), where the limit r_t^2 is ARCH(∞) model with

$$\sigma_t^2 = \sigma^2(1-\mu) + \mu\left(1-(1-L)^q\right)r_t^2 = \sigma^2(1-\mu) + \mu\sum_{k=1}^{\infty}\tilde{b}_k r_{t-k}^2.$$

Since the covariance function r_t^2 is absolutely summable, differently from AR(1) case the aggregation procedure does not lead to the long memory. On the other hand, a similar aggregation procedure of random coefficient GLARCH(1,1) is likely to result in a long memory limiting process, but this possibility was not yet studied.

Leipus and Viano (2002) also discussed the aggregation (averaging) procedure

$$r_{N,t}^2 = N^{-1}\sum_{i=1}^{N}(r_t^{(i)})^2,$$

of squared random coefficient ARCH models

$$r_t^{(i)} = \sigma_t^{(i)}\varepsilon_t,\ (\sigma_t^{(i)})^2 = b_0^{(i)} + \sum_{k=1}^{\infty} b_k^{(i)}(r_{t-k}^{(i)})^2,$$

where ε_t are an i.i.d. noise variables and the weights $(b_0^{(i)}, b_1^{(i)}, b_2^{(i)}...)$ are independent copies of the random sequence $(b_0, b_1, b_2...)$. They have shown that under additional restrictions on random weights $b_0, b_1, \ldots$ ensuring existence

---

[4] We say that $\xi_N \xrightarrow{L_2} \xi$ if $E|\xi_N - \xi|^2 \to 0$ as $N \to \infty$.

of stationary solution with a finite moment $Er_t^4 < \infty$, the aggregated model can be written as Volterra series

$$r_{N,t}^2 = \varepsilon_t^2 \Big(\beta_N(0) + \sum_{k=1}^{\infty} \sum_{j_1,\dots,j_k=1}^{\infty} \beta_N(j_1,\dots,j_k)\varepsilon_{t-j_1}^2 \cdots \varepsilon_{t-j_1-\dots-j_k}^2 \Big),$$

with random coefficients $\beta_N(j_1,\dots,j_k) = N^{-1}\sum_{i=1}^N b_0^{(i)} b_{j_1}^{(i)} \cdots b_{j_k}^{(i)}$. Since $\beta_N(j_1,\dots,j_k) \to \beta(j_1,\dots,j_k) = E[b_0 b_{j_1} \cdots b_{j_k}]$ a.s. by the Strong Law of Large Numbers, for all $j_1,\dots,j_k \geq 1$, the aggregated model, as $N \to \infty$, converges in L_2 norm to the ARCH model with deterministic coefficients,

$$r_{N,t}^2 \xrightarrow{L_2} \bar{r}_t^2 = \varepsilon_t^2 \Big(\beta(0) + \sum_{k=1}^{\infty} \sum_{j_1,\dots,j_k=1}^{\infty} \beta(j_1,\dots,j_k)\varepsilon_{t-j_1}^2 \cdots \varepsilon_{t-j_1-\dots-j_k}^2 \Big)$$

in view of general result by Kazakevičius *et al.* (2004) on stability of random coefficient ARCH models. The covariance function of the limit process $\bar{r}_t^2$ is absolutely summable, i.e. $\bar{r}_t^2$ has short memory, but the aggregated process $r_{N,t}^2$ and the limit process $\bar{r}_t^2$ are no longer ARCH(∞) processes.

Zaffaroni (2000) investigated contemporaneous aggregation of GARCH processes. A number of papers in the ARCH literature deal with the temporal aggregation, partially motivated by the need of bridging between low and high frequency ARCH models (including their continuous time counterparts), see Drost and Nijman (1993), Drost and Werker (1996), Corradi (2000) among others.

Kazakevičius *et al.* (2004) studied general properties of ARCH(∞) random coefficient model r_t defined by equations (2), in the case when the coefficients $\{b_0, b_1, \dots\} =: b$ form a sequence of non-negative random variables, and $\varepsilon_t, t \in \mathbf{Z}$ is and i.i.d zero mean sequence independent of b. Under similar conditions as in the non-random coefficient case, they showed the existence of a stationary solution r_t with a finite moment Er_t^4. However, differently from the deterministic coefficient case, in this case the covariance function

$$\mathrm{Cov}(r_h^2, r_0^2) \to \mathrm{Var}E^b r_0^2 > 0 \tag{53}$$

tends as $h \to \infty$ to a nonzero constant $\mathrm{Var}E^b r_0^2$, where E^b denotes a conditional expectation given b, which can be easily seen from the equality $\mathrm{Cov}(r_{k+h}^2, r_k^2) = E\mathrm{Cov}^b(r_{k+h}^2, r_k^2) + \mathrm{Var}E^b r_k^2$. Then

$$E(r_h^2 - E^b r_h^2)(r_0^2 - E^b r_0^2) \equiv E\mathrm{Cov}^b(r_h^2, r_0^2) \to 0,$$

which implies the convergence $N^{-1}\sum_{h=1}^N r_h^2 \xrightarrow{d} E^b r_0^2 \neq Er_0^2$ to the random limit $E^b r_0^2$, thus showing that the ARCH(∞) process $\{r_t^2\}$ with constant random coefficients is non-ergodic.

5 Statistical Inference

5.1 Estimation of Parameters

Earlier works on (quasi) maximum-likelihood estimation in GARCH(p, q) model used the assumption of Gaussianity of errors, which was later dropped. Asymptotic properties of such estimators were considered in several papers.

Engle (1982) and Bollerslev (1986) considered the maximum likelihood estimator (MLE) for conditionally Gaussian ARCH(p) and GARCH(p, q) models, which extends readily to ARCH(∞) of (2). Given observations r_t, $t = 1, ..., N$, the log-likelihood, apart from an additive constant, is equal to

$$\ell_N(\theta, b) = -\frac{1}{2}\sum_{t=1}^{N}\left\{\log \sigma_t^{*2}(\theta, b) + \frac{r_t^2}{\sigma_t^{*2}(\theta, b)}\right\}, \tag{54}$$

where

$$\sigma_t^{*2}(\theta, b) = b + \sum_{j=1}^{t-1} b_j(\theta) r_{t-j}^2,$$

b is any admissible value of b_0 and $b_j(\theta)$ depend on a parameter $\theta \in \mathbf{R}^p$. The likelihood (54) was used by Robinson (1991) for testing for long memory in ARCH(∞). Note that (54) is only approximate because $\sigma_t^{*2}(\theta, b)$ is not equal to $E_{\theta,b}(r_t^2|r_{t-1}, \ldots, r_1)$.

The MLE of (θ_0, b_0) is defined by

$$(\widetilde{\theta}, \widetilde{b}) = \arg\max \ell_N(\theta, b),$$

where the maximization is taken over a suitable subset of $\mathbf{R}^{p+1}$. For inference, the limiting distribution of $(\widetilde{\theta}, \widetilde{b})$ is of interest. Weiss (1986) showed that $(\widetilde{\theta}, \widetilde{b})$ is $\sqrt{N}$-consistent and asymptotically normal in the case of ARCH(p) and finite p, while Lee and Hansen (1994), Lumsdaine (1986) established similar properties of the MLE in the case of GARCH(1,1). The asymptotic theory developed by these authors makes significantly weaker assumptions than conditional Gaussianity, by considering $\ell_N(\theta, b)$ as a quasi-log-likelihood. The analysis of $\ell_N(\theta, b)$ becomes more complicated in the case of GARCH(p, q) and general p, q. Asymptotic properties of the corresponding estimators for general GARCH(p, q) models were established by Hall and Yao (2003), who used in (54) the following approximation of the conditional variance $\sigma_t^2 = \mathrm{Var}(r_t|r_1, ..., r_{t-1})$:

$$\begin{aligned}\tilde{\sigma}_t(\theta)^2 &= \frac{\alpha_0}{1 - \sum_{i=1}^{p}\beta_i} \\ &+ \sum_{i=1}^{\min(q,t-1)} \alpha_i\Big(r_{t-i}^2 + \sum_{k=1}^{t-i-1} \sum_{\substack{1\le j_1,\ldots,j_k\le p:\\ j_1+\ldots+j_k\le t-i}} \beta_{j_1}\ldots\beta_{j_k} r_{t-i-j_1-\ldots-j_k}^2\Big).\end{aligned}$$

Hall and Yao (2003) showed that if ε_t in (65) have finite fourth moment, then the quasi-maximum likelihood estimate of parameters of GARCH(p,q) is $\sqrt{N}$-consistent and the asymptotic distribution is normal. In the case of heavy-tailed errors with $E\varepsilon_0^4 < \infty$, the class of possible limit distributions is extremely large, the rate of convergence is slower than $N^{1/2}$ and the limit distributions of estimators are no longer Gaussian. Hall and Yao (2003) also devised a bootstrap procedure for estimation of parameters of the limit distribution.

Peng and Yao (2003) suggested the modified least absolute deviation estimator

$$\hat{\theta}_2 = \arg\min_\theta \sum_{t=1}^{n} |\log(r_t^2) - \log(\tilde{\sigma}_t^2(\theta))|$$

which is less sensitive to heavy-tailed errors. This estimator has the standard $\sqrt{N}$-rate of convergence and the asymptotically normal limit distribution regardless whether the errors are heavy-tailed or not. Their study suggests using of the least absolute deviation estimator $\hat{\theta}_2$ when the errors have very heavy tails as $E(|\varepsilon_t|^3) < \infty$, and using the quasi-maximum likelihood estimator $\tilde{\theta}$ as long as ε_t are not very heavy tailed. Peng and Yao (2003) also investigated asymptotic properties of the absolute deviation estimator $\hat{\theta}_1 = \arg\min_\theta \sum_{t=1}^{n} |r_t^2/\tilde{\sigma}_t^2(\theta) - 1|$, and by the regression motivated estimator $\hat{\theta}_3 = \arg\min_\theta \sum_{t=1}^{n} |r_t^2 - \tilde{\sigma}_t^2(\theta)|$. Let us note that asymptotic normality of quasi–maximum likelihood estimator of parameters of GARCH(p,q) under mild moment conditions was obtained in Berkes *et al.* (2003a), Berkes and Horváth (2004).

Straumann and Mikosch (2003) discussed quasi-maximum likelihood estimation in a general heteroscedastic time series model $X_t = \sigma_t\varepsilon_t$, where unobservable volatility σ_t is a parametric function of (X_{t-1}, σ_{t-1}), and ε_t is a standard i.i.d. noise. They showed the existence and uniqueness of a stationary solution X_t, established the consistency and asymptotic normality of the estimator and applied their results to GARCH$(1,1)$, asymmetric GARCH$(1,1)$ and EGARCH models. For some recent results, see also Berkes *et al.* (2003a), Francq and Zakoïan (2004), Ling and McAleer (2003b).

Giraitis and Robinson (2001) considered Gaussian or Whittle estimation based on $r_t^2, t = 1, \ldots, N$. The idea of such estimation in the GARCH case was pointed out by Bollerslev (1986) who noted that r_t^2 generated by (65) has an ARMA type representation, albeit with conditionally heteroscedastic innovations, and it was employed in Harvey (1998) and Robinson and Zaffaroni (1997, 1998) for certain class of stochastic volatility and non-linear moving average processes.

Rewrite ARCH(∞) model (2) as

$$r_t^2 = b_0 + \sum_{j=1}^{\infty} b_j(\theta) r_{t-j}^2 + \nu_t,$$

where $\nu_t = r_t^2 - \sigma_t^2$ are martingale differences $b_j(\theta)$ are depend on a parameter θ. Assuming r_t is a fourth-order stationary sequence, under some additional conditions r_t^2 has spectral density

$$f(\lambda) = \frac{\sigma^2}{2\pi}\, g(\lambda, \theta), \quad g(\lambda, \theta) = \left|1 - \sum_{j=1}^{\infty} b_j(\theta) e^{ij\lambda}\right|^{-2}, \quad -\pi < \lambda \leq \pi,$$

where $\sigma^2 := E\nu_t^2$. Consider the objective function

$$w_N(\theta) = \sum_{j=1}^{N-1} \frac{I(\lambda_j)}{g(\lambda_j; \theta)},$$

where $I(\lambda) = \frac{1}{2\pi N} \left|\sum_{t=1}^N r_t^2 e^{it\lambda}\right|^2$ is the periodogram of the r_t^2, $\lambda_j = 2\pi j/N$. The Whittle estimate is defined by

$$\widehat{\theta} = \arg\min_{\theta \in \Theta} w_N(\theta), \tag{55}$$

where Θ is a compact subset of $\mathbf{R}^p$. The normalized spectral density $g(\lambda; \theta)$ is often explicitly given, for example, for GARCH(p, q) models it equals (see Bollerslev, 1986)

$$g(\lambda; \theta) = \left|\frac{1 - \beta(e^{i\lambda})}{1 - \alpha(e^{i\lambda}) - \beta(e^{i\lambda})}\right|^2, \tag{56}$$

where $\alpha(z) = \sum_{j=1}^q \alpha_j z^j$, $\beta(z) = \sum_{j=1}^p \beta_j z^j$. Under some additional regularity and moment conditions and the normalizing assumption

$$\int_{-\pi}^{\pi} \log g(\lambda; \theta) d\lambda = 0, \tag{57}$$

Giraitis and Robinson (2001) showed that

$$N^{1/2}(\widehat{\theta} - \theta_0) \xrightarrow{d} N\left(0, 2W^{-1} + W^{-1} V W^{-1}\right),$$

where

$$W = \frac{1}{2\pi} \int_{-\pi}^{\pi} \frac{\partial \log g(\lambda; \theta_0)}{\partial \theta} \frac{\partial \log g(\lambda; \theta_0)}{\partial \theta'}\, d\lambda,$$

and

$$V = \frac{2\pi}{\sigma^2} \int_{-\pi}^{\pi} \int_{-\pi}^{\pi} \frac{\partial g(\lambda; \theta_0)^{-1}}{\partial \theta} \frac{\partial g(\omega; \theta_0)^{-1}}{\partial \theta'} f(\lambda, -\omega, \omega)\, d\lambda\, d\omega,$$

where $f(\lambda, \omega, \nu)$ is the fourth-order cumulant spectrum of r_t^2. Note that the matrices V and W can be consistently estimated.

The following two remarks should be made in this context. Firstly, the above Whittle estimation in the case of GARCH(p, q) processes requires that their spectral density (56) is sufficiently regular and satisfies condition (57).

Secondly, the above results apply to weights which decrease much more slowly than exponentially, e.g. as

$$b_j = c_1 j^{-\zeta},$$

where $\zeta > 1$ and $c_1 > 0$, in which case we have in the above example, $\theta = (\theta_1, \theta_2)$ and

$$g(\lambda;\theta) = \Big|1 - (\theta_1 - 1)\theta_2 \sum_{j=1}^{\infty} j^{-\theta_1} e^{ij\lambda}\Big|^{-2},$$

where the true θ_1 equals ζ, θ_2 is bounded by 1 from above, and $g(\lambda;\theta)$ satisfies (57). Of course, the lack of a closed form representation of the above spectral density is a practical disadvantage.

Although the Whittle estimation method is simple and easy to use, $\widehat{\theta}$ has a different limiting variance from $\widetilde{\theta}$ as shown in Lee and Hansen (1994), Lumsdaine (1996), and is asymptotically less efficient than $\widetilde{\theta}$ when the r_t's are conditionally Gaussian.

Let us note, finally, the result of Mikosch and Straumann (2002), who showed some non-standard properties of the Whittle estimator in a heavy-tailed GARCH(1, 1) model.

5.2 Change-point Problem

In order to avoid spurious inference, an important problem in financial data analysis is testing for parameter constancy against some kind of instability in the conditional variance, in particular for structural breaks of variance $\operatorname{Var} r_t$.

Kokoszka and Leipus (1999) studied the CUSUM type tests for a change-point in the parameters of the conditional variance of the ARCH(∞) model defined by equations (2), which results in the change of the variance $\operatorname{Var} r_t$. They assumed the following condition

$$(E\varepsilon_0^8)^{1/4} \sum_{j=1}^{\infty} b_j < 1 \tag{58}$$

guarantying the existence of an eighth order strictly stationary solution r_t and the validity of the Functional Limit Theorem for the squares r_t^2.

Denote by $b := (b_0, b_1, \dots)$ the parameter sequence of ARCH(∞) process r_t^2, (2), and write $\{r_t\} \in \mathcal{R}(b)$ if b satisfies condition (58).

Suppose we want to test the null hypothesis

$$H_0 : r_1^2, \dots, r_N^2 \text{ is a sample from } \{r_t\} \in \mathcal{R}(b) \text{ for some } b$$

against the alternative that the sample $r_1^2, \dots, r_N^2$ has the form

$$r_t^2 = \begin{cases} (r_t^{(1)})^2, & \text{if } 1 \le t \le k^*, \\ (r_t^{(2)})^2, & \text{if } k^* < t \le N, \end{cases} \tag{59}$$

where $k^* = [N\tau^*]$, $0 < \tau^* < 1$ is fixed, and sequences $\{r_t^{(1)}\} \in \mathcal{R}(b^{(1)})$, $\{r_t^{(2)}\} \in \mathcal{R}(b^{(2)})$ with $b^{(1)} \neq b^{(2)}$ have different variances $\operatorname{Var} r_t^{(1)} \neq \operatorname{Var} r_t^{(2)}$. The sequences $\{r_t^{(1)}\}$ and $\{r_t^{(2)}\}$ are generated by the same noise sequence $\{\varepsilon_t\}$.

Kokoszka and Leipus (1999) studied CUSUM type tests based on the process $\{U_N(\tau), \tau \in [0,1]\}$, where

$$U_N(\tau) = N^{1/2} \frac{[N\tau](N-[N\tau])}{N^2} \left(\frac{1}{[N\tau]} \sum_{j=1}^{[N\tau]} r_j^2 - \frac{1}{N-[N\tau]} \sum_{j=[N\tau]+1}^{N} r_j^2 \right). \tag{60}$$

The partial sums in (60) are empirical estimates of the variance of the r_t, so the tests are designed to detect a change in parameters which leads to a change in variance of returns.

Under the null hypothesis H_0,

$$\Big\{U_N(\tau),\, \tau \in [0,1]\Big\} \rightarrow_{D[0,1]} \Big\{\sigma W^0(\tau),\, \tau \in [0,1]\Big\},$$

where $\sigma^2 = \sum_{k=-\infty}^{\infty} \operatorname{Cov}(r_k^2, r_0^2)$ and $W^0(\tau) = W(\tau) - \tau W(1)$ is a Brownian bridge. Under H_0, this implies the convergence of the following standard statistics

$$\sup_{0\le\tau\le1} |U_N(\tau)| \xrightarrow{d} \sigma \sup_{0\le\tau\le1} |W^0(\tau)|, \qquad \int_0^1 U_N^2(\tau)d\tau \xrightarrow{d} \sigma^2 \int_0^1 (W^0(\tau))^2 d\tau. \tag{61}$$

The tests based on such continuous functionals of the process $\{U_N(\tau)\}$ have positive asymptotic power against the alternative H_1.

Note that under H_0, the process r_t^2 has short memory and $\sigma^2 < \infty$. To construct the asymptotic critical regions, the only unknown parameter σ^2 of the limit distributions in (61) can be estimated by

$$\hat{s}_{N,q}^2 = \sum_{|j|\le q} \left(1 - \frac{|j|}{q+1}\right) \hat{\gamma}(j),$$

where $q \to \infty$, $q/N \to 0$ is the bandwidth parameter, $\hat{\gamma}(j) = N^{-1}\sum_{i=1}^{N-|j|}(r_i^2 - \bar{r}^2)(r_{i+|j|}^2 - \bar{r}^2)$ are the sample covariances and $\bar{r}^2 = N^{-1}\sum_{j=1}^{N} r_j^2$ is the sample mean. Then, as $N \to \infty$,

$$\hat{s}_{N,q}^2 \xrightarrow{P} \sigma^2.$$

To estimate the change-point τ^* Kokoszka and Leipus (2000) have used the estimator $\hat{\tau} = \hat{k}/N$, which is based on the CUSUM statistic (60), setting

$$\hat{k} = \min\{k : |U_k| = \max_{1\le j<N} |U_j|\},$$

where $U_k = U_N(k/N)$. Then

$$|\hat{\tau} - \tau^*| = O_P\left(\frac{1}{N\Delta^4}\right),$$

where $\Delta = E(r_k^{(2)})^2 - E(r_k^{(1)})^2 \neq 0$. Here $E(r_k^{(i)})^2 = \lambda_1 b_0^{(i)}/(1 - \lambda_1 \sum_{j=1}^{\infty} b_j^{(i)})$, $i = 1, 2$.

Andreou and Ghysels (2002, 2003) applied these results to the stock and FX markets. Tests for changes in volatility, based on squared model residuals and applicable to the GARCH model were designed by Kokoszka and Teyssière (2002); for related research see Horvàth *et al.* (2001), Berkes and Horváth (2003a), Berkes *et al.* (2003b). Berkes *et al.* (2004) used a test based on the likelihood ratio. Sequential monitoring scheme for detection of changes in the parameters of GARCH sequence model was studied by Berkes, Gombay, Horváth and Kokoszka (2004).

References

1. Andersen, T. G., Bollerslev, T., Diebold, F. X. and Labys, P. (2001) The distribution of realized exchange rate volatility. *Journal of American Statistical Association*, **96,** 42–55.
2. Andreou, E. and Ghysels, E. (2002). Detecting multiple breaks in financial market volatility dynamics. *Journal of Empirical Econometrics*, **17,** 579–600.
3. Andreou, E. and Ghysels, E. (2003). Tests for breaks in the conditional comovements of asset returns. *Statistica Sinica*, **13,** 1045–1073.
4. Baillie, R. T., Bollerslev, T. and Mikkelsen, H. O. (1996a). Fractionally integrated generalized autoregressive conditional heteroskedasticity. *Journal of Econometrics*, **74,** 3–30.
5. Baillie, R. T., Chung, C.-F. and Tieslau, M. A. (1996b). Analysing inflation by the fractionally integrated ARFIMA-GARCH model. *Journal of Applied Econometrics*, **11,** 23–40.
6. Barlow, R. E. and Proschan, F. (1981). *Statistical Theory of Reliability and Life: Probability Models.* Silver Spring, MD.
7. Bera, A. and Higgins, M. L. (1993). ARCH models: properties, estimation and testing. *Journal of Economic Surveys*, **7,** 305–366.
8. Berkes, I., Gombay, E., Horvàth, L. and Kokoszka, P. S. (2004). Sequential change-point detection in GARCH(p, q) models. *Econometric Theory*, **20,** 1140–1167
9. Berkes, I., Horvàth, L. and Kokoszka, P. S. (2002b). Probabilistic and statistical properties of GARCH processes. *Fields Institute Communications*, forthcoming.
10. Berkes, I. and Horvàth, L. (2003a). Limit results for the empirical process of squared residuals in GARCH models. *Stochastic Processes and their Applications*, **105,** 271–298.
11. Berkes, I. and Horvàth, L. (2003b). Asymptotic results for long memory LARCH sequences. *Annals of Applied Probability*, **13,** 641–668.
12. Berkes, I. and Horvàth, L. (2004). The efficiency of the estimators of the parameters in GARCH processes. *Annals of Statistics*, **32,** 633–655.
13. Berkes, I., Horvàth, L. and Kokoszka, P. S. (2003a). GARCH processes: structure and estimation. *Bernoulli*, **9,** 201–227.

14. Berkes, I., Horvàth, L. and Kokoszka, P. S. (2003b). Asymptotics for GARCH squared residual correlations. *Econometric Theory*, **19,** 515–540.
15. Berkes, I., Horvàth, L. and Kokoszka, P. S. (2004). Testing for parameter constancy in GARCH(p, q) models. *Statistics and Probability Letters*, **70,** 263–273,
16. Black, F. (1976). Studies in stock price volatility changes. *Proceedings of the 1976 Business Meeting of the Business and Economic Statistics Section. American Statistical Association*, 177–181.
17. Bollerslev, T. (1986). Generalized autoregressive conditional heteroskedasticity. *Journal of Econometrics*, **31,** 307–327.
18. Bollerslev, T. (1988). On the correlation structure for the generalized autoregressive conditional heteroskedastic process. *Journal of Time Series Analysis*, **9,** 121–131.
19. Bollerslev, T., Chou, R.Y. and Kroner, K.F. (1992). ARCH modeling in finance: a review of the theory and empirical evidence. *Journal of Econometrics*, **52,** 5–59.
20. Bollerslev, T., Engle, R. F. and Nelson, D. B. (1994). ARCH models. In: *Handbook of Econometrics, New York* (eds R. F. Engle and D. L. McFadden), volume 4, pp. 2961–3031. Elsevier Science.
21. Bollerslev, T. and Mikkelsen, H. O. (1996). Modeling and pricing long memory in stock market volatility. *Journal of Econometrics*, **73,** 151–184.
22. Bougerol, P. and Picard, N. (1992). Stationarity of GARCH processes and of some nonnegative time series. *Journal of Econometrics*, **52,** 115–127.
23. Breidt, F. J., Crato, N. and de Lima, P. (1998). On the detection and estimation of long memory in stochastic volatility. *Journal of Econometrics*, **83,** 325–348.
24. Cai, J. (1994). A Markov model of switching-regime ARCH. *Journal of Business & Economic Statistics*, **12,** 309–316.
25. Chow, Y.-F. and Liu, M. (1999). Long swings with memory and stock market fluctuations. *Journal Financial and Quantitative Analysis*, **34,** 341–367.
26. Corradi, V. (2000). Reconsidering the continuous time limit of the GARCH(1,1) process. *Journal of Econometrics*, **96,** 145–153.
27. Cox, J. T. and Grimmett, G. (1984). Central limit theorems for associated random variables and the percolation models. *Annals of Probability*, **12,** 514–528.
28. Dacorogna, M. M., Müller, U. A., Nagler, R. J., Olsen, R. B. and Pictet, O. V. (1993). A geographical model for the daily and weekly seasonal volatility in the foreign exchange market. *Journal of International Money and Finance*, **12,** 413–438.
29. Dahlhaus, R. and Subba Rao, S. (2003). Statistical inference for time-varying ARCH processes. *Annals of Statistics*, forthcoming.
30. Davidson, J. (2004). Moment and memory properties of linear conditional heteroscedasticity models, and a new model. *Preprint.*
31. Davidson, J. and Sibbertsen, Ph. (2002). Generating schemes for long memory processes: regimes, aggregation and linearity. *Preprint.*
32. Diebold, F.X. and Inoue, A. (2001). Long memory and regime switching. *Journal of Econometrics*, **105,** 131–159.
33. Ding, Z. and Granger, C. W. J. (1996). Modeling volatility persistence of speculative returns: a new approach. *Journal of Econometrics*, **73,** 185–215.
34. Ding, Z., Granger, C. W. J. and Engle, R. F. (1993). A long memory property of stock market returns and a new model. *Journal of Empirical Finance*, **1,** 83–106.

35. Drost, F. C. and Nijman, T. E. (1993). Temporal aggregation of GARCH processes. *Econometrica*, **61,** 909–927.
36. Drost, F. C. and Werker, B. (1996). Closing the GARCH gap: continuous time GARCH modeling. *Journal of Econometrics*, **74,** 31–57.
37. Dueker, M. J. (1997). Markov switching in GARCH processes and mean-reverting stock-market volatility. *Journal of Business & Economic Statistics*, **15,** 26–34.
38. Engle, R. F. (1982). Autoregressive conditional heteroscedasticity with estimates of the variance of United Kingdom inflation. *Econometrica*, **50,** 987–1008.
39. Engle, R. F. (1990). Stock volatility and the crash of '87. discussion. *The Review of Financial Studies*, **3,** 103–106.
40. Engle, R. F. and Bollerslev, T. (1986). Modelling the persistence of conditional variances. *Econometric Reviews*, **27,** 1–50.
41. Engle, R. F. and Russell, J. R. (1998). Autoregressive conditional duration: a new model for irregularly spaced transaction data. *Econometrica*, **66,** 1127–1162.
42. Fan, J. and Yao, Q. (2002). *Non-linear Time Series: Nonparametric and Parametric Methods.* Springer, New York.
43. Francq, C. and Zakoïan, J.-M. (2004). Maximum likelihood estimation of pure GARCH and ARMA-GARCH processes. *Bernoulli*, **10,** 605–637,
44. Francq, C., Roussignol, M. and Zakoïan, J.-M. (2001). Conditional heteroskedasticity driven by hidden Markov chains. *Journal of Time Series Analysis*, **22,** 197–220.
45. Giraitis, L., Kokoszka, P. and Leipus, R. (2000a). Stationary ARCH models: dependence structure and Central Limit Theorem. *Econometric Theory*, **16,** 3–22.
46. Giraitis, L., Kokoszka, P., Leipus, R. and Teyssière, G. (2000b). Semiparametric estimation of the intensity of long memory in conditional heteroskedasticity. *Statistical Inference for Stochastic Processes*, **3,** 113–128.
47. Giraitis, L., Robinson, P. M. and Surgailis, D. (2000c). A model for long memory conditional heteroskedasticity. *Annals of Applied Probability*, **10,** 1002–1024.
48. Giraitis, L., Kokoszka, P., Leipus, R. and Teyssière, G. (2003). Rescaled variance an related tests for long memory in volatility and levels. *Journal of Econometrics*, **112,** 265–294.
49. Giraitis, L., Leipus, R., Robinson, P. M. and Surgailis, D. (2004). LARCH, leverage and long memory. *Journal of Financial Econometrics*, **2,** 177–210.
50. Giraitis, L. and Robinson, P. M. (2001). Whittle estimation of ARCH models. *Econometric Theory*, **17,** 608–631.
51. Giraitis, L. and Surgailis, D. (2002). ARCH-type bilinear models with double long memory. *Stochastic Processes and their Applications*, **100,** 275–300.
52. Gonçalves, E. and Gouriéroux, C. (1988). Aggrégation de processus autoregressifs d'ordre 1. *Annales d'Economie et de Statistique*, **12,** 127–149.
53. Gouriéroux, C. *ARCH Models and Financial Applications.* Springer, Berlin.
54. Gourieroux, C. and Jasiak, J. (2001). Memory and infrequent breaks. *Economics Letters*, **70,** 29–41.
55. Granger, C.W.J. (1980). Long memory relationships and the aggregation of dynamic models. *Journal of Econometrics*, **14,** 227–238.
56. Granger, C. W. J. and Andersen, A. P. (1978). *An Introduction to Bilinear Time Series Models.* Vandenhoek & Ruprecht, Göttingen.

57. Granger, C. W. J. and Hyung, N. (1999). Occasional structural breaks and long memory. *Discussion paper 99-14*, Department of Economics. University of California, San Diego.
58. Hall, P. and Yao, Q. (2003). Inference in ARCH and GARCH models with heavy-tailed errors. *Econometrica*, **71,** 285–317.
59. Hamilton, J. D. and Susmel, R. (1994). Autoregressive conditional heteroskedasticity and changes in regime. *Journal of Econometrics*, **64,** 307–333.
60. Harvey, A. (1998). Long memory in stochastic volatility. In: *Forecasting Volatility in the Financial Markets, Oxford* (eds J. Knight and S. Satchell), pp. 307–320. Butterworth & Heineman.
61. He, C. and Teräsvirta, T. (1999). Fourth moment structure of the GARCH(p, q) process. *Econometric Theory*, **15,** 824–846.
62. Horváth, L., Kokoszka, P. and Teyssière, G. (2001). Empirical process of the squared residuals of an ARCH sequence. *Annals of Statistics*, **29,** 445–469.
63. Jensen, M. J. and Liu, M. (2001). Do long swings in the business cycle lead to strong persistence in output? *Journal of Monetary Economics*, forthcoming.
64. Karanasos, M. (1999). The second moment and the autocovariance function of the squared errors of the GARCH model. *Journal of Econometrics*, **90,** 63–76.
65. Kazakevičius, V. and Leipus, R. (2002). On stationarity in the ARCH(∞) model. *Econometric Theory*, **18,** 1–16.
66. Kazakevičius, V. and Leipus, R. (2003). A new theorem on existence of invariant distributions with applications to ARCH processes. *Journal of Applied Probability*, **40,** 147–162.
67. Kazakevičius, V., Leipus, R. and Viano, M.-C. (2004). Stability of random coefficient autoregressive conditionally heteroskedastic models and aggregation schemes. *Journal of Econometrics*, **120,** 139–158.
68. Kokoszka, P. and Leipus, R. (1999). Testing for parameter changes in ARCH models. *Lietuvos Matematikos Rinkinys*, **39,** 231–247.
69. Kokoszka, P. and Leipus, R. (2000). Change-point estimation in ARCH models. *Bernoulli*, **6,** 513–539.
70. Kokoszka, P. and Teyssière, G. (2002). Change-point detection in GARCH models: asymptotic and bootstrap tests. *Preprint.*
71. Koulikov, D. (2003). Modeling sequences of long memory non-negative covariance stationary random variables. *Preprint.*
72. Lee, S.-W. and Hansen, B. E. (1994). Asymptotic theory for the GARCH(1,1) quasi-maximum likelihood estimator. *Econometric Theory*, **10,** 29–52.
73. Leipus, R., Paulauskas, V. and Surgailis, D. (2005). Renewal regime switching and stable limit laws. *Journal of Econometrics*, **129,** 299–327.
74. Leipus, R. and Surgailis, D. (2003). Random coefficient autoregression, regime switching and long memory. *Advances in Applied Probability*, **35,** 737–754.
75. Leipus, R. and Viano, M.-C. (2002). Aggregation in ARCH models. *Lietuvos Matematikos Rinkinys*, **42,** 68–89.
76. Leipus, R. and Viano, M.-C. (2003). Long memory and stochastic trend. *Statistics and Probability Letters*, **61,** 177–190.
77. Li, W. K., Ling, S. and McAleer, M. (2002). Recent theoretical results for time series models with GARCH errors. *Journal of Economic Surveys*, **16,** 245–269.
78. Ling, S. and Li, W. K. (1998). On fractionally integrated autoregressive moving average time series models with conditional heteroskedasticity. *Journal of the American Statistical Association*, **92,** 1184–1193.

79. Ling, S. and McAleer, M. (2002a). Stationarity and the existence of moments of a family of GARCH processes. *Journal of Econometrics*, **106,** 109–117.
80. Ling, S. and McAleer, M. (2002b). Necessary and sufficient moment conditions for the GARCH(r, s) and asymmetric power GARCH(r, s) models. *Econometric Theory*, **18,** 722–729.
81. Ling, S. and McAleer, M. (2003a). Asymptotic theory for a vector ARMA-GARCH model. *Econometric Theory*, **19,** 278–308.
82. Ling, S. and McAleer, M. (2003b). On adaptive estimation in nonstationary ARMA models with GARCH errors. *Annals of Statistics*, **31,** 642–674.
83. Lippi, M. and Zaffaroni, P. (1999). Contemporaneous aggregation of linear dynamic models in large economies. *Preprint.*
84. Liu, M. (2000). Modeling long memory in stock market volatility. *Journal of Econometrics*, **99,** 139–171.
85. Louichi, S. (1998). Théorèmes limites pour des suites positivement ou faiblement dépendantes. *PhD thesis*, Université de Paris Sud Orsay, 1998.
86. Lumsdaine, R. L. (1996). Consistency and asymptotic normality of the quasi-likelihood estimator in IGARCH(1,1) and covariance stationary GARCH(1,1) models. *Econometrica*, **64,** 575–596.
87. Mikosch, T. (2003). Modelling dependence and tails of financial time series. In: *Extreme Values in Finance, Telecommunications and the Environment* (eds B. Finkenstädt and H. Rootzén), pp. 185–286. Chapman & Hall.
88. Mikosch, T., Resnick, S., Rootzén, H. and Stegeman, A. (2002). Is network traffic approximated by stable Lévy motion or fractional Brownian motion? *Annals of Applied Probability*, **12,** 23–68.
89. Mikosch, T. and Stărică, C. (1999). Change of structure in financial time series, long range dependence and the GARCH model. *Preprint.*
90. Mikosch, T. and Stărică, C. (2000). Is it really long memory we see in financial returns? In: *Extremes and Integrated Risk Management, London* (ed. P. Embrechts), pp. 149–168. Risk Books.
91. Mikosch, T. and Stărică, C. (2003). Long-range dependence effects and ARCH modeling. In *Theory and Applications of Long-Range Dependence, Boston* (eds P. Doukhan, G. Oppenheim and M. S. Taqqu), pp. 439–459. Birkhäuser.
92. Mikosch, T. and Stărică, C. (2004). Non-stationarities in financial time series, the long-range dependence and the IGARCH effects. *Review of Economics and Statistics*, **86,** 378–390.
93. Mikosch, T. and Straumann, D. (2002). Whittle estimation in a heavy-tailed GARCH(1,1) model. *Stochastic Process and their Applications*, **100,** 187–222.
94. Nelson, D. B. (1990). Stationarity and persistence in the GARCH$(1, 1)$ model. *Econometric Theory*, **6,** 318–334.
95. Nelson, D. B. (1991). Conditional heteroskedasticity in asset returns: a new approach. *Econometrica*, **59,** 347–370.
96. Nelson, D. B. and Cao, C. Q. (1992). Inequality constraints in the univariate GARCH model. *Journal of Business & Economic Statistics*, **10,** 229–235.
97. Newman, C. M. (1984). Asymptotic independence and limit theorems for positively and negatively dependent random variables. In *Inequalities in Statistics and Probability* (ed. Tong, Y. L.), Hayward, CA: Inst. Math. Stat.
98. Newman, C. M. and Wright, A. L. (1981). An invariance principle for certain dependent sequences. *Annals of Probability*, **9,** 671–675.

99. Oppenheim, G. and Viano. M.-C. (1999). Obtaining long memory by aggregating random coefficients discrete and continuous time simple short memory processes. PUB. IRMA, LILLE, Vol. 49, No. V.
100. Pagan, A. (1996). The econometrics of financial markets. *Journal of Empirical Finance*, **3,** 15–102.
101. Parke, W. R. (1999). What is fractional integration? *The Review of Economics and Statistics*, **81,** 632 638.
102. Peng, L. and Yao, Q. (2003). Least absolute deviations estimation for ARCH and GARCH models. *Biometrika*, **90,** 967–975.
103. Pipiras, V., Taqqu, M. S. and Levy, J.B. (2004). Slow, fast and arbitrary growth conditions for the renewal reward processes when the renewals and the rewards are heavy-tailed. *Bernoulli*, **10,** 378–390.
104. Pourahmadi, M. (1988). Stationarity of the solution of $X_t = A_t X_{t-1} + \varepsilon_t$ and analysis of non-Gaussian dependent variables. *Journal of Time Series Analysis*, **9,** 225–239.
105. Robinson, P. M. (1991). Testing for strong serial correlation and dynamic conditional heteroskedasticity in multiple regression. *Journal of Econometrics*, **47,** 67–84.
106. Robinson, P. M. (2001). The memory of stochastic volatility models. *Journal of Econometrics*, **101,** 195–218.
107. Robinson, P. M. and Zaffaroni, P. (1997). Modelling nonlinearity and long memory in time series. *Fields Institute Communications*, **11,** 161–170.
108. Robinson, P. M. and Zaffaroni, P. (1998). Nonlinear time series with long memory: a model for stochastic volatility. *Journal of Statistical Planning and Inference*, **68,** 359–371.
109. Rydberg, T. (2000). Realistic statistical modelling of financial data. *International Statistical Review*, **68,** 233–258.
110. Sentana, E. (1995). Quadratic ARCH models. *Review of Economic Studies*, **62,** 639–661.
111. Shephard, N. (1996). Statistical aspects of ARCH and stochastic volatility. In *Time Series Models in Econometrics, Finance and Other Fields* (eds D. R. Cox, D. V. Hinkley and O. E. Barndorff-Nielsen), pp. 1–55. Chapman & Hall.
112. Straumann, D. and Mikosch, T. (2003). Quasi-maximum-likelihood estimation in heteroscedastic time series: a stochastic recurrence equations approach. *Annals of Statistics*, forthcoming.
113. Subba Rao, T. and Gabr, M. M. (1984). *An Introduction to Bispectral Analysis and Bilinear Time Series Models.* Lecture Notes in Statistics, volume 24. Springer Verlag, New York.
114. Surgailis, D. and Viano, M.-C. (2002). Long memory properties and covariance structure of the EGARCH model. *ESAIM: Probability and Statistics*, **6,** 311–329.
115. Taqqu, M.S. and Levy, J.B. (1986). Using renewal processes to generate long-range dependence and high variability. In *Dependence in Probability and Statistics* (eds E. Eberlein and M. S. Taqqu), pp. 73–89. Birkhäuser, Boston.
116. Taylor, S. (1986). *Modelling Financial Time Series.* Wiley, New York.
117. Teräsvirta, T. (1996). Two stylized facts and the GARCH(1,1) model. Stockholm School of Economics. SSE/EFI Working Paper Series in Economics and Finance, No. 96.

118. Terdik, G. (1999). *Bilinear Stochastic Models and Related Problems of Nonlinear Time Series Analysis: A Frequency Domain Approach.* Lecture Notes in Statistics, volume 142. Springer Verlag, New York.
119. Teyssière, G. (2000). Double long-memory financial time series. *Preprint.*
120. Tsay, R. S. (2002). *Analysis of Financial Time Series.* Wiley, New York.
121. Weiss, A. A. (1986). Asymptotic theory for ARCH models: estimation and testing. *Econometric Theory,* **2,** 107–131.
122. Zaffaroni, P. (2000). Contemporaneous aggregation of GARCH processes. Preprint.

Intermittency, Long–Memory and Financial Returns

Raj Bhansali[1], Mark P. Holland[2], and Piotr S. Kokoszka[3]

[1] The University of Liverpool, U.K. sa17@liverpool.ac.uk
[2] University of Surrey, U.K. Mark.Holland@eim.surrey.ac.uk
[3] Utah State University, U.S.A. piotr@math.usu.edu

1 Introduction

There is now a vast literature supporting the empirical stylized fact that the volatility of financial returns, as measured either by the absolute values or the squares of the returns series, exhibits long-memory and has correlations which remain positive for long lags and decay slowly to zero, and an associated stylized fact that the marginal distribution of the returns has heavy tails. There is also evidence to suggest that the returns display 'intermittency', often called 'volatility clustering', and the relatively benign periods of market activity are often interrupted by the occurrence of violent market movements. Reference may be made to Greene and Fielitz (1977), Akgirav and Booth (1988), Ding, Granger and Engle (1993), Kokoszka and Taqqu (1996), Podobnik *et al.* (2000), Mittnik and Rachev (2000), Kirman and Teyssière (2002), Cont (2001), among others, and which are just a few among hundreds of publications in this area. At the same time, it is also recognised that the standard ARCH and GARCH models, introduced originally to describe the dependence structure of the volatility, exhibit short range dependence and thus do not capture the long-memory property of the returns, though they do imply volatility clustering and a heavy-tailed marginal distribution. Consequently, several different modifications of the basic ARCH specification have been introduced so as to incorporate the slow decay of correlations, see Baillie *et al.* (1996), for a review. Many of these modifications are, however, rather ad-hoc and a mathematical theory behind some is not fully developed as yet and indeed a few appear in fact not to exhibit long memory, at least asymptotically, see Giraitis, Kokoszka and Leipus (2000).

In this chapter, we consider a new and an entirely different approach to modelling phenomena exhibiting long-memory, intermittency and heavy-tailed marginal distributions, namely by chaotic intermittency maps. This class of maps has witnessed much development in recent years and it marks an important emerging branch of the subject area of *Dynamical Systems Theory*. It should also be stressed that the idea of using deterministic maps as a candidate class of non-linear, non-stochastic models for Economic time series has an established pedigree by now and their use has previously been considered by several different authors; see, for example, Brock and Hommes (1997). For the purpose of the present discussion, three principal properties of these maps are relevant and which qualify them as a plausible class of models for financial returns. First, unlike some of the standard chaotic maps, for example, the Logistic, Tent and Bernoulli shift maps, the intermittency maps display long memory and have correlations decaying at a sub-exponential rate, meaning at a polynomial rate or even slower. Secondly, the invariant density of these maps can display 'Pareto' tails and thus go down to zero at a polynomial rate. Thirdly, as their name implies, these maps display intermittency and generate time series, called the orbit of the map, which display intermittent chaos, meaning the orbit of the map alternates between laminar and chaotic regions. A brief outline of the main theoretical properties of these maps is given in Section 2; for a more detailed discussion, reference may also be made to Bhansali, Holland and Kokoszka (2004). A further motivation for considering the use of intermittency maps for modelling financial data comes from the work of Mondragon (1999), who has successfully applied a sub-class of intermittency maps for modelling the internet traffic, which is a related yet different example of phenomena exhibiting long-memory and heavy-tailed marginal distributions, see, for example, Park and Willinger (2000).

The generic characteristic features of intermittency maps described above are, however, asymptotic and apply, for example, as the lag, u, of the correlation function tends to infinity. Moreover, the bounds on the correlations have been developed for some Hölder continuous function of the map time series, w_t, say, and it is as yet not known precisely for which functions this bound would actually hold in practice. In Section 3, therefore, we present results of a simulation study aimed at investigating the empirical behaviour of the estimated correlations for three different categories of intermittency maps, namely the Polynomial, Logarithmic and Cusp maps, and for a range of different parameter values defining these maps. In addition, we examine the behaviour of the estimated invariant density for these three categories of maps and also that of the partial correlations and the associated 'linear Gaussian' statistics but when these are computed from a simulated realization of the map.

The invariant distribution of the Polynomial and Logarithmic intermittency maps is concentrated on a compact interval, $[0, 1]$. On the other hand, however, the absolute returns on financial time series could in principle take values over the entire non-negative real line, $\mathbb{R}^+$. In Section 4, we accordingly consider transformations, $h(w_t)$, which have distributions defined over $[0, \infty)$ and investigate empirically the correlation structure and related properties of the transformed series by a simulation study.

Currently, there are very few theoretical results on the modelling of absolute returns as they are difficult to treat analytically, see, however, Granger and Ding (1996). The paper concludes in Section 5, where we examine the modelling potential of the intermittency maps for absolute returns and present concluding remarks for direction of future research.

2 Chaotic Intermittency Maps

2.1 General Properties

By a map time series we mean a deterministic sequence, $\{w_t,\, t = 0, 1, \ldots\}$, generated by iteratively applying a one-dimensional map of the following form:

$$w_{t+1} = \zeta(w_t) \ (t = 0, 1, \ldots), \tag{1}$$

where w_0 is real-valued and specifies the *initial condition* for the iterative scheme specified by (1), $\zeta : J \to J$ is a non-linear map (function), J denotes a closed interval of the real line, $\mathbb{R}$.

It is recognized, see Rasband (1990), that such maps can generate *chaotic dynamics* and although the successive values of w_t evolve in a deterministic way from the immediately preceeding one, they could still be *chaotic* and sensitive to the choice of the initial condition, w_0. This is in contrast to a *random system* in which the present observation does not entirely depend on the previous observation.

If, in addition, the map, $\zeta : J \to J$, is ergodic and admits an invariant density, $\chi(w)$, say, then provided the initial condition, w_0, is a random number from this invariant distribution, $\{w_t\}$ defines a strictly stationary stochastic process, see Berliner (1992), Bhansali, Holland and Kokoszka (2004), among others, and hence such maps qualify as alternative non-linear, non-stochastic models for an observed time series, and can be considered for describing phenomena which appear random.

As discussed in Section 1, the correlation functions for some of the popular and widely-studied examples of deterministic maps, for example the Logistic, Tent and Beta transformations, decay exponentially and thus these maps imply that the observed time series has short-memory; see Lawrance and Balakrishna (2001), Hall and Wolff (1995), among others. Lately, there has been much development on devising maps exhibiting slowly-decaying correlations and intermittency. Some of the early work on this class of maps, see Manneville and Pomeau (1980), was in connection with continuous-time Lorenz systems and the main motivation was to construct a mathematical model for intermittent chaos, that is, phenomena exhibiting long periods of laminar behaviour together with short bursts of erratic behaviour. The intermittency maps also differ from the more popular short-memory maps as regards their recurrence properties, and, in particular, in the behaviour of the probability distribution of the time taken by the orbit of the map to return back to a given fixed interval $C \subset J$, having started from within this interval; an important property of intermittency maps is that there may exist sub-intervals, C, for which the distribution function of this recurrence time decays at a sub-exponential rate; see Bhansali, Holland and Kokoszka (2004), for details.

Below, we briefly outline some of the known statistical properties of four categories of intermittency maps, namely the *Polynomial, Logarithmic, Generalised Polynomial-Logarithmic* and *Cusp* maps, focusing particularly on the results concerning their invariant densities and the rate of decay of correlations.

Most theoretical results concerning the rate of decay of correlations are however valid for *all* Hölder continuous functions of the map time series, $\{w_t\}$. Hence, we first give a formal mathematical definition of this class of functions: A function, φ, is said to be Hölder continuous if there exist a $\gamma \in (0,1)$ and a $C < \infty$, independent of x, y, such that

$$|\varphi(x) - \varphi(y)| \leq C|x-y|^{\gamma}, \quad \forall\, x, y \in J,$$

where J denotes the domain of the function, φ.

Moreover, on the assumption that the map, $\zeta : J \to J$, is ergodic and admits an invariant density, $\chi(w)$, it would also be useful to state at the outset a formal definition of the mean and variance functions of $\varphi(w_n)$ for each $n \in \mathbb{N}$, and also of the covariance between $\varphi(w_n)$ and $\psi(w_0)$, where $\varphi, \psi : J \to \mathbb{R}$ are two Hölder continuous functions. The expectation and variance of $\varphi(w_n)$ are given by

$$E\{\varphi(w_n)\} = \int_J \varphi(w)\chi(w)dw, \tag{2}$$

$$V\{\varphi(w_n)\} = E\{\varphi(w_n)^2\} - (E\{\varphi(w_n)\})^2. \tag{3}$$

Similarly, for each $n \in \mathbb{N}$, the covariance, $R_{\varphi,\psi}(n)$, between $\varphi(w_n)$ and $\psi(w_0)$ is defined as follows:

$$\begin{aligned} R_{\varphi,\psi}(n) &= E\{\varphi(w_n)\psi(w_0)\} - E\{\varphi(w_n)\}E\{\psi(w_0)\} \\ &= \int_J \varphi(\zeta^n(w))\psi(w)\chi(w)dw - \int_J \varphi(w)\chi(w)dw \int_J \psi(w)\chi(w)dw. \end{aligned} \tag{4}$$

If $\psi(w) = \varphi(w) = w$, then $R_{\varphi,\psi}(n) = R_w(n)$ is the autocovariance function of the weakly stationary process $\{w_t, t \in \mathbb{N}\}$. Ergodic theory however enables us to consider also more general functions of w_n and w_0; for example, $\varphi(w_n) = w_n^2$ and $\psi(w_0) = w_0^2$ (Lawrance, 2001).

2.2 Polynomial Maps

A celebrated example, with $J = [0,1]$, of this category of maps is:

$$\zeta_\alpha(w) = \begin{cases} w(1 + 2^\alpha w^\alpha) & \text{if } 0 \leq w \leq 1/2, \\ 2w - 1 & \text{if } 1/2 < w \leq 1, \end{cases} \tag{5}$$

where $\alpha > 0$; see Liverani *et al.* (1999), Young (1999).

When $\alpha \in (0,1)$, an explicit expression for the invariant density, $\chi_\alpha(w)$, is as yet unknown, but it is known that, see Thaler (1980), $\chi_\alpha(w)$ is of the following form:

$$\chi_\alpha(w) = \frac{V_\alpha(w)}{w^\alpha}, \tag{6}$$

where $V_\alpha(w)$ depends on the value of α and, for each fixed α, it is a piecewise continuous, uniformly bounded function of w which is also bounded away from zero.

For the map (5), the asymptotic behaviour of the correlation function for large values of the lag, n, has been studied by several authors, and, as discussed in Section 2.1, the results obtained on the rate of decay of correlations are quite general and apply to all Hölder continuous functions of the map time series, $\{w_t\}$. The most general results are due to Sarig (2001) and Gouëzel (2004), who show that

$$\lim_{n\to\infty} n^{(1/\alpha-1)} R_{\varphi,\psi}(n) = U, \tag{7}$$

where φ and ψ are Hölder continuous and U is a constant.
By contrast, the rate of decay of correlations for a standard $ARFIMA(p,d,q)$ process is, Hosking (1981):

$$R_{\varphi,\psi}(n) \sim Bn^{2d-1}, \tag{8}$$

where $d \in (-0.5, 0.5)$ is the fractional-differencing or the memory parameter; the process is said to have long-memory if $d \in (0, 0.5)$ and an intermediate memory if $d \in (-0.5, 0.0)$. On comparing (7) with (8), the following equivalence relationship between d and α may be established:

$$d = 1 - (2\alpha)^{-1}. \tag{9}$$

Hence, if $\alpha \in (0.5, 1.0)$, the result, (7), implies that the correlations for some Hölder continuous functions, φ and ψ, of the map time series, $\{w_t\}$, behave like that of a long-memory time series, and if $\alpha \in (0.0, 0.5)$, they behave like that of a time series with intermediate memory. In Section 3, we examine by simulations the extent to which this result in fact holds for the map time series itself.

2.3 Logarithmic Maps

A second category of intermittency maps has recently been introduced by Holland (2002), and is defined on $J = [0,1]$ by:

$$\zeta_\beta(w) = \begin{cases} w\left[1 + Y(\beta) w(-\log w)^{1+\beta}\right] & \text{if } 0 \le w \le 1/2, \\ 2w - 1 & \text{if } 1/2 < w \le 1, \end{cases} \tag{10}$$

where $Y(\beta) = 2(\log 2)^{-(1+\beta)}$ is chosen to ensure that $\lim_{w\to 1/2^-} \zeta_\beta(w) = 1$.

The invariant density, $\chi_\beta(w)$, for the logarithmic map (10) exists for all $\beta \in (0, 2\log 2 - 1)$, and takes the following form on $(0, 1/2)$:

$$\chi_\beta(w) = \frac{V_\beta(w)}{w \log(1/w)^{\beta+1}} \quad \forall\, w \in J, \tag{11}$$

where $V_\beta(w)$ depends on β and for each fixed β it is a uniformly bounded piecewise continuous function, that is also bounded away from zero. On $[1/2, 1]$, the invariant density is a bounded piecewise continuous function. By contrast, for the polynomial map, (5), the invariant density exists only if $\alpha \in (0, 1)$, and it diverges to infinity as $w \to 0$ at a much faster rate than $\chi_\beta(w)$.

For the logarithmic map, Holland (2002) also examined the rate of decay of the correlation function of $\{w_t : t \in \mathbb{N}\}$ and shows that

$$R_{\varphi,\psi}(n) \le B(\log n)^{-\beta}, \tag{12}$$

where B is a uniform constant. Thus, within the class of all Hölder continuous functions, the correlations for this map could decay at a slower rate than for the polynomial map. It should be emphasized, nevertheless that (12) only provides an upper bound on the rate of decay of the correlations; a lower bound analogous to (7) has recently been obtained by Gouëzel (2004).

2.4 Generalised Polynomial-Logarithmic Maps

Holland (2002) also considers the possibility of combining and generalising the polynomial and logarithmic maps by introducing an extended category of maps of the following general form, with $J = [0, 1]$:

$$\zeta_\beta(w) = \begin{cases} w\,[1 + Y(\beta) w^\alpha \Delta(w)] & \text{if } 0 \le w \le 1/2, \\ 2w - 1 & \text{if } 1/2 < w \le 1, \end{cases} \tag{13}$$

where $\alpha > 0, Y(\beta) = 2^\alpha/\Delta(1/2)$ and $\Delta(1/w)$ is a slowly-varying function at infinity and such that it is twice differentiable for all $w \in [M, \infty]$ and M denotes a bounded constant. Holland (2002) has also investigated the rate of decay of correlations for this extended category of maps and shows that

$$R_{\varphi,\psi}(n) \le B\Delta^*(n) n^{1-1/\alpha}, \tag{14}$$

where $B > 0$ is a uniform constant and $\Delta^*(w)$ is another slowly-varying function at infinity and its precise functional form may be derived from that of $\Delta(w)$. The result (14) shows that by choosing an appropriate functional form for $\Delta(w)$, it is possible to construct intermittency maps belonging to this category whose rate of decay of correlations is bounded above by a function that has a prescribed decay rate.

2.5 Cusp Maps

This category of intermittency maps has been investigated by Balakrishnan *et al.* (1997, 2001), Balakrishnan and Theunissen (2001). A specific example of the maps they consider, with $J = [-1, 1]$, is as follows:

$$\zeta(w) = \begin{cases} 1 - 2\sqrt{(-w)} & \text{if } -1 \leq w \leq 0, \\ 2\sqrt{(w)} - 1 & \text{if } 0 < w \leq 1. \end{cases} \tag{15}$$

The invariant density, $\chi_C(w)$, say, for the cusp map exists and it coincides with that of a Uniform distribution on $J = [-1, 1]$.

Although an explicit expression for the rate of decay of the correlation function for the cusp map is as yet not available, from the results of Young (1999), it may be hypothesized that

$$R_{\varphi,\psi}(n) = O(n^{-1}) \quad (n \to \infty). \tag{16}$$

It should be noted that the rate of decay of correlations specified by (16) is at the boundary of that given in (8) for distinguishing between a long-memory and an intermediate process, and it is not possible to attain this particular rate of decay if only the standard $ARFIMA(p, d, q)$ models are considered.

Indeed, a comparison of the results (16), (14), (12) and (7) with (8) reveals that the rate of decay of correlations for the class of intermittency maps could either equal or be slower than that for a standard $ARFIMA(p, d, q)$ process, and in this sense the class of intermittency maps offers a wider choice of possible models for describing phenomena exhibiting slowly-decaying correlations; in addition, these maps mostly imply a heavy-tailed invariant density, which is also not possible within the standard $ARMA$ framework, though there have been developments on extending this framework to allow for innovations with a stable distribution, see Kokoszka and Taqqu (1995), among others. In Section 3, we examine by simulations the extent to which these theoretical results hold empirically, and especially if the correlation function is estimated by a standard 'non-parametric' method.

3 Simulation Results for the Intermittency Maps

3.1 The Invariant Density

The Polynomial Map

Since a closed-form analytic expression for the invariant density of the polynomial map is as yet not available, we decided to simulate this density for three different values of α, namely $\alpha = 0.25, 0.5$ and 0.75. To this end, a total of $N = 10^7$ iterations, w_t, $0 \leq t \leq N$, of the map (5) were obtained, starting with an initial value, w_0, generated from a uniform distribution on $[0, 1]$. A kernel estimate of the density of the last $N-m$ observations, w_t, $m < t \leq N$, with $m = 10^3$, was then computed by adopting a rectangular kernel of width 0.02. Other kernels of similar lengths were also applied and gave comparable results. To avoid edge effects, the kernel estimate was computed at 350 equispaced points over the interval $[0.01, 0.99]$. We found that with $N = 10^7$, the choice of the initial value, w_0, and the "burn in" period, m, did not materially influence the shape of the estimated densities. A plot of the estimated densities is shown in Figure 1.

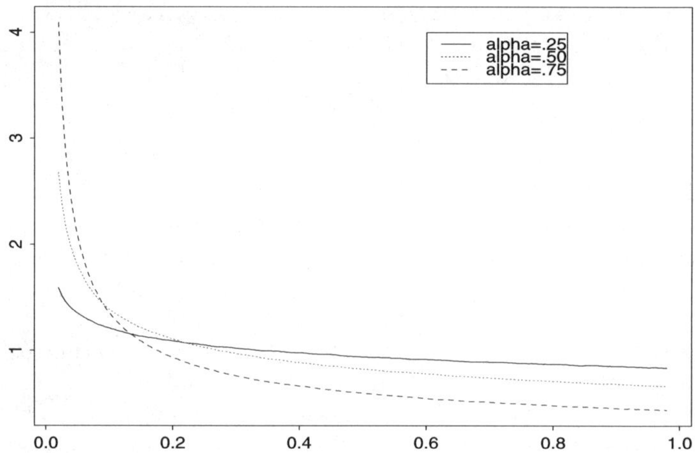

Fig. 1. Estimated invariant densities for the Polynomial intermittency maps

In a neighbourhood of $w = 0$, the estimated invariant densities take values close to $w^{-\alpha}$, and their shape accords with the asymptotic theoretical result, (6). Moreover, in this neighbourhood, the unknown function, $V_\alpha(w)$, may be approximated by a constant, namely $C_\alpha = (1-\alpha)$, whose value depends on that of α. But, for values of w outside this neighbourhood, the estimated density assumes values which are much smaller than $w^{-\alpha}$ and this asymptotic result does not provide a good approximation to the actual shape of the simulated densities. A more detailed analysis, which is not reported here, nevertheless showed that now the simulated densities could in fact be approximated by a function of the form $\chi_\alpha^*(w) = K_\alpha^* w^{-\alpha^*}$ with some $\alpha^* < \alpha$ and $K_\alpha^* > 0$.

The Logarithmic Map

Figure 2 shows the estimated invariant densities for the map (10), with $\beta = 0.1, 0.2, 0.3$. The same method of kernel density estimation as used in Section

3.1 was applied, but we generated $N = 10^7 + 10^4$ iterations of the map (10), starting with an initial value of $w_0 = 0.2$, and discarded the initial $m = 10^4$ points.

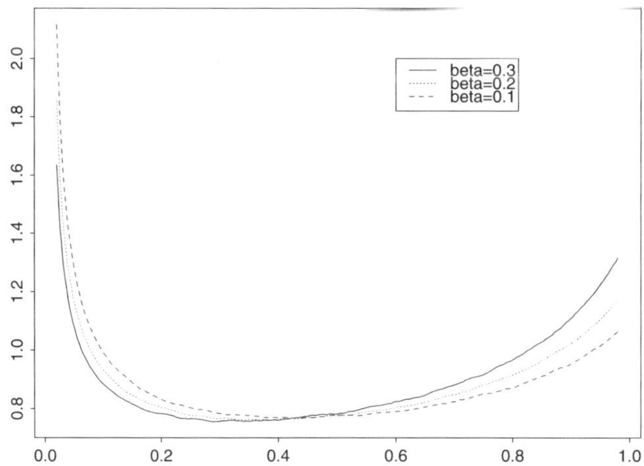

Fig. 2. Estimated invariant densities for the Logarithmic intermittency maps

For $w \in (0.0, 0.5)$, the estimated density is seen to decrease monotonically as w increases; by contrast, for $w \in [0.5, 1.0)$, the estimated density increases monotonically with w. This behaviour does however accord with the asymptotic results described in Section 2.3, since the result (11) holds only for $w \in (0.0, 0.5)$ and for $w \in [0.5, 1.0]$, an explicit analytic form for the invariant density is not as yet available.

3.2 The Correlations, Partial Correlations and the Autoregressive Order Selected

Plan of the Study

For studying the behaviour of the estimated correlations and the associated model selection procedures used in a linear Gaussian analysis of time series, a slightly different approach to that described in Section 3.1 was adopted. A stretch of $N = 10^7$ iterations of the Polynomial, Logarithmic and Cusp maps was generated, but only the last $T = 10^4$ observations, $\{w_t, t = m+1, ..., m+T\}$, with $m = 10^7 - 10^4$, were actually used for computing the estimated correlations and the related statistics, and thus the first m generated values of w_t were discarded so as to avoid the possible 'transient' effects. Two higher values of N, namely $N = 10^8$ and 10^9, but with the value of $T = 10^4$ remaining unchanged, were also considered. A principal effect was to reduce the simulated variances of the estimated correlations, without noticeably changing their means; at the same time, however, there was a considerable, more than five-fold, increase in the computational costs and it was therefore decided to adopt $N = 10^7$ as a suitable compromise between the computational effort and a suitably large 'burn-in' period.

The initial value, w_0, for the Polynomial and Logarithmic maps was generated from a truncated Exponential distribution with mean 0.2, the truncation point being set equal to 1.0, the upper limit of the interval, J, over which these two maps are defined. The initial value for the Cusp map was, however, generated from the known invariant density for this map, namely that for the Uniform distribution over $[-1, 1]$. Several different values of the parameter, α, for the Polynomial map were considered, namely $\alpha = 0.05, 0.1, 0.2, 0.3, 0.45, 0.5, 0.55, 0.6, 0.7, 0.8, 0.85, 0.9, 0.95$; similarly, we set $\beta = 0.05, 0.1, 0.15, 0.2, 0.25, 0.3, 0.35$ for the Logarithmic map. For each generated realization of the corresponding maps, the standard 'non-parametric' estimates of correlations, $r^{(T)}(u)$ $(u = 0, 1, \ldots, 100)$, were computed as follows:

$$r^{(T)}(u) = R_T(u)/R_T(0), \tag{17}$$

where

$$R_T(u) = T^{-1} \sum_{t=1}^{T-u} (w_t - \bar{w})(w_{t+u} - \bar{w}) \tag{18}$$

denotes a 'positive-definite' estimate of the covariance function of $\{w_t\}$ and

$$\bar{w} = T^{-1} \sum_{t=1}^{T} w_t$$

denotes the estimated mean of the generated sequence.

The associated values of the estimated partial correlations, $\hat{\pi}(k)$ $(k = 0, 1, \ldots, 50)$, were then computed by applying the NAG library routine $G13ACF$; in this routine, the partial correlations are computed by implementing the Durbin-Levinson algorithm.

For examining the behaviour of the standard linear Gaussian model selection procedures when these are applied to the generated sequence, the order of an appropriate linear autoregressive approximation was also determined by applying the BIC criterion:

$$BIC(k) = T \log \hat{\sigma}^2(k) + (\log T)k \qquad (k = 0, 1, \ldots, 50), \tag{19}$$

where $\hat{\sigma}^2(k) = \hat{\sigma}^2(k-1)\{1 - \hat{\pi}^2(k)\}$, was computed recursively by setting $\hat{\sigma}^2(0) = R_T(0)$; see Bhansali (1993) for a further discussion of this and the related autoregressive order selection procedures.

As discussed in Section 2, the intermittency maps that we consider are ergodic and hence the estimated correlations, $r^{(T)}(u)$, converge with probability one to the corresponding autocorrelation function of these maps, where the convergence is pointwise, for each fixed u, and hence so do the partial correlations and related statistics used for model selection and described above.

The computations described above were replicated 1000 times, but with a different initial condition each time. The simulated sampling means, variances and standard deviations, SD, of the various estimates over these replications were determined. In addition, the simulated sampling distributions of $r^{(T)}(u), u = 1, 2, 3$ and also that of $\bar{w}$ and the autoregressive order selected by the BIC criterion were obtained. For convenience, however, the discussion of the results given below is based largely on an analysis of the simulated sampling means and variances of the various estimates, though we do consider the simulated sampling distributions of $r^{(T)}(1)$ and $\bar{w}$.

The Polynomial Map

A plot of the simulated means of the $r^{(T)}(u)$ $(u = 1, \ldots, 100)$ for $\alpha = 0.95, 0.8, 0.45, 0.3$ is shown in Figure 3 together with two curves representing mean+SD and mean-SD. A reason why we use the simulated standard deviations is that an explicit expression for the correlation function of the Polynomial map is as yet unknown and the standard asymptotic results on the sampling variance of the $r^{(T)}(u)$ involve the unknown correlation function; an additional difficulty, which we further discuss below, is that for $\alpha \in (0.5, 1.0)$ these asymptotic results do not apply. Furthermore, some preliminary calculations, not reported here, suggested that the mean-SD curve provides a rough approximation to the simulated lower 90% confidence limits for these estimates when the limits are calculated from their simulated sampling distributions.

Although, in the interest of space, the results for other values of α are not shown, they are nevertheless consistent with those presented in Figure 3.

Consider first the behaviour of the mean-correlations for $\alpha = 0.95$, equivalent $d = 0.474$. The mean-correlations are positive for all 100 lags and show persistence in the sense that their numerical magnitudes remain noticeably large even at large values of the lag, u, and decrease very slowly as the value of u increases. Although the simulated standard deviations, SD, of the $r^{(T)}(u)$ increase steadily as u increases, reflecting the greater variability of the estimated correlations at high lags, the curve corresponding to mean - SD lies above the 0 value even when u is as high as 100. Thus, the empirical behaviour of the mean-correlations is incompatible with a rapid exponential decay of correlations observed with a standard 'short-memory' model, and accords with the theoretical 'long memory' property (7) of the Polynomial maps discussed in Section 2.2. The simulation results thus indicate that this property could in fact hold when $\psi(w) = \varphi(w) = w$.

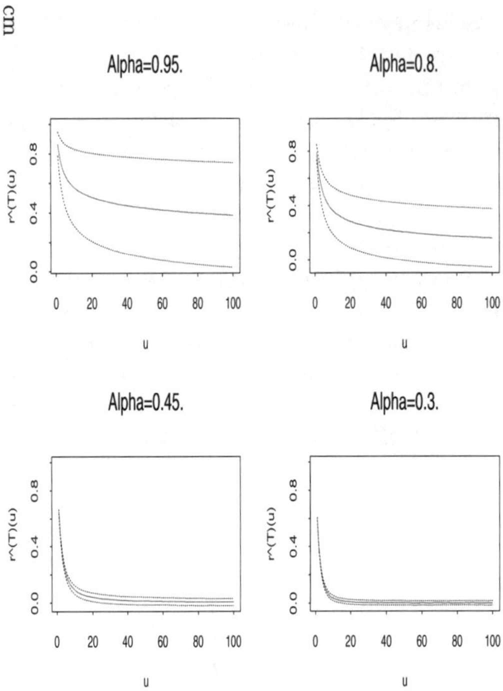

Fig. 3. Plots of means of $r^{(T)}(u)$ versus u, with solid curve representing $r^{(T)}(u)$ and the dotted curves $r^{(T)}(u)\pm$ SD

The behaviour of the mean-correlations for $\alpha = 0.8$ and 0.6 is similar to that with $\alpha = 0.95$. The numerical magnitudes of the mean-correlations, and also of the simulated standard deviations of these estimates decrease monotonically, however, as α decreases and the rate at which they approach 0 increases, and the mean-corelations with $\alpha = 0.3$, and to a slightly lesser extent with $\alpha = 0.45$ behave in a manner that is akin to that of a 'short-memory' time series. The simulation results for these two values of α are nevertheless in agreement with the asymptotic result (7), which shows that for $\alpha < 0.5$, the map exhibits 'intermediate memory' and the theoretical rate of decay of correlations with $\alpha = 0.3$, in particular, is quite rapid.

The asymptotic result (7) suggests that a plot of $log[r^{(T)}(u)]$ against $log(u)$ may be useful for constructing an initial estimate of the parameter, α, especially if the bound given by this result were to hold actually for $\psi(w) = \varphi(w) = w$. Figure 4 shows a plot of the logarithm of the mean of $r^{(T)}(u)$ against $\log(u)$ for $\alpha = 0.95$. The log-log plot of the mean correlations appears linear. The line of best fit was determined by an Ordinary Least Squares regression. Note, however, that the result (7) only applies asymptotically as the lag $u \to \infty$. Hence, the regression line was estimated only for the values of $u > n$, where we allowed n to vary by setting $n = 3, 4, 5, 6, 8, 26$. We found that the estimated slope of this line varies considerably as the value of n changes and only the plot for $n = 6$ is shown in Figure 4. The results however do not support the use of such a graphical procedure for estimating α from an observed realization of the map. On the other hand, however, this finding is not surprising since (7) provides only an upper bound on the rate of decrease of correlations and it shows that this rate is not any slower as $n \to \infty$. Similar plots for all other values of α considered in the paper were also obtained but the results are omitted in the interest of space.

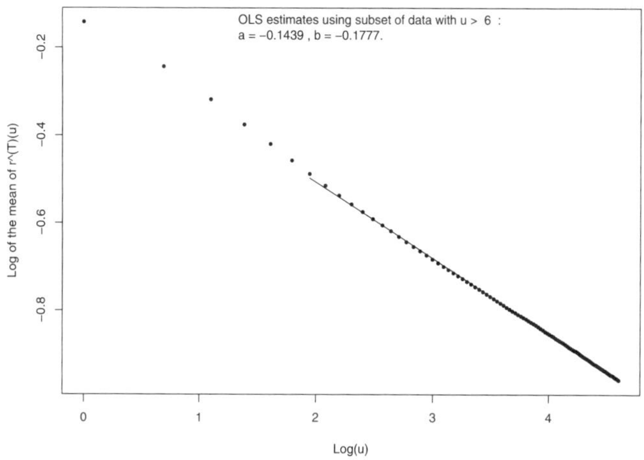

Fig. 4. Plot of the logarithm of the mean of $r^{(T)}(u)$ against $\log(u)$ for $\alpha = 0.95$

The corresponding plots of the simulated means of the estimated partial correlations are shown in Figure 11 together with two curves representing mean+SD and mean-SD. To save space, the results are only presented for $\alpha = 0.95, 0.8, 0.45$ and 0.3. It is again convenient to first consider the detailed results for $\alpha = 0.95$. The mean partial correlations, like the mean correlations, are positive at all lags, u, and decrease steadily to zero as u increases. Moreover, the first partial correlation substantially dominates all other partial correlations, in the sense that its numerical magnitude is very much bigger than that of all others. On the other hand, however, the curve corresponding to mean - SD lies above the value of 0 for all $u < 10$ and in this sense the corresponding mean partial correlations are significantly different from 0. A similar pattern of results is also observed for the other three values of α considered in Figure 11, but now the values of the partial correlations at each lag decrease monotonically as the value α decreases and for $\alpha = 0.3$, in particular, the mean partial correlations for all $u \geq 2$ are close to the 0 value, though the curve corresponding mean - SD falls below 0 only for $u > 4$.

An examination of the plots of the mean correlation and mean partial correlations for the Polynomial maps in accordance with the standard graphical approach to model identification suggests that a low-order autoregressive model could well be fitted to a typical realization of this map for approximating its second-order behaviour, especially for values of α not too close to 1.0. We investigated this hypothesis further by determining the order of a linear autoregressive model selected when the BIC-criterion, (19), is applied, albeit uncritically, to simulated realizations of a Polynomial map. To save space, the detailed results are omitted and Figure 5 only shows the average autoregressive order selected by this criterion in 1000 replications of the Polynomial maps with different values of α.

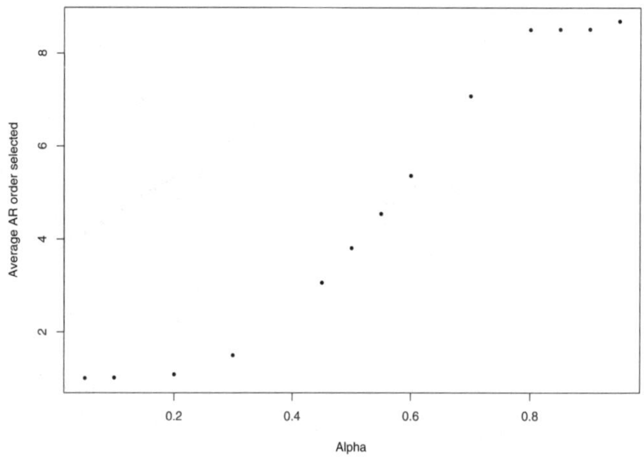

Fig. 5. Plot of average autoregressive order selected for different values of α

For $\alpha \leq 0.3$, the average order selected is close to 1.0 and this model does describe the estimated correlations, as judged by a comparison of the mean correlations with the corresponding theoretical value implied by an autoregressive process of order 1, but with the parameter set equal to the value of the first mean correlation coefficient. A further check was also carried out by separately fitting an AR(1) model at each simulation and applying the *Box-Ljung* diagnostic test, see Ljung and Box (1978), for checking the adequacy of the fitted model; this test also did not reject the $AR(1)$ model in all 1000 simulations. By contrast, for $\alpha > 0.5$, higher order autoregressions are selected, and the average order selected increases noticeably as α increases from 0.55 to 0.8, and thereafter it increases rather slowly and reaches a plateau. For

$\alpha \geq 0.8$, the equivalent value of the long-memory parameter, d, is quite close to its upper limit and, by (7), the asymptotic rate of decay of correlations is rather slow and in this sense the simulation results do accord with this theoretical result.

An additional feature of the results for $\alpha \geq 0.8$ was that the actual frequency distribution of the autoregressive order selected is not adequately described by its mean and increasing α also increases the frequency of selecting the more extreme orders. This point is illustrated in Table 1, where the frequency of selecting an $AR(1)$ model and that of selecting a very long model, that is, an $AR(k)$ model with $k \geq 40$ is shown as α varies over this interval.

Table 1. Frequency of selecting extreme autoregressive orders in 1000 realizations of the Polynomial Maps with $\alpha \in [0.8, 1.0)$

α	$AR(1)$	$\geq AR(40)$
0.8	55	9
0.85	117	6
0.9	201	11
0.95	258	17

A further feature of the results for $\alpha \geq 0.8$ was that there were simulations for which the orbit of the map remained close to 0 in the laminar region and changed very slowly and hence the estimated correlations were close to 1 for many lags, and even for lags as large $u = 50$. An example of this phenomenon is shown in Figure 6, where the orbit of the map for a particular simulation with $\alpha = 0.95$ and $w_{m+1} = 0.96403486 \times 10^{-6}$ is shown. The orbit of the map for the entire stretch of $T = 10,000$ observations remains in the laminar region, close to the neutral fixed point of 0, and increases monotonically as t increases. As discussed by Bhansali *et al.* (2003), this behaviour is congruent with the characteristic behaviour of all intermittency maps. On the other hand, however, for such orbits, the estimated correlations approximately equal 1, at lags as large as $u = 100$ and lie along a straight line, while the value of the partial correlations for all $u \geq 2$ is very close to 0 and that of the first partial correlation is close to unity; in addition, the mean, $\bar{w}$, takes a value very close to 0.

To understand these results, observe that, see Aaronson (1997), by definition, (5), of the Polynomial map,

$$w_t \approx \frac{1}{2}\left((2\delta)^{-\alpha} - \alpha t\right)^{-\frac{1}{\alpha}}, w_0 = \delta \quad \text{and} \quad w_t \ll 1, \tag{20}$$

where $w_0 = \delta$ and this approximation holds for all w_t such that $w_t \ll 1$. Hence, if, for some $t \geq 0$, $w_t = O(10^{-6})$, then the subsequent orbit of the map will

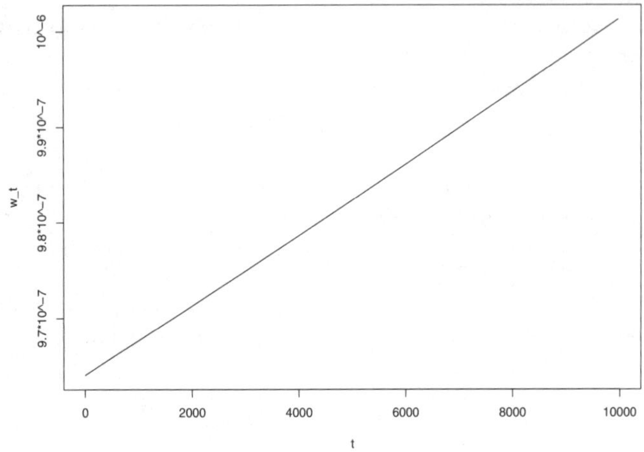

Fig. 6. An orbit of the Polynomial map with $\alpha = 0.95$ and $w_{m+1} = 0.96403486 \times 10^{-6}$

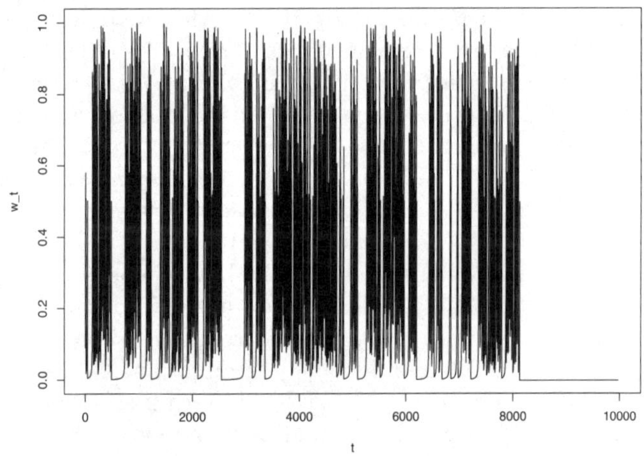

Fig. 7. A different orbit of the Polynomial map with $\alpha = 0.95$ and $w_{m+1} = 0.54495027$

take values of this magnitude for the next $N = 10^{-5}$ iterates before its magnitude increases to 10^{-5}. To compare the results for this particular simulation with the corresponding results for a simulation in which the orbit of the map visits the chaotic and laminar regions more frequently, in Figure 7 we show the orbit for another simulation with $\alpha = 0.95$ and $w_{m+1} = 0.54495027$. Although the orbit of the map remains in the laminar region and approximately equals a constant for the last 2000 observations, the graph of the estimated correlations and partial correlations for this orbit was similar in appearance to that of the mean correlations and mean partial correlations in Figures 3 and 11 and it is not shown here. We also plotted the orbits of the maps with several other values of α but with the same starting values as used in the Figure 7. To save space, the detailed results are not shown here. Suffice to say, however, that the behaviour of the orbit is consistent with the theoretical results described in Section 2 and as α increases, the time spent by the orbit in the laminar region increases also, but for small values of α the orbit appears rather chaotic and it rarely displays the laminar behaviour shown in Figures 6 and 7.

To further provide a perspective on the unusual behaviour of the estimated correlations for the Polynomial map when α takes a value close to 1.0, Figure 8 shows the estimated kernel density of the simulated sampling distribution of $r^{(T)}(1)$.

In 255 simulations, the simulated values of $r^{(T)}(1)$ exceeded 0.999 and thus approximately equaled 1; these values were omitted when constructing the kernel density estimate. One reason for excluding this discrete mass was that its inclusion gave rise to a bimodal density. It is also pertinent to note that since $\{w_t\}$ is not a linear process, the classical results on the asymptotic sampling distribution of the estimated correlations, see Anderson (1971), do not apply to the $r^{(T)}(u)$ as defined by (17), and nor do the results of Brillinger (1969), who does not require that the observed process is linear but nevertheless assumes that its joint cumulants of all orders are summable. The simulations therefore provide an insight into the empirical behaviour of the sampling distributions of these estimates and provide a warning against an unjustified application of the standard asymptotic results to the present context. For a visual comparison of the shape of the truncated density estimate with that of a Normal distribution, see Figure 8, we have also superimposed the density curve of a Normal distribution with the same mean and variance as that of the truncated distribution. The kernel estimate even after truncation is seen to be skewed with a long right hand tail, indicating that the sampling distribution has many more estimated correlation values greater than 0.9 than implied by the fitted Normal distribution. Similar results were also obtained for $\alpha = 0.8, 0.9$. As may be gleaned from Figure 9, for $\alpha = 0.3$, however, for smaller values of α, the fitted Normal distribution provides a good approximation for the estimated density.

A similar set of somewhat unusual results for α close to 1.0, arising mainly because the orbit of the corresponding map may remain in the laminar region close to 0, the neutral fixed point, for a very long time, was also evident in the simulated sampling distributions of the mean, $\bar{w}$, but now the discrete mass occurred for values close to 0. Thus, for $\alpha = 0.95$, there were 256 simulations in which $\bar{w} < 0.0001$ and these were omitted before constructing a kernel

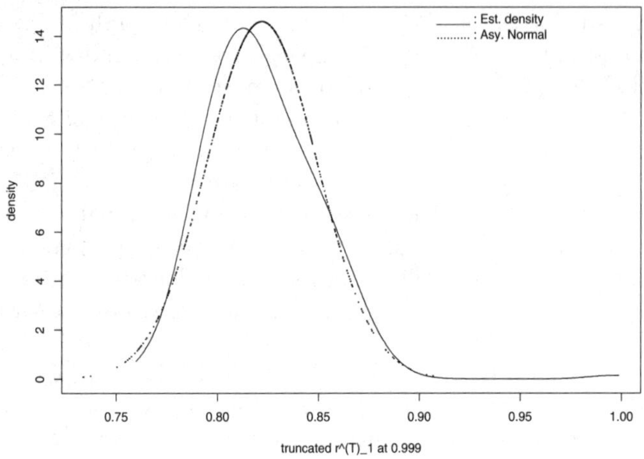

Fig. 8. Estimated kernel density of $r^{(T)}(1)$ with the asymptotic Normal curve superimposed for $\alpha = 0.95$

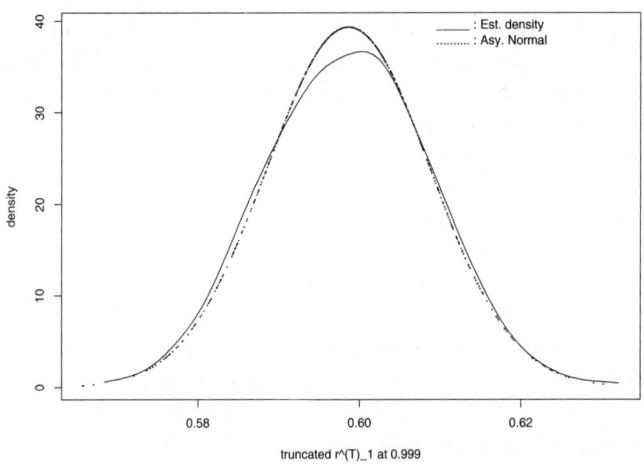

Fig. 9. Estimated kernel density of $r^{(T)}(1)$ with the asymptotic Normal curve superimposed for $\alpha = 0.3$

estimate of its density, shown in Figure 10. The kernel estimate, even after excluding the discrete mass close to 0, is skewed to the left and not well approximated by the fitted Normal distribution. As α decreases, however, the shape of the estimated Kernel density does get closer to that of a Normal distribution, and for $\alpha < 0.5$, the fitted Normal distribution does provide a good approximation to the estimated density. It is also relevant to note that, see Bhansali *et al.* (2003) and Young (1999), the known central limit theorem for $\bar{w}$ is only valid for $\alpha \in (0, 1/2)$ and the simulation results thus provide a caution against presuming that a result of this type would also hold for $\alpha \in [0.5, 1.0)$ and suggest that a more delicate theoretical analysis is needed for the latter case.

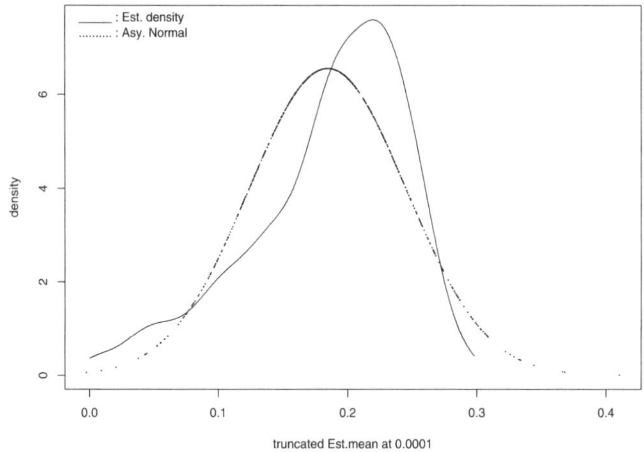

Fig. 10. Estimated kernel density of $\bar{\omega}$ with the asymptotic Normal curve superimposed for $\alpha = 0.95$

The Logarithmic Map

Although results similar to those described above in Section 3.2 were also obtained for the Logarithmic map with several different values of β, to save space, only the results for $\beta = 0.05$ are discussed in detail.

Figure 12 shows a plot of the simulated means of the $r^{(T)}(u)$ $(u = 1, \ldots, 100)$ for $\beta = 0.05$ together with two curves representing mean+SD and mean-SD. The graph of the mean correlations is similar in shape to that for the Polynomial map with $\alpha = 0.8$ but now the mean correlations decrease to 0 somewhat more slowly and for all $u > 16$ the curve of mean - SD takes negative values. In the interest of space, the corresponding results for the mean partial correlations are not shown, but their behaviour is also somewhat similar to that for the Polynomial maps. Increasing the value of β reduces the numerical magnitudes of the correlations, but it does not alter the overall appearance of the mean correlation curves. Indeed, the behaviour of

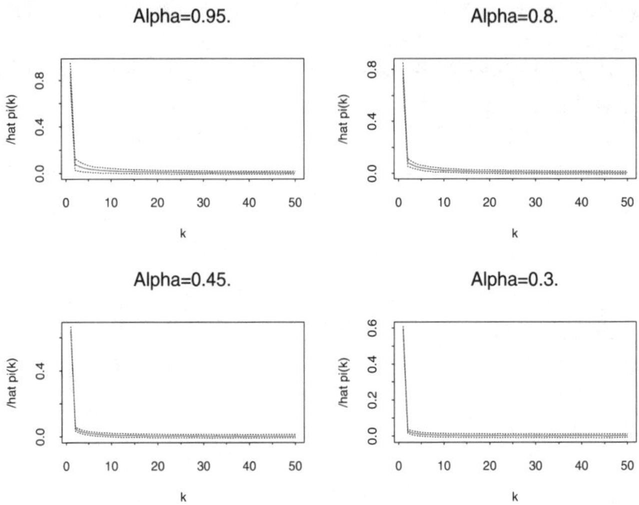

Fig. 11. Plots of means of partial correlation $\hat{\pi}(k)$ against k, with solid curve representing $\hat{\pi}(k)$ and dotted curve $\hat{\pi}(k)\pm$SD

the mean-correlations for the Logarithmic maps was broadly consistent with the asymptotic result (14), which shows that unlike the Polynomial maps, the asymptotic rate of decrease of the correlations for the Logarithmic maps is directly related to that of the map parameter, β, implying that this rate decreases as β decreases; moreover, this rate of decrease is logarithmic rather than polynomial and hence much slower.

The result (14) also suggests that a plot of the $\log[r^{(T)}(u)]$ against $\log[\log(u)]$ may be useful for estimating the value of β and if the bound given by this result were to hold for $\psi(w) = \varphi(w) = w$, then this plot would be linear and an estimate of β would be provided by the slope of the resulting line. This suggestion was investigated but with somewhat mixed results. Thus, the resulting plot for $\beta = 0.05$ is shown in Figure 13 and which is almost quadratic in appearance and far from being linear. Although as β increases, this plot does get closer to being linear, the estimated slope provides a somewhat poor approximation to the actual value of β and which moreover varies considerably depending on how many initial values of the mean correlations are ignored before fitting a straight line.

A useful way of quantifying the slower rate of decay of the correlations for the Logarithmic maps relative to the Polynomial maps is by an inspection of Figure 14, which shows the average autoregressive order selected by the *BIC* criterion for the former map. The average autoregressive order decreases, almost linearly, as the value of β increases, but very slowly and even for values of β close to the upper limit of the admissible range, the selected values of the autoregressive order are generally high. On the other hand, however, as with

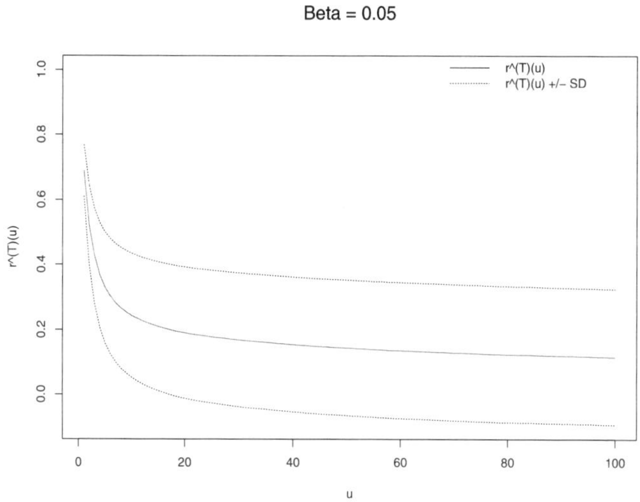

Fig. 12. The simulated means of $r^{(T)}(u)$ for the Logarithmic map; the solid curve shows the means and the dotted curves show mean $\pm$SD

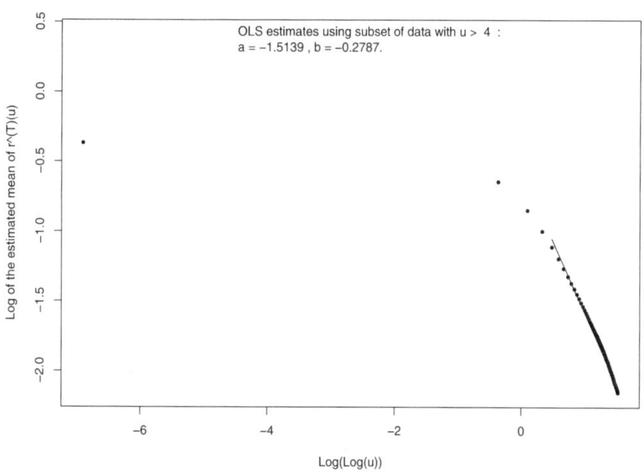

Fig. 13. Plot of the logarithm of the mean of $r^{(T)}(u)$ against $\log(\log(u))$ for $\beta = 0.05$

the Polynomial maps, the average autoregressive order selected does not adequately describe the actual frequency distributions of the autoregressive order selected with small values of β and, in particular, with $\beta = 0.05$ there were several simulations, 42 out of 1000, for which the orbit of the map remained close to the value of 0 in the laminar region for all $T = 10,000$ values and for which the selected autoregressive order equaled 1, and the estimated correlations even at very high lags were close to unity. Although not unexpected, this behavior also gave rise to a discrete mass in the simulated sampling distributions of the estimated correlations. Moreover, the latter, which to save space is not shown here, even after removing this discrete mass, is skewed to the right and its appearance is similar to that of the corresponding sampling distribution shown in Figure 8 for the Polynomial map.

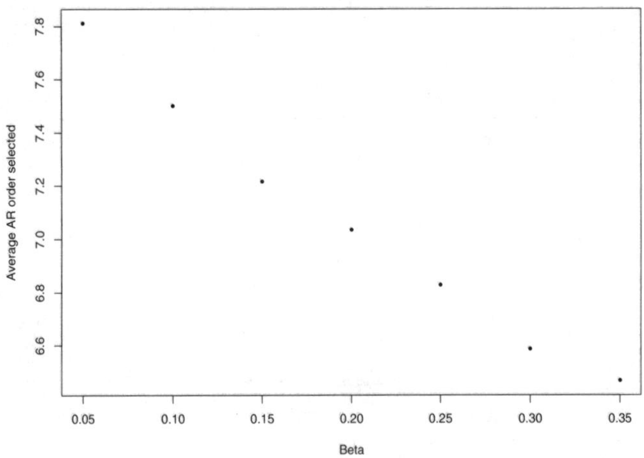

Fig. 14. Plot of the average autoregressive order selected for the Logarithmic map with different values of β

The Cusp Map

A typical orbit of the Cusp map, with $w_1 = 0.79971443$, is shown in Figure 15. The orbit displays the characteristic intermittent behaviour of all intermittency maps: for values close to the two neutral fixed points, $\tilde{w} = \pm 1$, the orbit is in the laminar region and displays monotonic behaviour and away from these points it displays chaotic behaviour. However, the length of time the orbit stays in the laminar region is now much smaller than for the Polynomial maps with α close to 1 or the Logarithmic maps with β close to 0.

Figure 16 shows the means of the estimated correlations for the Cusp map together with the associated mean $\pm$SD curves. The mean correlations typically take quite large values for small values of the lag, u, but decrease more rapidly to zero than for either of the other two categories of maps considered earlier. Although a plot of the mean partial correlations is not shown here,

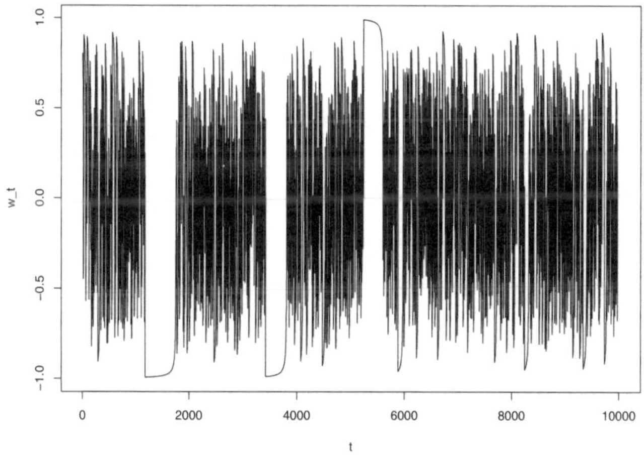

Fig. 15. Orbit of the Cusp map with $\omega_1 = 0.79971443$

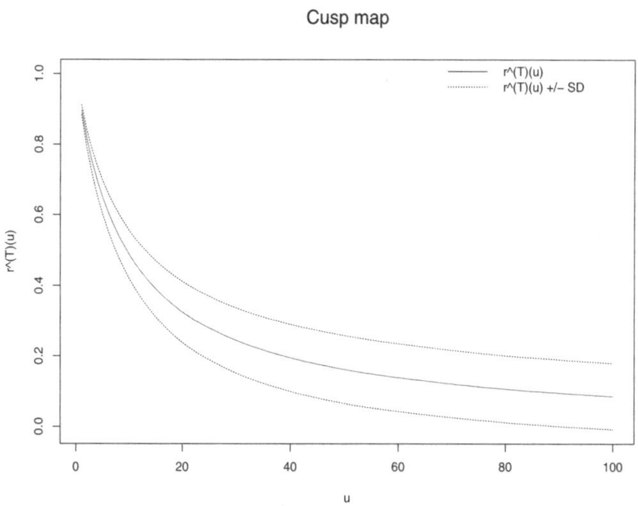

Fig. 16. The simulated means of $r^{(T)}(u)$ for the Cusp map; the solid curve shows the means and the dotted curves show mean $\pm$SD

their behaviour is also analogous to that of the correlations and the simulation results broadly accord with the asymptotic results described in Section 2.5.

The simulated sampling distributions of the first three estimated correlations for the Cusp map were also examined and in contrast to the corresponding distributions for the Polynomial and Logarithmic maps, these were closer to Normality. While pleasant, this result is not surprising and in agreement with the known asymptotic results concerning this map and the estimated correlations. Thus, as discussed in Section 2.5, the rate of decay of correlations for this category of maps is on the boundary of the rate distinguishing a long-memory and an intermediate memory process and, as discussed above in Section 3.2, even for the Polynomial maps with $\alpha < 0.5$, the simulated sampling distributions of the estimated correlations are close to Normality.

4 Transformed Polynomial Map Sequences

The map time series, $\{w_t\}$, produced by the Polynomial or Logarithmic Maps is restricted to take values over the unit interval, $[0, 1]$. On the other hand, many financial series such as asset prices and absolute returns could in principle take values over the entire positive real line, $[0, \infty)$ and admit a heavy-tailed marginal distribution. Indeed, there is much empirical evidence, see Gopikrishnan *et al.* (1998), among others, to suggest that the probability distribution for many financial returns may have 'Pareto' tails with tail index, $\tau \approx 3.1$, meaning that

$$P(r_t > g) \sim g^{-\tau} \quad (g \to \infty), \tag{21}$$

with $\tau \approx 3.1$ and thus the probability that the returns take a value equal to or greater than a threshold value, g, say, is well-approximated, as $g \to \infty$, by an inverse cubic law. For modelling such series using the Polynomial or Logarithmic maps, we consider an instantaneous family of transformations, $y_t = h(w_t)$, of the following form:

$$h(w) = \delta w^p + \gamma(1 - w)^{-q}, \tag{22}$$

where $-\infty < p < \infty$, $q > 0$, $\delta \geq 0$, $\gamma \geq 0$ are some fixed constants such that y_t is distributed on $[0, \infty)$.

To study the effect of such a transformation on the invariant distribution of y_t, consider the Polynomial maps and suppose first that $\delta = 0$. As discussed in Section 3.1, Figure 1 suggests that for $w \sim 0$ the invariant density, $\chi_\alpha(w) \sim C_\alpha w^{-\alpha}$ where C_α is a constant that does not depend on w. It then readily follows from (6) that

$$P(\gamma(1 - w_t)^{-q} > x) = P(1 - w_t > (\gamma/x)^{1/q}) \approx C_\alpha(\gamma/x)^{1/q}.$$

Thus, the tail index of $y_t = \gamma(1 - w_t)^{-q}$ is $\tau = 1/q$. Similarly, if $\gamma = 0$, the tail index of $y_t = \delta w^p$ is $\tau = p(1 - \alpha)$. Moreover, the relative sizes of γ and δ would also influence the tail index of y_t,

In the simulations reported here, we consider the following three different sets of transformations, two of which are special cases of the transformation (22), while the third is closely related to this transformation:

In the first set of transformations, we set $p = 1$, $\gamma = \delta = 1.0$, and only vary q, and thus the transformed series of the following form are considered:

$$\text{Transformation A}: \quad y_t = w_t + (1 - w_t)^{-q}, \tag{23}$$

with $q = 0.1, 1/3$ and 4.3.

The computations described in Section 3.2 for the Polynomial and Logarithmic maps were repeated for the series obtained by applying Transformation A, but for the Polynomial map only three different values of α, namely $\alpha = 0.3, 0.6, 0.95$ are considered and for the Logarithmic map only one value, $\beta = 0.05$, of β is considered. Detailed results are too extensive to be presented here. Suffice to say, however, that for $q = 0.1$, the shape of the mean correlation curve and that of the associated mean $\pm$SD curves is quite similar in appearance to that for the untransformed series, and so is the behaviour of the autoregressive order selected by the BIC criterion. Increasing q to 1/3 reduces the numerical magnitudes of the correlations and thus pulls these curves downwards without essentially altering their respective shapes; moreover, it also leads to selecting lower order autoregressions. An example of the former effect is shown in Figure 17 where the mean correlations for the Polynomial map with $\alpha = 0.95, q = 1/3$ are shown.

By contrast, taking a very large value of q, namely $q = 4.3$ gave rise to mean correlations whose appearance is qualitatively different from that of the untransformed maps and resembled that of an MA(1) process with a very small parameter value. Moreover, in this case, the variability of the estimates also increases considerably as α increases and indeed it seemed plausible that the transformed maps may behave quite differently than the untransformed map and in particular the sampling properties of the estimated correlations could have changed substantially as a consequence of the very heavy tails induced by a large value of q. Analogous set of results were also obtained for the Logarithmic map; the details are omitted.

A second set of transformations we consider is of the following form:

$$\text{Transformation B}: \quad y_t = w_t + w_t^{-q}, \tag{24}$$

and w_t is generated by the Polynomial map. Although several different values of α and q were considered, we only present the results for $\alpha = 0.3, q = 1/3$. Figure 18 shows a typical orbit of the transformed map, and Figure 19 shows the mean correlations in 1000 simulations. The orbit does display the presence of many large values interrupted by medium to small values. In addition, as compared with the untransformed maps, the mean correlations for the transformed map are much larger and decrease slightly more slowly. The variability of the correlations is however also much greater than that for the untransformed map, as is evident by the fact that the mean + SD and mean - SD curves take rather large values for values of u close to 20. Indeed, the shape of the mean correlation curve is consistent with that of an $AR(1)$ process with a parameter value close to 0.9, and which was also the model

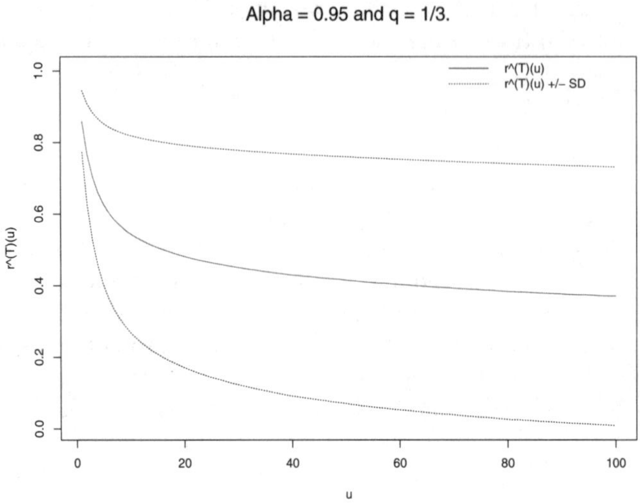

Fig. 17. The simulated means of $r^{(T)}(u)$ for Transformation A; the solid curve shows the means and the dotted curves show mean $\pm$SD

selected by the BIC criterion in 99.3 per cent of all simulations. Moreover, an application of the Box-Ljung diagnostic test did not reject the fitted $AR(1)$ model in all these simulations. It should be noted that although an $AR(1)$ model is also selected quite frequently under Transformation A and for the untransformed series, the value of the parameter value is now much greater and almost close to the unit root.

A third set of transformations we consider is of the following form:

$$\text{Transformation C}: \quad y_t = \delta w_t + \gamma(1 - w_t)^{-1/3}, \tag{25}$$

and w_t is generated by the Polynomial map with $\alpha = 0.625$. This transformation is thus related to Transformation A, but now the potential effects of varying the constants (γ, δ) are studied by considering two different sets of values of these constants, namely

i): $(\gamma = 0.01, \delta = 0.001)$;
ii):$(\gamma = 0.004, \delta = 0.007)$.

A stretch of $N = 10^7$ values of w_t were generated and only the last $T = 3000$, were retained. The corresponding transformed sequences, y_t, according to the choice of constants given in i) and ii) above were obtained from this stretch. Their time series plots, which are not shown here, indicated that the transformed series still display intermittent behaviour and for values of y_t close to 0 the orbit remain in the laminar region and increases slowly but monotonically and away from this region it appeared chaotic. The estimated correlations basically remain unchanged as the values of the constants, γ and

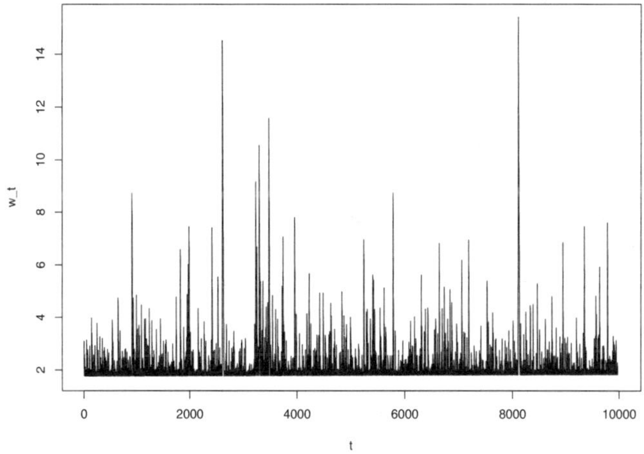

Fig. 18. A realization of the series transformed according to transformation B

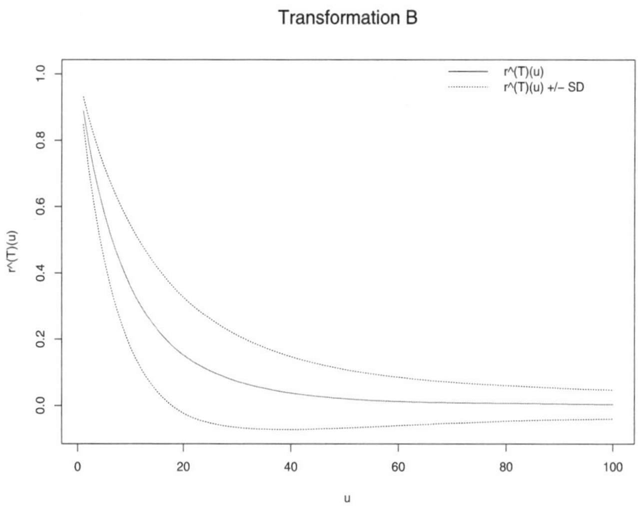

Fig. 19. The simulated means of $r^{(T)}(u)$ for Transformation B; the solid curve shows the means and the dotted curves show mean $\pm$SD

δ are varied and indeed the appearance of the graph is also consistent with the correlation function of a low-order autoregressive process. A kernel estimate of the density function for the transformed sequence was also obtained. This confirmed that the transformation does produce a sequence with a heavy-tailed marginal distribution; the details are omitted to save space.

5 Modelling Absolute Returns

For examining how effective the intermittency maps considered in this paper might be for modelling the absolute returns of financial data, we considered $T = 3,000$ consecutive daily returns on the *SP500* and *NASDAQ* indices for the period ending in November 2002. We also considered the same length of consecutive daily returns on the Pound-Dollar and Deutschmark-Dollar exchange rates for a period ending in mid 1990, well before the introduction of the Euro, but well after the introduction of the European Monetary System in 1979. The returns for the two share index series thus include a period of marked change in volatility which occurred in late 1990, whereas the returns on FX rates correspond to a relatively uneventful period.

The sample paths, the dependence structure, that is, the estimated correlation functions and the GPH estimate of the long-memory parameter of these four series were computed and compared with those of the intermittency maps discussed earlier in Sections 3 and 4.

The results are omitted to save space. Suffice to say, however, that there are three features of intermittency maps which could make them unsuitable for modelling absolute returns. First, their correlations are generally greater than those usually observed with absolute returns. Secondly, for certain values of parameters, the orbit of the Polynomial and Logarithmic maps could remain in the laminar region for a very long time and increase monotonically but still take values very close to 0, which are features not usually present in observed absolute returns. Thirdly, although the transformed maps considered in this paper could produce time series with a heavy-tailed marginal distribution, the behaviour of the resulting densities does not accord with that of absolute returns. As the intermittency maps are not specially designed for modelling financial data, this finding is perhaps altogether not surprising.

On the other hand, however, the simulation results for the three intermittency maps presented in this paper show that the map time series, w_t, itself could possess long-memory and have slowly-decaying correlations. Moreover, the results for Transformations A and C show that it is possible to define map time series which could take values over the entire positive real line and yet retain their long-memory property; Transformation B, by contrast, produces time series which behave like a non-Gaussian $AR(1)$ process with a parameter value close to the unit circle. Moreover, it is possible that time series with some random component may be constructed from these maps, for example, by adding a suitable random error term or allowing the parameter defining the maps to be a random variable itself, which could prove more useful for financial and related Economic time series and this possibility should be investigated further.

Acknowledgment. Thanks are due to Mr. Aonan Zhang, Utah State University, for his help in producing the graphs included in this paper.

References

1. Aaronson, J. (1997). *An Introduction to Infinite Ergodic Theory.* Vol. 50, Mathematical Surveys and Monographs, AMS.
2. Akgirav, V. and Booth, G.G. (1988). The stable law model of stock returns. *Journal of Business and Economic Statistics*, **6,** 51–57.
3. Anderson, T.W. (1971). *The Statistical Analysis of Time Series.* Wiley and Sons, New York.
4. Baillie, R. T., Bollerslev, T. and Mikkelsen, H. O. (1996a). Fractionally integrated generalized autoregressive conditional heteroskedasticity. *Journal of Econometrics*, **74,** 3–30.
5. Balakrishnan, V., Nicolis, G. and Nicolis, C. (1997). Recurrence time statistics in chaotic dynamics. I. Discrete time maps. *Journal of Statistical Physics*, **86,** 191–212.
6. Balakrishnan, V., Nicolis, G. and Nicolis, C. (2001). Recurrence time statistics in chaotic dynamics: multiple recurrences in intermittent chaos. *Stochastics and Dynamics*, **1,** 345–359.
7. Balakrishnan, V. and Theunissen, M. (2001). Power–law tails and limit laws for recurrence time distributions. *Stochastics and Dynamics*, **1,** 339–343.
8. Berliner, L.M. (1992). Statistics, probability and Chaos. *Statistical Science*, **7,** 69–122.
9. Bhansali, R.J. (1993). Order selection for linear time series models: a review. In: *Developments in Time Series Analysis*, T. Subba Rao editor, pp. 50–56. Chapman & Hall, London.
10. Bhansali, R.J., Holland, M.P., and Kokoszka, P.S. (2004). Chaotic maps with slowly decaying correlations and intermittency. *Fields Institute Communications*, **44,** 99–126.
11. Brilinger, D.R. (1969). Asymptotic properties of spectral estimates of second–order. *Biometrika*, **56,** 375–390.
12. Brock, W.A. and Hommes, C.H. (1997). A rational route to randomness. *Econometrica*, **65,** 1059–1095.
13. Cont, R. (2001). Empirical properties of asset returns: stylized facts and statistical issues. *Quantitative Finance*, **1**, 223–236.
14. Ding, Z., Granger, C.W.J. and Engle, R.F. (1993). A long memory property of stock market returns and a new model. *Journal of Empirical Finance*, **1,** 83–106.
15. Giraitis, L., Kokoszka, P. and Leipus, R. (2000). Stationary ARCH models: dependence structure and Central Limit Theorem. *Econometric Theory*, **16,** 3–22.
16. Gopikrishnan, P., Meyer, M., Amaralan, M. and Stanley, H. (1998). Inverse cubic law for the distribution of stock price variations. *The European Physical Journal*, B **3,** 139–140.
17. Gouëzel, S. (2004). Sharp polynomial estimates for the decay of autocorrelations. *Inventiones Mathematicae*, **150,** 629–653.
18. Granger, C.W.J. and Ding, Z. (1996). Varieties of long memory models. *Journal of Econometrics*, **73,** 61–77.
19. Greene, M.T. and Fielitz, B.D. (1977). Long–term dependence in common stock returns. *Journal of Financial Economics*, **4,** 339–349.
20. Hall, P. and Wolff, R.C.L. (1995). On the strength of dependence of a time series generated by a chaotic map. *Journal of Time Series Analysis*, **16,** 571–583.

21. Holland, M.P. (2002). Slowly mixing systems and intermittency maps. *Preprint.*
22. Hosking, J.R.M. (1981). Fractional differencing. *Biometrika*, **68,** 165–176.
23. Kirman, A. and Teyssière, G. (2002). Microeconomic Models for long memory in the volatility of financial time series. *Studies in Nonlinear Dynamics and Econometrics*, **5,** 281–302.
24. Kokoszka, P.S. and Taqqu, M.S. (1995). Fractional ARIMA with stables innovations. *Stochastic Processes and their Applications*, **60,** 19–47.
25. Kokoszka, P.S. and Taqqu, M.S. (1996). Infinite variance stable moving averages with long memory. *Journal of Econometrics*, **73,** 79–99.
26. Lawrance, A.J. and Balakrishna, N. (2001). Statistical aspects of chaotic maps with negative dependence in communication setting. *Journal of the Royal Statistical Society*, B **63,**, 843–853.
27. Lawrance, A.J. (2001). Statistical aspects of chaos. A series of 7 lectures for MSM4S2. *Technical report* School of Mathematics and Statistics, University of Birmingham.
28. Liverani, C., Saussol, B. and Vaienti, S. (1999). A probabilistic approach to intermittency. *Ergodic Theory Dynamical Systems*, **19,** 671–685.
29. Ljung, G. and Box, G.E.P. (1978). On a measure of lack of fit in time series models. *Biometrika*, **66,**, 67–72.
30. Manneville, P. and Pomeau, M. (1980). Intermittent transition to turbulence in dissipative dynamical systems. *Communication in Mathematical Physics*, **74**, 189–197.
31. Mittnik, S. and Rachev, S. (2000). *Stable Paretian Models in Finance.* Wiley, New York.
32. Mondragon, R.J. (1999). A model of packet traffic using a random wall model. *International Journal of Bifurcation and Chaos*, **9,** 1381–1392.
33. Park, K. and Willinger, W. (2000) (eds.) *Self-similar Network Traffic and Performance Evaluation.* John Wiley & Sons, New York.
34. Podobnik, B., Ivanov, P., Lee, Y., Chessa, A. and Stanley, H. (2000). Systems with correlations in the variance: Generating power law tails in probability distribution. *Europhysics Letters*, **50,** 711–717.
35. Rasband, S.N. (1990). *Chaotic Dynamics of Nonlinear Systems.* Wiley, Chichester New York.
36. Sarig, O. (2001). Subexponential decay of correlations. *Preprint.*
37. Thaler, M. (1980). Estimates of invariant densities of endomorphisms with indifferent fixed points. *Israel Journal of Mathematics*, **37,** 303–314.
38. Young, L.S. (1999). Recurrence times and rates of mixing. *Israel Journal of Mathematics*, **110,** 153–188.

The Spectrum of Euro–Dollar*

Vincent Brousseau

European Central Bank, Directorate General Operations.
vincent.brousseau@ecb.int

Summary. The purpose of this study is to analyse the time series of the euro–dollar exchange rate. This exchange rate is studied as if it was a continuous–time physical process, which implies that we systematically make use of different degrees of time resolution. The analysis takes into account various statistical indicators, but puts a special emphasis on the spectrum of the process. In other words, we consider the process of the returns of the euro–dollar exchange rate as produced by a superposition of oscillations with different frequencies and we try to determine the relative weight of those frequencies within the returns process. The spectrum of this process is the function giving those relative weights. We find that this spectrum has an identifiable pattern and we claim that this pattern is a core characteristic of the process. We simulate then a process having the same spectrum and compare the behaviours of actual process and of the simulated process in terms of various statistical indicators. We find that the simulated process provides a good, but not perfect, replication of the behaviour of the actual euro–dollar exchange rate.

1 Introduction

The euro–dollar exchange rate is the archetype of the liquid market price. Trades and quotations occur so frequently that it can be thought as a continuous–time process. Probably no other economic or financial time-series can produce, or has ever produced, databases comparable both in term of size and of accuracy. In short, the euro–dollar exchange rate is one of the rare economic phenomena, if not the only one, that can be studied as if it was a physical phenomenon. Physical phenomena, indeed, can be described by data

* I wish to thank Sandrine Corvoisier, Simone Manganelli, Andres Manzanares, Ben Weller and Flemming Würtz for their help in the difficult process of the capture of the data. I am especially indebted to Gilles Teyssière for his very careful reading of this chapter. I accept sole responsibility for any remaining errors. The views expressed therein are those of the author and not necessarily those of the European Central Bank.

whose precision and whose quantity are virtually unbounded, and physical phenomena are frequently thought in continuous time.

This being said, it is tempting to examine whether or not the phenomenon "euro–dollar exchange rate" obeys to some analogue of physical laws. In concrete terms, this would mean to identify statistical regularities in the sample path of the exchange rate process, which could be simply described in mathematical terms and which would remain present over time. Answering to this question is in principle feasible owing to the enormous number of available data, and to the considerable accuracy of those data.

The idea that the dynamics of liquid prices could be regarded as physical phenomena are regarded by physicists is probably due to Benoît Mandelbrot. In 1973, Mandelbrot gave three conferences at the Collège de France, in which he was in essence introducing a methodology applicable to "self–similar" phenomena. Those phenomena were mostly physical ones, and the methodology aimed at discovering in those phenomena useful empirical regularities that could lead to the formulation of law: This is nothing else, of course, that the most classical physicist approach. However, Mandelbrot was also considering, among those physical phenomena, a class of more economics ones, namely the dynamics of some market prices. His point was that the proposed methodology was equally valid for those economical phenomena.

In 1973, nonetheless, this was not much more than an appealing idea. The data that would have permit to explore the question with precision were simply not existing. The best price data that were available had daily frequencies. Mandelbrot proudly presented an astonishingly long series of cotton prices, more than one century of daily data. The daily records of the Dow Jones index were also available up to the First World War. Even with a history of one or two centuries, daily data only make 10,000 or 20,000 records. This is a bit insufficient to start a physics–like study.

Things changed when a few core market instruments became liquid, around the mid–eighties or the beginning of the nineties. Those few core market instruments "main exchange rates, main future contracts" were traded and quoted with high frequency, and the traded or quoted prices started to be recorded. At that point, the Mandelbrot vision became feasible.

The availability of tick–data has created a new domain of research. We will not here attempt to give an exhaustive description of this domain, but rather briefly evoke some of its sub-area.

- Let us mention first the direct application of Mandelbrot concepts to tick–data sets. In this area, Mandelbrot himself of course played a major role, but he was followed by many others. This type of literature assesses the presence or the absence of "Noah effect" and "Joseph effect". It researches in the time series the property of self–similarity, it applies the Hurst test to them. It makes uses of the notion of fractal, the notion of dimension. This sort of research is undoubtedly the one that is the closer to what could be called "a physics of finance".

- Let us then mention the volatility trend. This type of literature is concerned with the notion, the definition, and the practical measurement, of volatility (This means here the historical or empirical volatility, by opposition to the implied one). The potential applications of this literature lie primarily in the area of risk management.
- Let us also mention the market microstructure area. Those papers tend to derive, to prove or to calibrate market microstructure models from tick data sets.

This classification is obviously not clear–cut, it is easy to find examples of papers falling into two of those categories. Furthermore, this classification is not complete. It is only intended to, say give the reader a flavour of what exists and of what can be done. The present study would probably fall more in the first above–mentioned category, the "physics of finance".

Tick–data have granted many empirical findings of heterogeneous meaning, heterogeneous precision, and heterogeneous simplicity. Only two of them are generally valid, i.e., valid for all liquid prices, abundantly documented and relatively simple to express or to explain. Those two findings can therefore be regarded as the fundaments of our empirical knowledge about the topic. It is interesting to note that any of those two findings can also be found from daily data, instead of tick–data or of high frequencies data, albeit with less precision. Those two findings are the followings:

1. Leptokurtosis: A probability distribution is leptokurtic when it gives a relatively high weight to high (in absolute value) outcomes, to small outcomes, and thus a relatively low weigh to medium–sized outcomes. For example, a uniform distribution is less leptokurtic than the Gaussian distribution, while a Levy–stable distribution, other than the Gaussian one itself, is more leptokurtic. It is well established and well documented that the returns of liquid market prices have leptokurtic distributions. How leptokurtic, this may depend of the market instrument, of the time–scaled considered, of the measurement method. Leptokurtosis, however, is a general feature of the distributions of the returns.
2. Correlation structures: It is well established and well documented that the subsequent returns of liquid market prices are not correlated, while their squares or their absolute values are positively correlated. More precisely, subsequent returns might exhibit a negative correlation at very fine time scale but this memory vanishes anyway very quickly, after at most some minutes. On the contrary, the positive correlation among the absolute values, or the squares, of returns, persists in time and decays only slowly at a hyperbolic rate. Squared returns or absolute values of returns are said to have "long–memory".

The present study is focused on one market instrument, the euro–dollar exchange rate, because of its tremendous liquidity. In addition, it is specially focused on one mathematical property of time series, namely their spectral decomposition, or their decomposition as the sum of oscillations (of fixed frequency). This particular focus involves a technical difficulty. In short, we used tick data of quoted prices extracted from Reuters. In those data, quotes of heterogeneous quality and accuracy succeed one another, triggering a noise with low amplitude but high frequency. Such a noise must be eliminated if what we are interested in is the measurement of the spectrum, i.e., the frequency decomposition, of the process.

This technicality is treated in the Section 2, whose theme is the available data, the way to understand them, and the way to filter them. The analysis of the empirical regularities of the euro–dollar takes place in Section 3. To start with, we examine several statistical indicators with the idea to cross–check the output of those indicators and of the spectral analysis. Those several indicators are the probability densities, the autocorrelation functions, henceforth ACF, and the realised variations of the returns of the euro–dollar at different time–scale. Then we tackle the spectral analysis itself, and try to reproduce this spectral structure and other statistical properties in a simulated random process. Section 4 concludes.

2 The Methodology

In order to enable the reader to reproduce, or simply to judge, the results, a precise description of the methodology is useful. The present part has a technical scope, and it is subdivided in two headings: The data, and the various algorithms of data filtering. A reader who does not intend to reproduce the experiment may nevertheless find those technical sections useful. The first section clarifies things about Reuters data, what they are, what they mean, what they don't mean. The second section does not only expose the details of the methodology that has been applied to the data. It also makes the point that once this methodology has been used, the Reuter data actually reveal a precise description of the history of the market price.

2.1 The Data

About Reuters Quotes

The data set that we will be using contains real–time quotes contributed on Reuters. It is therefore necessary to examine what are exactly those Reuters quotes, what information they are likely to carry, and what noise is likely to affect them.

The system, which works according to the same principles since several years, can be described as follows. First, a certain number of banks place quotes, i.e., indicative bid and ask quoted prices, on one of their so–called pages. Those pages, which are readable by any other user, unless otherwise specified, are made of x lines of y ASCII characters, typically 25 lines of

81 characters. The quotes appear in this rectangles of ASCII characters as ASCII strings of various formats, such as "1.0768-73", or "1.0768 1.0773", or "1.0768/73", or other formats. Those strings of characters have of course an unambiguous meaning for the human reader to whom they are destined. However, the problem of converting them towards numerical figures in an automated and reliable way is not completely easy. It implies to locate in the page where the quote is written, to recognise the format, and to avoid being fooled by a variety of things, the enumeration of which would be tedious.

The system has this ability to automatically perform conversion. This conversion is usually referred to with the word "parsing", and the information obtained after conversion is usually referred to as the "parsed" or "logicised" information. Whenever a page is updated, the system fishes the updated quotes, converts them from ASCII strings to numerical figures and update an appropriate structure, called "RIC", with the converted bid price, the converted ask price, the time stamp, the identifiers of the contributing bank, the page from which the data have been read (the so–called "background page"), and possibly other items of informations. Of course, some items of the parsed information can be redundant, for example the knowledge of the background page implies the knowledge of the quoting bank. This produces an uninterrupted flow of numbers that are readable and understandable not only by human readers, but also by software packages such as spreadsheets or other dedicated applications. For example, the RIC of the euro–dollar exchange rate may receive some 10,000 or even 20,000 updates a day. During the period 2001–2002, this has represented for this instrument slightly more than 10 million of quotes.

Those quotes are of diverse nature or significance. At least for the foreign exchange market, they are never legally binding and when calling a dealer at the very moment when he contributes the quote, this dealer is not legally forced to trade on that bid price or on that ask price. For major market players, the quote nevertheless reflects, at the precise time when it is quoted, the price that a dealer would agree to trade with another bank of the same importance (not with a customer). For others banks, the price is more indicative because the incentives to hold the quotes are less strong. Finally, other banks use the contribution of quotes purely as an advertising medium, to be seen on the screens. It should also be noted that some banks produce their quotes in an automated way, and the input needed by the automated process can be, purely and simply, the content of the RIC itself.[2] This introduces some circularity in the quotes–generating process. In short: the quality of the informational content of raw data is heterogeneous.

[2] This can be observed when two or three of them remain by mistake active during the weekend, when all the others are silent. Then the two or three automated quoted react on their own output and on nothing else. After a while, it becomes even possible to disentangle the algorithm(s) of quote production that they are using, as those algorithms can be rather simple.

It is therefore clear that the sequence of quoted price, despite its huge size and impressive accuracy, does not in itself provide a perfect continuous–time description of the euro–dollar price process. As will be shown below, however, there is a way to recover such a description from the sequence of quoted prices and quoting banks. The way to do so is called in this paper the "dynamical filtering".

The Data Set

The three items of information we are interested in are then the quoted bid price, the quoted ask price and some key identifying the quoting bank. This corresponds, in the Reuters terminology, to three "fields", namely the bid field, the ask field and, for example, the field indicating the background page. The paper uses a data set containing records of those three fields for the RIC of the euro–dollar exchange rate.[3] On the top of those three fields, of course, the records contain the time stamp and the identity of the quoted instrument. For what concerns the time stamp, it is not the one that can be found in the Reuters RIC because in the case of quoted exchange rate this information gives only the minute, but not the second.[4] Thus, the time stamp of the recording computer has been used. For what concerns the identity of the quoted instrument, several choices would have been possible but the simplest one, of course, was the RIC name itself, which is, for the exchange rate we are interested in, "EUR=". The following table summarises the descriptive statistics of the data set:

Table 1. Descriptive statistics of the data. (13,642,666 observations)

	Mid Price	Time Stamp
First	0.86605	07/09/2000 10:28:53
Last	1.07635	22/02/2003 08:03:58
Low	0.8227	26/10/2000 09:37:31
High	1.0937	05/02/2003 09:36:22

In short, this is slightly more than two years, encompassing the whole year 2001 and the whole year 2002.

Contributions have been provided by circa 200 banks, some of them may have shifted their background page at some point within the period under

[3] The reader wishing to reproduce the experiment may find useful to know the following: The name of the RIC having been used is "EUR=". The names of the three fields recorded within this RIC are "BID", "ASK" and "BACKGROUND PAGE". The numbers of those three fields are 22, 25 and 105.

[4] This is likely a consequence of the early implementation of those RICs. For instruments that have been introduced more recently in the system, the time stamp gives also the second.

study, while others have contributed anonymously and then cannot be distinguished.

2.2 Data Filtering

It is clear that in a database of this size, erroneous data are inevitably present. Moreover, the quality of the informational content of raw data is heterogeneous, as was explained above. Finally, owing to the absence of legal binding, no isolated raw data can be a priori deemed as carrying the information we are really interested in, namely: At a given instant, where stands the market.

For all those reasons, it is necessary to manipulate the raw data so as to eliminate errors and to reconstruct the hidden reality of the market price. For those purposes, we will apply three filtering algorithms to the raw data. We refer to those three algorithms as to the simple filtering, the dynamical filtering and the retroactive filtering. The simple filtering is mainly about the elimination of obvious errors. The dynamical filtering can be seen as the reconstruction of the reality hiding behind the sequence of raw quotes, but it has also the effect of filtering out bad quality data. Applying only those two filters would nonetheless allow a certain class of isolated erroneous data to survive, as will be explained below. The so–called retroactive filtering is specially designed to eliminate those ones.

The three algorithms are to be applied to the data in the following order: Firstly, the retroactive filtering, secondly, the simple filtering, thirdly, the dynamical filtering. One may hope that after having submitted the raw data to those three procedures, one obtains something more accurate. But it is not necessary to rely on hope only: There exist a possibility to check and measure the gain of accuracy. This can be done by comparing the filtered data to traded data, which contain (anonymously posted) bid and ask prices that can actually be hit. Such tradable prices happen to exist, since an important proportion of the euro–dollar market is traded on an electronic market, namely EBS^{TM}. We will compare the EBS mid–price, the mid price of the raw quoted data and the mid price of the filtered data over three sample days. This will give a sense of the accuracy of the information that can be extracted from the raw data when filtering them.

Simple Filtering

The so–called simple filtering is, as the name indicates, the most rudimentary of the three filtering algorithms. It simply rejects any data differing from the previous data from more than twenty pips. More precisely, the rejection occurs when the absolute value of the logarithm of the ratio of the two subsequent mid prices exceeds 0.002. After 1 hour, however, this rejection rule ceases to operate. The value of twenty pips has no scientific “raison d'être”. It has been chosen because of the two following motives:

- It eliminates an anomaly in the chart of a statistical indicator to be presented below, see Figure 5. This anomaly, strong enough to trigger a visible effect of the chart of the indicator, was generated by only two data. It is impressive to think that the twelve millions other data were correct enough to avoid triggering visible anomaly.
- It does it at a low cost, i.e., rejecting a extremely small number of data.

The so–called simple filtering is certainly an elementary precaution, but clearly it cannot be seen as sufficient to eliminate all errors, and it does not do any job of extracting the market price from a bunch of quotes of mixed quality.

Dynamical Filtering

The principle of the dynamical filtering is simple. The idea is the reconstruction of the set of "currently valid" quotes "around" a given point in time. It is thinkable to try something in this direction, provided only that one has at one's disposal, not only the quoted bid and ask prices, but also some key identifying the quoting bank, which can be its name, the name of its background page or something else. As it has been said, the database used for this paper contained records of the background page, bid prices, and ask prices.

Now how can this principle be implemented in practise? No perfect method exists, as shows the following observation. Even if a quote is valid, i.e., represents the true intentions of the quoting bank, at the second where it is posted, there can be no guarantee that it is still valid a few seconds after, because the market may have shifted. The only thing that can be hoped for is that, if the market actually shifts during those few seconds, then a more recent quote will tell about this shift within those few seconds. It is reasonable to hope it because liquid markets have frequent quotes. For the euro–dollar, the average frequency of quotation lies around two seconds.

The problem becomes then to design an algorithm that tentatively reconstructs the relevant quotations of banks and performs this reconstruction at a low cost in term of complexity. Indeed, given that the reconstruction can never be absolutely sure, it is useless to design a too complex algorithm. Something which works reasonably well reasonably often and which is reasonably simple is the best that can be achieved. Once the problem is stated in these terms, its solution is easy to find. We construct the dynamically filtered bid–ask spread by:

1. Associating to each bank its most recent quoted bid–ask price,
2. Sorting the quotations, and thus the banks, in chronological order starting from the most recent,
3. Computing the best bid and the best ask for the n most recent banks, starting for $n = 1$, and incrementing n.
4. Stopping when the best bid is equal to or bigger than the best ask price,

5. Eliminating the bank on which we have stopped, this is bank number n, as well as all the less recent ones,
6. Taking the best bid and the best ask of the remaining banks, which are the $n-1$ most recent banks.

The following figure shows the output of this filtering method.

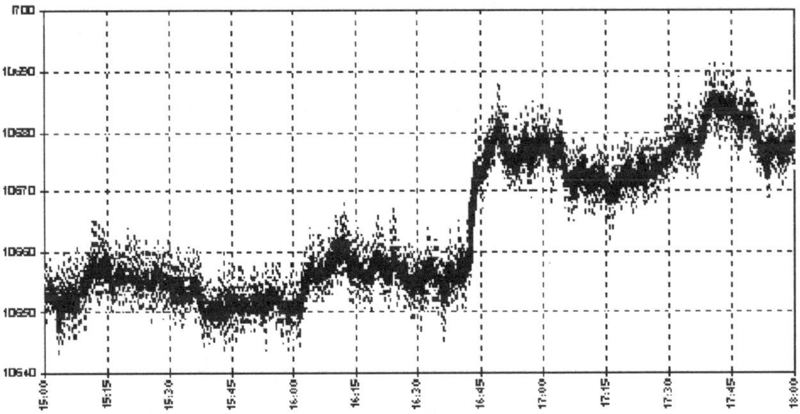

Fig. 1. Comparison between raw data and filtered data. Thin dotted line: raw data, bold line: filtered data (Source of raw data: Reuters)

The system is quite robust. For example, a quotation with large bid–ask spread encompassing the filtered bid–ask spread will not modify this filtered bid–ask spread. This is a good thing, since this quotation with large bid–ask spread is actually bringing no valuable information. The reader is invited to imagine other configurations and should find that the system produces always the reasonable reaction, except in one case. This is the case of a plainly false quote, which would be arbitrageable if it were to be really hold. Either its bid is higher than the market price, or its ask is lower than the market price. In this case, the response of the algorithm is to select this false quote only. To cope with those arbitrageable quotes, it will be necessary to rely on a complementary filtering, to be described in the next paragraph. However, this situation occurs rarely.

The dynamical filtering has a graphical effect, which is noteworthy. To expose this, let us consider some charts of tick data ranging over periods of circa 10–20 minutes. One can face two sorts of graphical patterns. In the first case, the chart represents the bid, or the ask, that are posted on some electronic trading system for some liquid instrument. It can be for example the Bund, the Bobl, the Schatz, on the trading system of Eurex, or the dollar–yen, the euro–dollar, the cable, on the EBS system. In the second case, the chart represents the quoted bid of some liquid instrument, having its liquidity, or having some liquidity, on the OTC market. It can be for example the dollar–yen, the euro–dollar, the cable, from the Reuters RIC EUR=, JPY=, GBP=, or some OAT or Bund from their corresponding RIC. The charts of the first type will look similar to each others, and the charts of the second type also will look similar to each other, but there will be a clear difference between the charts of two different types. The charts of the second type will indeed exhibit a kind of high–frequency oscillation which will be completely absent of the charts of the first type.

This, of course, is the result of the quick succession of banks producing the quotes. The narrowly quoting banks and of the largely quoting banks alternatively generate the price contained in the RIC and this produces the graphical impression of some high frequency shivering. Naturally, this high frequency shivering is not really there and should be eliminated. This is especially relevant for a study that focuses on the spectral decomposition of the returns process.

When plotting a chart of quoted data coming from RICs, but having been dynamically filtered in accordance to the method described above, one gets a chart of the first type instead of a chart of the second type. This small graphical experience, in itself, does not prove that the noise has been correctly killed: Only the test described in the second next paragraph below can do this job. However, it is quickly done and already suggests that, at least for what is regarding the spectral structure of the time series, dynamically filtered data are more advisable than raw data.

Retroactive Filtering

As was said, there is a type of erroneous data that are resistant against the dynamical filtering. This is the case of a quoted spread lying plainly out of the market. If the bank issuing such a spread was actually willing to trade on it, it could be arbitraged. The way to eliminate those data relies on the fact that they are usually isolated and surrounded by plenty of good data. In such a case, the sequence of prices should present a high price increase immediately followed by a price decrease of a comparable magnitude (or conversely). The method for taking advantage of the likely existence of this signal is therefore, before accepting a data, to read a few of the following data, and to validate the data only if this kind of statistical symptom is not found. One must thus implement a data buffer, in which a record extracted from the database must stay a while before being actually read (or rejected). This test implies to read a little bit of the future and to use the information found in the near future to assess the validity of a quote, consequently, this test could not be implemented in real time. This is why we will call it the "retroactive" filtering.

So much for the principle of it. There are two difficulties, which are immediately springing to the mind.

1. This retroactive test could possibly eliminate also good data,

2. This retroactive test could modify the chronological order of data in the case of several instruments.

Let us start by devoting some attention to the second point. The present study, of course, is concerned with just one instrument, which is the euro–dollar exchange rate. However, developing an algorithm of retroactive filtering which would require that only one instrument is considered at the same time would not be an excellent idea. It could not be re–used in the situations for which the simultaneous knowledge of several instruments prices is needed, and those situations are by far the most common ones. Here are two examples of it:

1. Studying the FX market for more of two currencies obliges of course to consider at least two exchange rates. Those two exchange rates continuously react on each other, and therefore it is not advisable to loose their common chronological structure.
2. The second example, on which we will not insist, is the case of yield curves.

Hence, there is a good case for building a retroactive filtering algorithm that can be consistently applied to more than one instrument at the same time. The way of doing so is to implement a cut–off time and to decide that whenever a data has not been rejected after this cut–off time, it is deemed to be accepted. One just need to read from the buffer only when the time stamp of the last data included in the buffer is more than the time stamp of the data to be read from the buffer, plus the cut–off time. Of course, when the end of the file is reached, the rule does not have to apply any more. This way of filling and emptying the buffer does not allow for a chronological distortion, even when the data pertain to several instruments.

This system of the cut–off time provides us with a solution for the other problem, which is that the retroactive filtering could very well also reject good data. The solution, actually, is not to preclude such unjustified rejections, but rather is to make those unjustified rejections unimportant. This can be easily done by choosing a small cut–off time. In this case, a "good" data can be unduly rejected only when it is surrounded by other good data, which are chronologically very closed to it, and which are not rejected. For example, choosing a cut–off time of two seconds ensures that in any circumstance, non–rejected data are not older than two seconds. This is fairly enough for practical purposes and as a matter of fact we will apply a cut–off time of two seconds.

The choice of the cut–off time does not entirely specify the algorithm, one must moreover specify the criterion that triggers the rejection. Let us consider the returns, i.e., the difference of the logarithms of the mid prices of two subsequent quotes. When the product of two subsequent returns is less than some (negative) threshold constant, then the intermediate quote is rejected if one of the four following inequalities hold true:

- First inequality: Its bid is bigger or equal to the ask of the proceeding quote and strictly superior to the ask of the next quote.

- Second inequality: Same condition, but exchanging the role of the next quote and of the previous quote.
- Third and fourth inequalities: Same conditions, but exchanging the role of the bid and the ask.

In short: When one of those four inequalities is true and moreover the product of the returns is below the threshold and moreover the next quote does not occur two seconds or more than the tested quote, then the tested quote is rejected.

The threshold value that has been retained is -4e-8, which corresponds to minus the square of two pips. This value and the value of the cut-off time, which is two seconds, entirely specify the algorithm. Concretely, out of 13,642,666 data, 13,525,729 data pass the simple filtering *and* the retroactive filtering, and therefore 0.86% of the data are rejected. The retroactive filtering accounts for nearly all of those rejections.

Testing the Validity of the Filtering Algorithms

As was said, the three algorithms are to be applied to the data in the following order: Firstly, the retroactive filtering, secondly, the simple filtering, thirdly, the dynamical filtering. This sequence of transformations represents a relatively complex task. This task aims at improving the quality of the knowledge of the price sample path. One can check that it actually does so, if one also possesses data of traded prices. In the case of the euro–dollar exchange rates, those data are to be found on the EBS system.

We have used three days of EBS tick–data, namely the 3rd of January, the 24th of January and the 28th of January 2003, starting from 8 hours, Frankfurt time, and finishing at 20 hours. For each of those three test days we have considered the period between 8 hours and 16 hours, and the period between 8 hours and 20 hours. The rationale for those two test periods is the following: Between 8 hours and 16 hours, the average EBS spread remains remarkably steady, around 1.3 pip. After 16 hours, it increases, reaches a peak around 1.6 pip at 18 hours and evolves between 1.5 and 1.6 afterwards. This is at least what suggests the data of our three test days. It suggests that the behaviour of the market is different before and after 16 hours, and it suggests that typical behaviour of EBS prices is the one that can be observed between 8 hours and 16 hours

We have then six test periods. For each of them, we examine the proportion of (physical) time during which the difference between the EBS mid price and the OTC quoted mid price, was at some given levels. For the OTC quoted mid price, we take first the non–filtered data (the raw tick data), and then the filtered data (with the three filtering presented above). Those proportions, for the levels comprised between -2 pips and 2 pips, are reported in the following table:

From the proportions reported in Table 2, it is easy to derive the proportions of time in which the difference between the quoted mid and the EBS bid

Table 2. Non filtered data (raw tick data)

	03-Jan 8h-16h	24-Jan 8h-16h	28-Jan 8h-16h	03-Jan 8h-20h	24-Jan 8h-20h	28-Jan 8h-20h
-2 pips	2%	1%	1%	2%	1%	1%
-1 1/2 pip	2%	2%	3%	2%	2%	2%
-1 pip	6%	6%	6%	6%	5%	5%
- 1/2 pip	10%	12%	11%	10%	10%	10%
0 pip	29%	33%	25%	30%	30%	23%
1/2 pip	25%	19%	18%	24%	19%	20%
1 pip	13%	13%	18%	14%	16%	19%
1 1/2 pip	7%	5%	7%	6%	6%	8%
2 pips	2%	2%	4%	3%	3%	4%

Filtered data (simple, dynamical and retroactive filtering)

	03-Jan 8h-16h	24-Jan 8h-16h	28-Jan 8h-16h	03-Jan 8h-20h	24-Jan 8h-20h	28-Jan 8h-20h
-2 pips	1%	1%	1%	2%	2%	1%
-1 1/2 pip	1%	1%	1%	1%	1%	1%
-1 pip	10%	15%	8%	10%	12%	7%
- 1/2 pip	7%	7%	5%	8%	7%	5%
0 pip	47%	47%	46%	42%	39%	39%
1/2 pip	17%	8%	10%	17%	11%	13%
1 pip	15%	17%	20%	16%	18%	21%
1 1/2 pip	1%	1%	2%	2%	4%	5%
2 pips	1%	2%	4%	2%	4%	6%

have been majored by two pips, by one pip, or have been zero. Those proportions summarise quite well the result of the experiment; they are reported in the following table:

The proportions reported in Table 3 show that the filtering procedures have bettered the informational content of the quoted data, and they also show by how much. The proportion of time where the difference EBS-OTC is zero is nearly doubled when applying the filters, and one reaches then a confidence level of 2% that the difference does not exceed two pips. Two pips is something quite small for this market, it corresponds to the width of a good price.

We conclude this exercise with two remarks:

Remark 1. The first remark is about the traded data used for the test. Those data have the clear advantage that they are made of bids and ask that can actually be traded, while the "tradability" of the quoted data is not enforced by law or by technical reasons. One should however not forget that the traded data have also a disadvantage: They concern small sizes, which can be as low as

Table 3. Non filtered data (raw tick data)

	03-Jan 8h-16h	24-Jan 8h-16h	28-Jan 8h-16h	03-Jan 8h-20h	24-Jan 8h-20h	28-Jan 8h-20h
$= 0$ pip	29%	33%	25%	30%	30%	23%
≤ 1 pip	83%	84%	77%	83%	80%	76%
≤ 2 pips	95%	94%	93%	95%	93%	92%

Filtered data (simple, dynamical and retroactive filtering)

	03-Jan 8h-16h	24-Jan 8h-16h	28-Jan 8h-16h	03-Jan 8h-20h	24-Jan 8h-20h	28-Jan 8h-20h
$= 0$ pip	47%	47%	46%	42%	39%	39%
≤ 1 pip	96%	93%	89%	93%	87%	85%
≤ 2 pips	99%	98%	98%	99%	97%	97%

1 million euro. The quoted data are meant for standard size that are of a bigger magnitude, at least 10 millions. It is thus only normal if they exhibit higher spreads and it is not completely obvious that their informational content is poorer than the one of the traded data, once the wrong quotations have been taken out.

Remark 2. The second remarks pertain to the general usefulness of the dynamical filtering. As the Figure 1 suggests, the impact of filtering on bids and on asks is much stronger than its impact on mid prices. This is because a non–negligible proportion of the large quotes are nevertheless correctly centred or nearly correctly centred. Consequently they cause much less noise on the mid than on the ask or on the bid. If one computes a table similar to Figure 1, but taking the asks or bids instead of mids, one finds the same level of quality for the filtered data, but a much poorer quality for the non–filtered data. As the present study is only concerned by the mid price, we do not further elaborate.

3 The Results

3.1 Discretisation of Time

All the results pertain to the time series of the euro–dollar exchange rates discretised in *physical time*. This means that the euro–dollar exchange rates are considered at regular intervals of constant length in physical time. This constant length is called the period of discretisation.

We have been considering 11 periods of discretisation, namely 15 seconds, 30 seconds, 1 minute, 2 minutes, 4 minutes, 8 minutes, 15 minutes, half an hour, 1 hour, 2 hours and 4 hours. Each period is thus the double of the immediately smaller period, except the period of 15 minutes.

All the results have been obtained after exclusion of nights and weekends. Weekends mean Saturdays and Sundays. Non working days other than Saturdays or Sundays are not excluded. Nights mean the period between 20 hours, Frankfurt time, and 8 hours, Frankfurt time. The length of the day is thus 12 hours, which is convenient because it is an integer multiple of all the periods of discretisation. To each period of discretisation corresponds thus a discretisation grid, with equally spaced points, starting at 8 hours and finishing at 20 hours.

3.2 General Statistical Indicators

Before tackling the spectral analysis of the euro–dollar, it is useful to consider some simpler indicators. We will examine the distribution of returns, the quadratic variation, and similarly defined variations, and the autocorrelation function. We will briefly say word of a less classical indicator, namely the Hausdorff–Besicovitch dimension, nowadays known as the fractal dimension, of the trajectory of the process. In all case, we will investigate what would have been the answer of the statistical indicator in the case of a standard Wiener process.

Distribution of Returns

Returns are defined as the first difference of the logarithm of the mid price of the discretised series. However, given the small order of magnitude of those returns, the concavity of the logarithm function has hardly a chance to play a role. Consequently, returns are either approximately an integer number of pips, or a half–integer number of pips.

Regarding the unconditional distribution of the returns, two observations should be made. The first one is that, quite surprisingly, the unconditional distribution of short period returns can be deemed constant. The second one is that, less surprisingly, the unconditional distribution of returns is leptokurtic and that the distribution of short–period returns is more leptokurtic than the distribution of long–period returns. The longest period taken into consideration is 4 hours.

The invariance of the distribution of short–term returns has been tested only on the 15–second returns because, as those returns are the more numerous, the results of the test will be more conclusive. We consider the quarters that are covered by our sample of data. They range from the fourth quarter of 2000 to the fourth quarter of 2002. We compute the empirical density function of the 15–second returns for each of those 9 quarters, as well as for the sample as a whole. The sample size of a quarter lies between 180,000 and 190,000 15–second returns, while the total sample size is 1,848,347 15–second returns,

Table 4. Distribution of the 15–second euro–dollar returns computed for the entire sample and for several sub–samples

Pips	All	00 Q4	01 Q1	01 Q2	01 Q3	01 Q4	02 Q1	02 Q2	02 Q3	02 Q4
-5 pips	0.1%	0.1%	0.1%	0.1%	0.1%	0.1%	0.0%	0.1%	0.1%	0.1%
-4 1/2 pips	0.2%	0.3%	0.4%	0.4%	0.3%	0.2%	0.2%	0.2%	0.1%	0.0%
-4 pips	0.2%	0.3%	0.3%	0.2%	0.2%	0.1%	0.1%	0.1%	0.3%	0.2%
-3 1/2 pips	0.6%	1.0%	0.7%	1.0%	0.9%	0.7%	0.5%	0.4%	0.1%	0.1%
-3 pips	0.6%	0.6%	0.9%	0.5%	0.4%	0.3%	0.3%	0.4%	0.9%	0.5%
-2 1/2 pips	1.3%	2.8%	0.7%	2.4%	1.8%	0.7%	1.7%	0.6%	0.4%	0.3%
-2 pips	2.1%	1.3%	4.0%	1.4%	2.0%	1.8%	0.4%	2.2%	3.4%	2.3%
-1 1/2 pip	1.5%	1.4%	2.0%	1.3%	1.5%	1.5%	1.2%	1.2%	1.9%	1.5%
-1 pip	11.8%	10.6%	13.0%	11.6%	11.8%	10.9%	10.5%	12.6%	13.2%	11.4%
- 1/2 pip	6.5%	6.5%	6.3%	6.4%	6.3%	6.4%	5.5%	5.2%	7.5%	7.7%
0 pip	49.7%	48.8%	42.4%	48.2%	48.4%	53.9%	58.9%	53.5%	44.1%	52.0%
1/2 pip	6.2%	6.2%	6.1%	6.1%	6.2%	6.1%	5.1%	5.0%	7.0%	7.2%
1 pip	12.0%	11.3%	13.5%	12.4%	11.8%	10.8%	10.7%	12.9%	13.1%	11.3%
1 1/2 pip	1.6%	1.4%	2.0%	1.3%	1.6%	1.7%	1.3%	1.3%	1.9%	1.6%
2 pips	2.1%	1.3%	3.9%	1.4%	2.0%	1.8%	0.4%	2.1%	3.4%	2.3%
2 1/2 pips	1.2%	2.6%	0.6%	2.3%	1.8%	0.8%	1.7%	0.5%	0.4%	0.3%
3 pips	0.6%	0.6%	0.8%	0.5%	0.4%	0.3%	0.2%	0.5%	0.9%	0.6%
3 1/2 pips	0.6%	0.9%	0.7%	0.9%	0.9%	0.7%	0.6%	0.4%	0.1%	0.1%
4 pips	0.2%	0.2%	0.2%	0.2%	0.2%	0.1%	0.1%	0.2%	0.3%	0.2%
4 1/2 pips	0.2%	0.3%	0.3%	0.3%	0.3%	0.2%	0.2%	0.2%	0.1%	0.0%
5 pips	0.1%	0.1%	0.1%	0.1%	0.1%	0.1%	0.0%	0.1%	0.1%	0.1%

excluding those of more of 25 pips. The results are shown in the following table:

In short, this shows that all the distributions look similar. The visual equivalent of this table is the chart plotting the densities:

The densities, without being strictly identical, appear difficult to distinguish. Another way to emphasise this resemblance is to consider the cumulative probability distribution functions corresponding to the various quarters as functions of the global cumulative probability distribution function. Such a scatter plot is called a "Q–Q plot".

The Figure 3 reveals tiny differences affecting the area beyond 1.5 pip: 2002Q1 exhibits relatively less 2 pips 15–second returns than the global distribution, while 2002Q3 exhibits relatively more of them. All in all, the quarter distributions are similar to the global one. They show therefore some common features, which are:

- Symmetry,
- Strong leptokurtosis,
- Overweight of the returns of an integer number of pips, underweight of the returns of a half–integer number of pips.

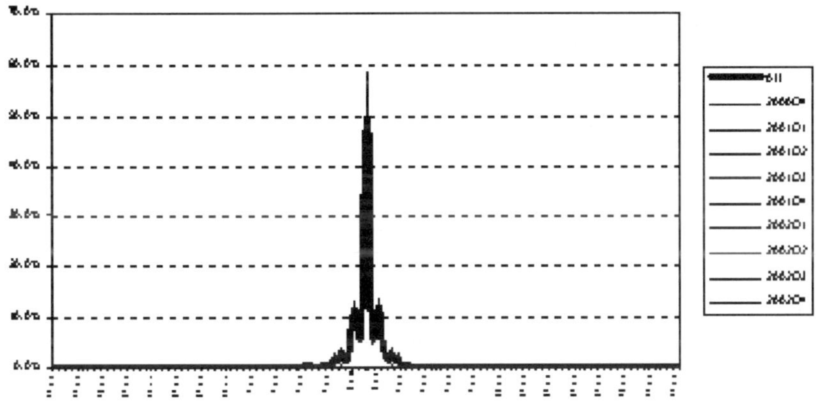

Fig. 2. Unconditional density functions of 15–second returns. Bold line: over all pdf, thin lines: pdf of quarters (Source of raw data: Reuters)

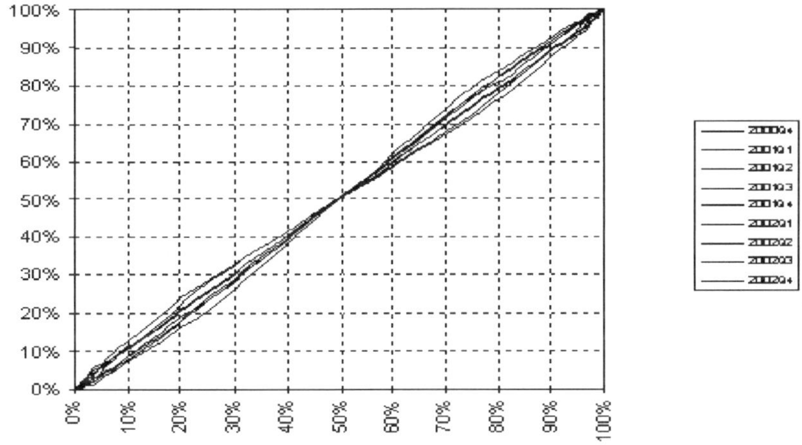

Fig. 3. Unconditional distribution functions of 15–second returns. Horizontal line: overall pdf, vertical axis: pdf of quarters (Source of raw data: Reuters)

The last feature is explained by the fact that the bid–ask spread, after dynamical filtering, is frequently 1 pip, thus the mid–price is frequently a half–integer number of pips. When two consecutive mid prices are a half–integer, number of pips, then the return is an integer number of pips. It is clear that no satisfactory continuous–time model can accommodate this third feature, but there would be no need for that, as it is only an artefact. Below, when we will construct a simulated process aiming at replicate the statistical behaviour of the euro–dollar, we will eliminate this feature. The distributions of returns for different period of discretisation, however, are not the same, and cannot be made the same by a simple change of scale. All of those are symmetrical and leptokurtic, but the shorter is the period of discretisation, the more pronounced is the leptokurtosis. This is immediately apparent when one plots the unconditional distribution functions of returns for the eleven periods of discretisation, from 15 seconds to 4 hours:

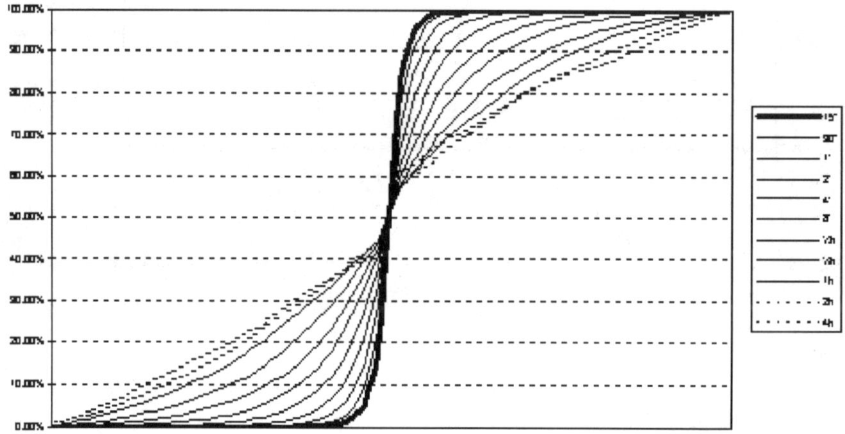

Fig. 4. Unconditional distribution function of returns (Source of raw data: Reuters)

Apart from the ones corresponding to the two longer periods, in dot lines on the graph, 2–hour and 4–hour, the curves are smooth, symmetrical and nicely ordered. Leptokurtosis is indicated by on this graph by the curvature of the curves around the level of 0% and around the level of 100%. The distribution function of the 15–second return, in bold on the graph, exhibits the stronger one, followed by the one of the 30–second returns, etc.

To get a more precise sense of the structure underlying those distributions of returns, it is useful to take recourse to the notion of cumulants. Briefly, the cumulant of a random variable is defined as the logarithm of the characteristic function of this random variable, which is in turn defined as the Fourier transform of the density function of this variable. We will denote with p its argument variable: the cumulant is a function $f(p)$ of a variable p. The justification for such a definition is as follows. When two random variables are independent, the density function of their sum is the convolution product of their density functions. The Fourier transform changes convolution products into usual products. Hence, the characteristic function of their sum is the product of their characteristic functions. The logarithm transforms products into sums. Hence, the cumulant of the sum of the two independent random variables is simply the sum of their cumulant.

This notion has a natural field of application in the case of random processes with independent increments. Indeed, the increment over a period of length t is the sum of two subsequent increments over periods of length $t/2$, so the cumulant of the t–increment must be twice the cumulant of the $t/2$–increment. We do not elaborate further here. Let us simply say that for a Wiener process, the cumulant of the increment over a time interval t is necessarily proportional to t, and proportional to the square of p. In other words, it takes the form tp^2. We are now going to glance at the cumulants of the returns of the euro–dollar.

The increments of the euro–dollar are always an integer or a half–integer number of pips, up to a small correction due to the concavity of the logarithm function, but as was said above it can be neglected. Therefore, the density functions of the returns are concentrated on a small number of points. Consequently, the cumulants can be computed only for a small number of points, and it is not possible to exactly specify which functional form they may have. It appears nonetheless that they are not proportional to the square of p. If a power function fits them, then this power function has a slightly lower exponent, lying around 1.6.

Another thing is that the cumulants do not add up. The cumulant of the 30–second return is smaller than twice the cumulant of the 15–second return, and so forth. The ratio, which would have been 2 for a process with independent increments, seems to lie around 1.6 or 1.7.

To conclude this paragraph, let us compare the euro–dollar log–exchange rate with a Wiener process:

- In the case of a Wiener process, we would also have found that the distributions of the first differences are the same for different time sub intervals. This is a similarity between the log euro–dollar and the Wiener process.
- We would have found that distributions corresponding to different periods of discretisation are the same, up to a scale factor, and in particular have the same level of leptokurtosis, i.e., a mild leptokurtosis. This is a difference between the log euro–dollar and the Wiener process.

The examination of the cumulants leads to precise this. For a Wiener process, the cumulants would have been a power function of exponent 2 (because of the normality of the increments). In addition, for a Wiener process, the cumulants of increments over $2t$ should have been the double of the increments of returns over t (because of the independence of the increments). Instead of that, we find that the cumulants of the returns of the euro–dollar are fitted by a power function of exponent less than 2. Furthermore, the cumulants of the returns of the euro–dollar less than double when the period of discretisation is doubled. This second feature suggests the existence of short–term negative correlations. It gives already some idea of what should be examined in the spectral decomposition of the exchange rate.

Quadratic Variations and Similar Objects

The next elementary indicator we will consider is the quadratic variation (QV). This one is simply defined as the cumulative sum of the squared increments of the process, along some discretisation grid. This is therefore an increasing function of time, starting from zero. It is a priori defined only on the points of the discretisation grid, but one can always extend it definition by interpolating, for example.

We discretise time with regularly spaced points, skipping only nights and weekends. Hence, we have as many quadratic variations as periods of discretisation, and namely eleven. All of them are well–defined (without interpolation) at 8 hours, 12 hours, 16 hours and 20 hours of any non–weekend day. It is then possible to compare them to each others:

Fig. 5. Euro: spot and realized quadratic variations (Source of raw data: Reuters)

Figure 5 shows the eleven quadratic variations. Their ranking on the graph corresponds to the ranking of their discretisation period. The higher one on the graph is the 15–second quadratic variation and the lower one is the 4–hour quadratic variation. The slopes of those eleven increasing functions are roughly constant. The smaller the discretisation period, the smoother the quadratic variation, and the less varying its slope.

Having looked at the graph of those eleven quadratic variations, let us remind which behaviour they should have had in the case of a Wiener process.

In the case of the Wiener process, the behaviour to be expected is simple. The quadratic variations with increasingly low periods of discretisation are the increasingly good approximations of a limit, and the form of this limit is extremely simple: It is a linear function of time. For a Wiener process, Figure 5 should theoretically show eleven confounded straight lines. A simulation shows that even with a smaller size sample, a simulated Wiener process actually produces those eleven confounded straight lines. So, Figure 5 can be seen as a (complicated) test that the log euro–dollar is not a Wiener process. It tells however a little more than simply that:

1. First, it questions the very notion of empirical volatility. If the discretised quadratic variations displayed in Figure 5 were converging towards some limit, this limit would be the conceptual or theoretical quadratic variation of the process. This would provide a natural definition of the volatility, which would simply be the square root of the slope of this limit quadratic variation. Unfortunately, the data do not speak in favour of the plausibility of such a limit. One can at most define "empirical volatilities of discretisation period of 15–second", "of 30–second", etc. Those several empirical volatilities do not happen to be the increasingly good approximation of something that could be named "the" empirical volatility. In essence, Figure 5 tells us that there is no such thing like "the" empirical volatility of the euro–dollar exchange rate. In other words, the empirical volatility does not appear to be intrinsically defined.
2. Notwithstanding the ambiguity about the right definition of the volatility, Figure 5 suggests the existence of something like general level of volatility that is slowly fluctuating over months. Indeed, the fluctuations of the eleven quadratic variations appear to be in line with each other. There are times when all the different quadratic variations increase more rapidly, and times where they all increase less rapidly. This creates, on the graphic, the appearance of slow undulations of the quadratic variations. Those undulations suggest the existence of persistent positive correlations of the squared returns. Of course, a more direct way to see that would be to consider the ACF of those squared returns. This will be done in the next paragraph.
3. Figure 5 suggests the existence of short–term negative correlations. This message was already brought by the study of the cumulants. Here, it arises from the increasingly high slopes of the quadratic variations of increasingly short period of discretisation. Assuming that those returns would have a finite variance, this patterns means in essence that the variance of the return of period $2t$ is less than twice the variance of period t. Hence, the correlation between two subsequent returns of period t must be negative. A more direct way to see that would be to consider the ACF of the returns. This will be done in the next paragraph.

Those undulations will be also observable on two (similar) statistical indicators, the absolute variations and the squared root variations. The absolute variation (AV) is the cumulative sum of the absolute values of the returns. The squared root variation (SV) is the cumulative sum of the absolute values of the squared roots of the returns. Just as the quadratic variations, the absolute and squared root variations are defined with respect to a period of discretisation and we consider therefore eleven representative of each sort, corresponding to the eleven periods of discretisation specified above. Unlike the quadratic variations, the absolute and squared root variations must be normalised by some power of the period of discretisation, otherwise they could not be compared with each other. Specifically, the cumulative sum of the absolute values of the returns must be multiplied by the squared root of the period of discretisation. The cumulative sum of the absolute values of the squared roots of the returns must be multiplied by the power of the period of discretisation. When so defined, the AV and the SV converge, for a Wiener process, towards a linear function of the time. Considering the AV and the SV of the log euro–dollar will show, once more, that it definitely differs from a Wiener process.

Fig. 6. Euro: spot and realized absolute variations (Source of raw data: Reuters)

The figures 6 and 7 exhibit several similarities with the Figure 5 . In every case, the variations have the appearance of smooth functions. There are differences too: the variation with shorter discretisation periods do not appear to be smoother than the other ones, contrarily to what was the case with quadratic variations. Also the ranking in term of slopes is different. It is reversed for the square root variations. The SV with longer discretisation period grow quicker that the ones with shorter discretisation period. For what regards the AV, no clear ranking emerges.

Autocorrelation Functions

The ACF of the returns r_t, denoted by $\rho(k)$, is defined as

$$\rho(k) = \frac{E(r_t - \bar{r})(r_{t-k} - \bar{r})}{E(r_t - \bar{r})^2} \tag{1}$$

The following table gives the ACF computed for the seven shorter periods of discretisation.

Fig. 7. Euro: spot and realized squared root variations (Source of raw data: Reuters)

Table 5. Autocorrelation function of the euro–dollar returns computed for several discretisation steps

	15 sec.	30 sec.	1 min.	2 min.	4 min.	8 min.	1/4 h
0	100.0%	100.0%	100.0%	100.0%	100.0%	100.0%	100.0%
1	-14.4%	-6.5%	-0.1%	0.6%	-2.0%	-4.0%	-2.2%
2	1.3%	2.8%	0.8%	-1.1%	-2.1%	0.2%	0.8%
3	0.9%	0.6%	-0.3%	-1.3%	-1.0%	-1.1%	1.3%
4	1.6%	0.6%	-0.6%	-1.0%	0.0%	0.2%	-0.3%
5	0.5%	0.1%	-0.6%	-1.0%	-0.8%	1.7%	1.3%
6	0.3%	-0.2%	-0.7%	-0.3%	-0.1%	0.3%	1.3%
7	0.0%	-0.5%	-0.7%	-0.2%	0.3%	0.0%	-0.8%
8	0.5%	-0.1%	-0.4%	0.1%	0.4%	0.3%	0.5%
9	0.0%	-0.6%	-0.5%	-0.2%	0.0%	0.0%	0.3%
10	0.1%	0.0%	-0.6%	-0.5%	0.9%	1.5%	-1.0%

The 1–minute returns appear to be perfectly non–correlated, they behave absolutely like a white noise. The 15–second and 30–second returns exhibit rather important negative correlations at the first lag, but after a few lags any trace of correlation disappear. Finally, the longer returns seem to show weakly negative correlations at the first lag.

The overall message of Table 5 is rather ambiguous. As it was the case with the indicators previously considered, it suggests the existence of short–term negative correlations. However, it does not suggest clearly what could be the structure of those negative correlations. To get some idea about it, one has to take a direct look at the spectral structure of the process. It is also worthwhile to examine the ACF of the squared returns or of the absolute value of returns.

We will focus on the first ones. The equivalent table for squared returns reads as follows:

Table 6. Autocorrelation function of the squared euro–dollar returns computed for several discretisation steps

	15 sec.	30 sec.	1 min.	2 min.	4 min.	8 min.	1/4 h
0	100.0%	100.0%	100.0%	100.0%	100.0%	100.0%	100.0%
1	1.1%	1.0%	1.3%	2.3%	3.1%	4.5%	6.6%
2	0.6%	0.7%	0.8%	1.3%	1.9%	2.7%	5.8%
3	0.5%	0.6%	0.8%	1.3%	1.3%	2.5%	3.0%
4	0.4%	0.4%	0.6%	0.9%	1.7%	1.9%	4.3%
5	0.4%	0.4%	0.5%	1.0%	1.5%	2.4%	4.5%
6	0.4%	0.5%	0.5%	0.7%	1.1%	1.9%	4.9%
7	0.3%	0.4%	0.5%	0.8%	1.2%	1.9%	3.4%
8	0.4%	0.3%	0.4%	1.0%	1.1%	1.6%	2.2%
9	0.4%	0.3%	0.4%	1.1%	0.8%	2.4%	2.6%
10	0.3%	0.3%	0.4%	0.7%	1.0%	2.1%	0.9%

The picture is here quite different. All the correlations are positive, all ACF are slowly and smoothly decreasing, ACF or shorter discretisation periods are dominated by ACF of longer ones. Those are the symptoms of the well–known phenomenon of the long-term memory of the squared returns.

In the case of a theoretical continuous–time Wiener process, both tables 5 and 6 would have given a correlation of 100% at row 0, and of 0 at other rows, independently of the period of discretisation. The actual log euro–dollar does not extremely differ from the Wiener process for what regards returns, with the exception of the short–term negative correlations observable for discretisation periods below 1 minute. For what regards squared returns, though, the difference is clear–cut: The phenomenon of long–memory is a well–defined and well–observable empirical regularity that should be entirely absent in the case of a Wiener process.

The long–memory of squared returns, or of absolute returns, which is well documented and which is observable also from daily data, has a connection to the so–called persistence of volatility.

The persistence of volatility refers to the fact that the empirical volatility defined as the standard deviation of returns over a given time–length, e.g. daily returns, tends to remain high (low) after having been high (low).[5]

[5] The persistence of volatility may also apply to implied volatility, instead of empirical volatility. The persistence of the one can cause the persistence of the other, and conversely, because they exist channels by which the empirical volatility can influence the implied volatilities, and channels by which the implied volatility can influence the empirical volatility.

The long–memory means in essence that high returns, in absolute value, appeal subsequent high returns, and thus that high returns are clustered. The long–memory is thus made graphically visible in the aforementioned undulations affecting the variations displayed on Figures 5–7.

To summarise, it seems legitimate to say that the undulations of the variations shown on Figures 5–7, the long–memory of squared returns shown by Table 6 and the persistence of the volatility are three aspects of the same empirical pattern of the log–price process.

Hausdorff–Besicovitch Dimension

So far, we have found only reasons to distinguish the sample path of the log euro–dollar from a Wiener process. Is this to say that the representation of (logarithms of) prices with Wiener processes is at all a wrong idea?

The history of the Wiener representation starts in 1900 with Bachelier. It has been the central intuition (rather than the central "finding") of his dissertation. At that occasion, Bachelier invented the Wiener process without grounding it in a totally mathematically rigorous way. Bachelier also gave intuitive, but logically not rigorous reasons to believe that speculative prices should behave according to a Wiener process (he did not mention about the logarithmic transformation, but that is not essential). In the opinion of Poincaré, those reasons were not convincing.

The idea that (log-) prices are Wiener remained. This idea has been at the origin of two important chapters of the history of finance (in both cases, the grounding paper has been published in the year 1973). First, Mandelbrot noticed that Wiener processes were only a special case of a more general family of processes possessing the important characteristic of being self–similar. Mandelbrot launched the idea that one might obtain a more accurate and more realistic representation of the log prices processes without loosing this property of self–similarity. Second, Black and Scholes (1973) invented the concept of pricing by replication, that was applicable to options under the hypothesis that the log price was following a Wiener process (or more precisely, was probabilistically equivalent to it). The Black and Scholes concept, model and formula were later applied by Garman and Kohlhagen (1983) to the specific case were the price is an exchange rate.

The vision of log prices as Wiener processes has a rich and fecund history. This suggests that it is neither inappropriate nor unrealistic. Likely, it represents some sort of optimal compromise between realism, mathematical tractability and conceptual richness. To remain aware of it, it is useful to underline also non–trivial resemblances between Wiener processes and the log euro–dollar. This is why we will conclude this section with a short glance at a statistical indicator of a special sort, namely the fractal dimension of the sample path or trajectory of the log euro–dollar. This will also be the opportunity to present the contrary situation of the case of the quadratic variations: They did not seem to be a quadratic variation of the log euro–dollar because its discretised approximations did not seem to converge towards an identifiable limit. In the case of the fractal dimension, the approximations will exhibit the required convergence.

We will not say a lot about what is the fractal dimension. In short, one should keep in mind the following points:

- There exist an intuitive notion of dimension, which is clear to everybody. According to this notion, a line has dimension 1, a surface has dimension 2, and a volume has dimension 3.
- There exists a technical way to define this dimension in terms of the limits of something. Actually, there exists several variants, but their technical differences do not matter here.
- Once so defined, the notion of dimension applies not only to lines, surfaces and volumes, but also to structures that are more complex.
- Among those structures, there is the sample path of a Wiener process.

The Hausdorff–Besicovitch dimension of a shape $\mathcal{S}$ is the real value D such that the Hausdorff measure d is infinite for $d < D$ and vanishes for $d > D$. D is also called a *critical dimension.*

The technical way to define the dimension that we will refer to is the so–called Hausdorff–Besicovitch dimension; see Hausdorff (1918), Besicovitch (1929), Ito and McKean (1974) Chap. 2 §5 note 2 pp. 53–54. Not any geometrical structure has a dimension: as it is defined of the limit of something, it may be not defined. The sample path of a Wiener process, however, has a dimension of 1.5: Therefore, it lies, from the point of view of dimensionality, between the lines and the surfaces.

One may wonder whether the log euro–dollar sample path have a dimension and if yes which one. Numerical estimations performed on the data set seem to indicate that there is one, and moreover that this one is equal to 1.5, as for a Wiener process.

3.3 The Spectral Analysis

The four families of indicators considered so far have delivered several messages about the path of the log euro–dollar:

- This path shows statistical regularities,
- Also, it exhibits some likelihood with a Wiener process,
- Even so, it exhibits clear differences from a Wiener process,
- An important difference is that returns are too leptokurtic (the returns of a Wiener process being normal),
- Another important difference is that returns exhibit short term negative correlations (the returns of a Wiener process being independent).

The focus, hereafter, will be put on that last difference. The most systematic way to tackle its study is probably to decompose the returns as a superposition of oscillations of various frequencies. In other words, this is to determine the spectrum of the process of returns, by computing the Fourier transform of the returns sequence.

Theoretical Definition of the Spectrum

The Fourier transform of the sequence of returns converts this sequence of returns into a superposition of sinusoidal oscillations. This transform conveys, in principle, information about the relative weights of frequencies and about the phases of the oscillations. We are here interested in the relative weights of frequencies, but not in the phases. Because of that, it is appropriate to define the spectrum as the modulus of the Fourier transform.

The Fourier transform applies to functions defined on the real axis and taking values in the complex plane. The result of the transform is also a function defined on the real axis and taking values in the complex plane. When moreover the function f takes real values, then its transform $g = F(f)$ is such that its real part is symmetrical (or even) and that its imaginary part is antisymmetrical (or odd). This amounts as saying that $\mathbf{Re}g(x)= \mathbf{Re}g(-x)$ and $\mathbf{Im}\ (g)(x)= -\mathbf{Im}(g)(-x)$ for any x, where $\mathbf{Re}(x)$ and $\mathbf{Im}(x)$ respectively stand for the real and the complex part of x. Hence, the modulus of g is also symmetrical (or even): $|g|(x) = |g|(-x)$. As we define the spectrum as the modulus of the Fourier transform of the returns, it results that this spectrum is symmetrical and therefore that its graph is symmetrical around the vertical axis.

The spectrum is an even function whose argument is a frequency and whose value is the weight with which this frequency is present in the returns process.

So defined, the spectrum of a Gaussian white noise would be a constant function. The spectrum of a theoretically defined random process would be some even real function. We are interested in the process of the returns of the euro–dollar, this process is an empirical one and not a theoretical one. It is not necessarily legitimate to speak of the spectrum of an empirical process. If it is legitimate, then this spectrum is a even real function.

One may speak of the spectrum (of the dimension, of the quadratic variation, etc.) of some empirical process if the discretised equivalent of this spectrum (of this dimension, of this quadratic variation) seems to converge towards some limit when the discretisation becomes thinner. This may or not be the case, and we have so far encountered examples of the two situations. As we have seen, there is no quadratic variation of the log euro–dollar, while there is one of a simulated Wiener process. In contrast, there is a Hausdorff–Besicovitch dimension of the path of the log euro–dollar, as well as of a simulated Wiener process. A study about the spectrum of some empirical process is not only about the computation of some Fourier transform, but also about the check that those transforms, computed for increasingly thinner discretisations, do converge towards something. Therefore, we will have to know how to consider jointly the spectrum of, say, the 15–second and the 30–second discretisations. We deal with this issue in the next paragraph.

Applied Construction of the Spectrum

Our problem is an applied one and not a purely mathematical one: we are obliged to perform numerical calculations to estimate an empirical spectrum. Thus, we have to proxy the mathematical objects (the functions from and to the real axis) by numerical objects, a finite sequence of numbers, and to replace the mathematical Fourier transform by the relevant numerical algorithm. This one, of course, is the standard discrete Fourier transform. When the finite sequence of numbers has length 2^n for some integer n, then the celebrated Fast Fourier Transform (FFT) algorithm applies.

We encounter two practical difficulties:

1. The number of returns is not of the form 2^n,
2. The number of returns is too big to be plotted, or the graph would be difficult to understand.

The solutions we apply are the following:

- We transform the series of returns into a series of 2^n returns, for the highest possible n, as follows. We consider the sequence of returns as a stepwise constant function, defined on some continuous time interval. We take its primitive, which is thus a stepwise linear function on the same continuous time interval. We cut the interval in 2^n subintervals of equal length. We consider the (left side) derivative of the stepwise linear function at the ends of each of those subintervals. Those are our 2^n returns, to which we can apply the FFT. The procedure has no effect if the number of returns has already the 2^n form, which is obviously quite unlikely. For example, the number of 15–second returns is 1,848,364, whose logarithm of base 2 lies between 20 and 21. The transformed series counts then 2^{20} returns.
- Regarding the reduction of the number of elements of the spectrum, we proceed as follows. We will divide the size of the spectrum by 2^{11}, so for example the spectrum of the 15–second returns should contain only $512 = 2^9$ points instead of 2^{20} points. The construction of this "aggregated spectrum" is done by the way of quadratic averages. The value of range k of the aggregated spectrum is the square root of the average of the squares of the values of range $2^{11}k, 2^{11}k+1, \ldots, 2^{11}(k+1)$, where the first and the last one enter with half weight only. One easily sees that this results into a symmetrical aggregated spectrum, where the kth first value is equal to the kth last value.

Plotting the Spectra

We plot those Fourier transform and the resulting spectra according to the convention that the zero frequency is at the centre, the highest positive frequency is on the right and the highest negative frequency on the left of the graph.

The following graph shows the Fourier transform (two thin dotted lines) and the resulting aggregated spectrum (one bold line) of the 15–second returns. (The scale of the spectrum has been doubled.)

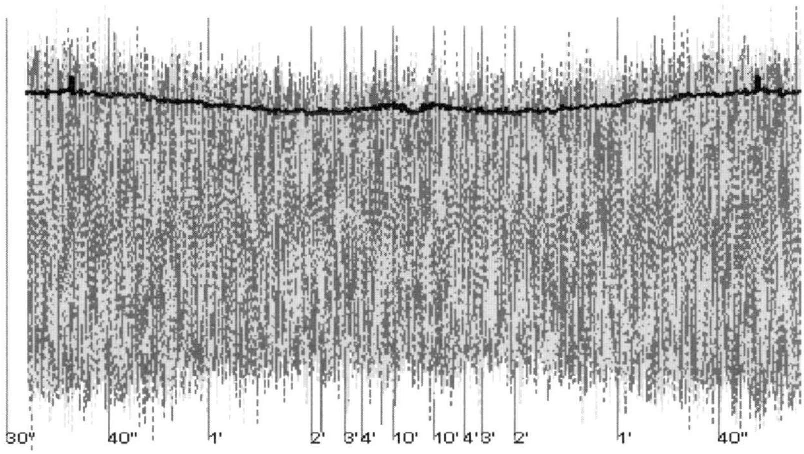

Fig. 8. Euro: Fourier transform of returns (Source of raw data: Reuters)

The two thin dotted lines of the Fourier transform correspond to its real part and its imaginary part. Actually, they look confounded in some cloud of points, due to the high number of data. Even so, the Figure 8 already delivers a message, namely that the high frequencies have slightly, but clearly, more weight than the low frequencies.

The spectrum captures the shape of this "cloud" of frequencies. The overweight of high frequencies gives it some convexity. The information that it carries is nonetheless not limited to this convexity. To get it fully, one needs to zoom on the vertical axis. One obtains the following figure:

On that figure, not only the 15–second spectrum, but also several spectra have been represented. The other ones, which occupy central sub-parts of the figure, overlap with the 15–second spectrum and are practically confounded with it. (Frequencies are designated by their reciprocal period.)

Figure 9 reveals that the actual form of the spectrum of the returns is not completely elementary. One can describe this form by three observations

1. Between zero and 10 minutes, frequencies have increasing weights,
2. Frequencies lying in the range 10 minutes to 2 minutes have decreasing weights,

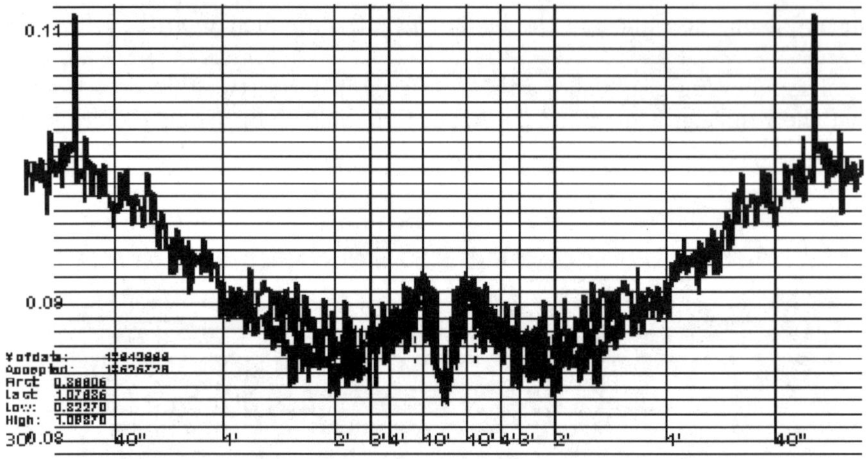

Fig. 9. Euro: spectra of returns (Source of raw data: Reuters)

3. Frequencies higher than 2 minutes have increasing weights, those weights seem to indefinitely increase and the highest weight overall are obtained for the highest frequencies.

The third observation, which was already suggested by Figure 8, points towards the spectrum of a fractionally derived white noise or of an Ohrstein–Uhlenbeck process. In the words of Mandelbrot, this is a "negative Joseph effect". This effect is clearly responsible for the negative correlations we found when studying the ACF, and for the symptoms of negative correlations we found by studying the repartitions of returns and the quadratic variations.

All together, those three observations capture the pattern, or the form, of the spectrum. Details that are more precise could be artefacts or the effect of randomness. However, the two peaks in the area of 30–40 seconds are striking, but we do not propose any explanation for them: They could be an artefact or have a real meaning. In any case, we do not account for them in the important remaining task of getting a stylised representation of the spectrum. A good stylised representation would be a depiction of the spectrum that simultaneously achieves an acceptable accuracy and keeps an acceptable level of simplicity. Figure 9 suggests that correct stylised representation of the spectrum describes it as the juxtaposition of three linear parts. We will rely on this muster when we will simulate a process whose spectral structure replicates the one of the euro–dollar.

The spectrum computation has been also conduced on subsets of the total data sample. It is remarkable that the form revealed by Figure 9 is already recognisable for time intervals of only some months. This form seems to be some constant characteristic of the process, just as the forms of the histograms, figures 11 and 12, in sub–section 3.4. As far as one can judge from the examination of our data sample, the ternary scheme of Figure 9 is the spectral signature of the euro–dollar.

3.4 Reconciling the Results

So far, we have examined various statistical indicators and characteristics of the log-price process of the euro–dollar. A successful description of this process would be a mathematical specification of this process such that the process must fit with this statistical behaviour. This means that

- The distributions of returns, variations, ACF and spectra should obey to some mathematical forms (preferably explicit ones),
- Those mathematical forms are recognisable on the tables and graphs shown so far,
- For a sample whose size is the size we have used, between 2^{20} and 2^{21} 15–second returns, those forms are obtained with the precision that is shown on those tables or graphs,
- A simulation of the mathematical process whose sample size is the same provides comparable tables and graphs.

This would be the criteria of a complete success, and their achievement would mean that the "physics" of the euro–dollar have been found. To see how far we are from this ambitious goal, the simplest is actually to rely on a simulation experiment. Specifically, one can build in a simulated process both the distribution of the 15–second returns and their spectrum, achieving so a partial reconciliation of our results. Then, we have to see how the process behaves regarding other statistical indicators.

Construction of the Simulated Process

We construct of the simulated process in two phases.

First, we generate N independent random variables whose distribution coincides with the empirical distribution of the 15–second returns.

The algorithm used for the underlying generation of N uniformly distributed variables follows Knuth's suggestion and has been taken from Press *et al.* (1988) (chap. 7, §1, p. 212). As it has an effective period of 714,025, we have to choose N below this limit (by comparison, the size of the actual sample of 15–second returns is 1,848,364). Actually, we have to choose N below or equal to the highest integer power of 2 which is less than 714,025, and that is $2^{19} = 524,288$.

We want to reproduce the conditions in which we studied the empirical process with the maximum of fidelity. We choose therefore a number which is in the same proportion to the immediately proceeding integer power of two than 1,848,364 is to 2^{20}. This leads us to take $N = 458,488$.

We generate $2^{19} = 524,288$ uniformly distributed deviates. Those are then transformed into as many euro–dollar-return-like distributed deviates. To do that,

1. We symmetrise the empirical distribution of those returns,
2. We fit the symmetrised distribution function, between 0 and 25 pips, with the ratio of two polynomials (each of degree four),
3. And we apply the reciprocal of this rational function to the uniform deviates.

This sequence of 524,288 deviates has a flat spectrum because the deviates are independents: it is a white noise, albeit a non–Gaussian one. Then, all what remains to do is to modify the spectral structure of this sequence. This is done by applying a direct Fast Fourier Transform, by multiplying the result with the desired spectrum, and by applying a reverse Fast Fourier Transform. The spectrum we have used was not directly the empirical spectrum, but its three–part piecewise–linear muster. We have then a sequence of 524,288 deviates having the required spectrum. Out of those ones, we select only the 458,488 first ones, in order to reproduce the case of the empirical sample, whose cardinal was not a power of two. Those 458,488 simulated returns constitute our new sample and we apply to it exactly the same algorithms that we applied to the sample of empirical 15–seconds returns.

In particular, we compute their spectrum, which looks as follows:

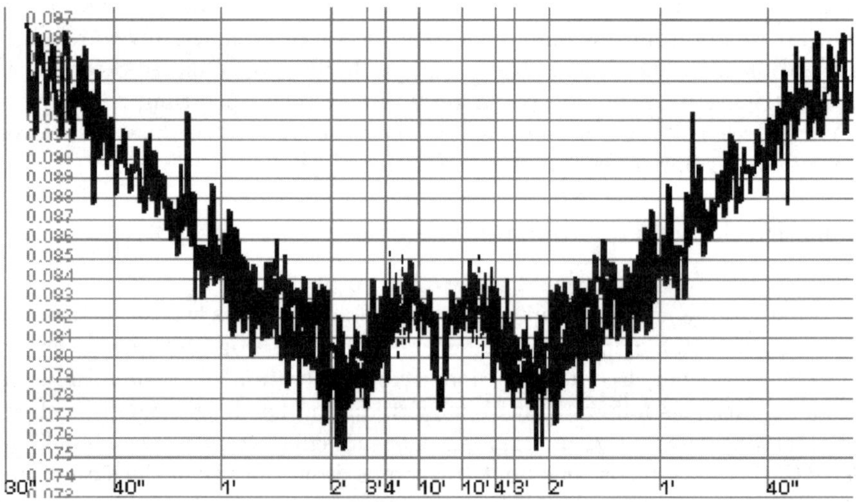

Fig. 10. Simulation: spectra of returns

The two peaks in the area of 30–40 seconds that were visible on Figure 9 are absent from Figure 10 because we did not take them on board of the stylised representation of the spectrum. Apart from that, the simulated spectrum reproduces the empirical spectrum.

Similarly, the actual distribution of the 15–second returns, see Figure 11, is well reproduced by the distribution of the simulated returns displayed by Figure 12.

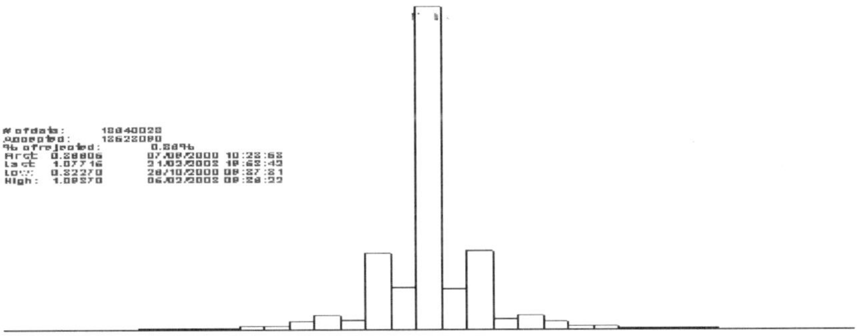

Fig. 11. Euro: Histogram of 15–second returns (Source of raw data: Reuters)

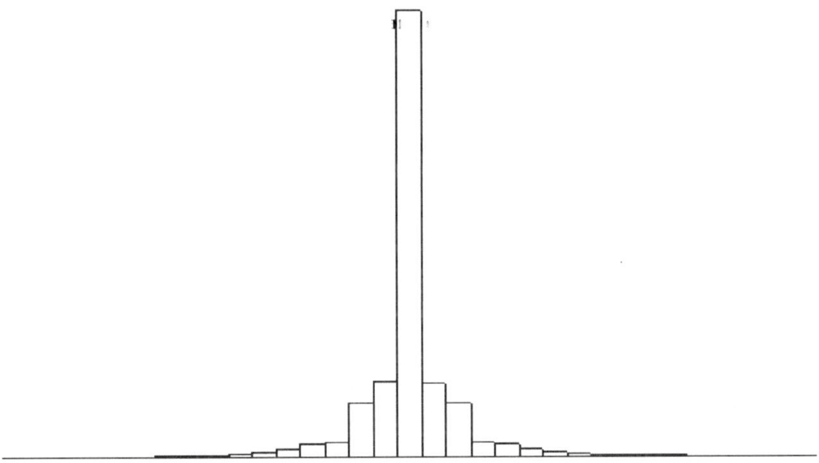

Fig. 12. Simulation: Histogram of 15–second returns

The only apparent difference concerns the slight over–representation of the returns with integer number of pips and under–representation of the returns with half integer number of pips. This empirical pattern, explained above in 3.2.1, is visible on Figure 11 but not on Figure 12. (The elimination of this empirical pattern was wishable. It was obtained when we fitted the probability distribution of the returns with a smooth function, namely a rational fraction, which was too rigid to account for it.)

The ACF, the Variations and the Dimension of the Simulated Process

The simulated process has the correct distribution of 15–second returns and the correct spectrum. It replicates the behaviour of the actual process with respect to those two statistical indicators. The remaining question is whether it also replicates the behaviour of the actual process with respect to the others statistical indicators that have been considered so far. We shall restrict our attention to the case of the ACF and to the cases of the quadratic variation (QV), absolute variation (AV) and squared root variation (SV).

We consider the ACF of the simulated process for the same discretisation periods than for the actual process. We obtain the following results:

Table 7. Autocorrelation function of the returns of the simulated process computed for several discretisation steps

	15 sec.	30 sec.	1 min.	2 min.	4 min.	8 min.	1/4 h
0	100.0%	100.0%	100.0%	100.0%	100.0%	100.0%	100.0%
1	-17.5%	-8.7%	-2.9%	0.1%	-0.8%	-3.3%	-3.5%
2	2.1%	0.5%	1.4%	-0.4%	-2.8%	-2.0%	-2.2%
3	-1.1%	1.4%	0.1%	-0.9%	-1.4%	-1.5%	3.2%
4	0.9%	0.6%	-0.2%	-1.6%	-0.3%	-0.2%	-1.4%
5	0.2%	0.4%	-0.3%	-1.2%	-0.3%	0.6%	-0.6%
6	0.7%	0.2%	-0.8%	-0.6%	-0.3%	1.3%	0.0%
7	0.4%	-0.1%	-0.3%	-0.6%	-1.5%	-0.2%	1.4%
8	0.1%	0.0%	-1.1%	0.2%	-0.2%	-0.8%	0.6%
9	0.3%	-0.6%	-0.4%	-0.4%	0.5%	-0.6%	-1.3%
10	0.2%	0.1%	-0.5%	-0.3%	0.3%	1.0%	-1.1%

Those results are to be compared with the ones contained in Table 5. The structures revealed by Table 5 and by Table are quite similar. There is only one noticeable difference: The 1–minute ACF of the actual process was practically identical to the one of a white noise, namely a Dirac, i.e., 1 for lag zero, 0 for other lags. This is not so precisely done by the 1–minute ACF of the simulated process. This might be because the simulated sample is 4 times shorter than the actual one.

Apart from that, the comparison of the two tables displays the same quantitative order of magnitudes of the individual correlations and the same overall picture.

Let us consider now the ACF of the squared returns. For the simulated process, it is given by the following table:

Table 8. Autocorrelation function of the squared returns of the simulated process computed for several discretisation steps

	15 sec.	30 sec.	1 min.	2 min.	4 min.	8 min.	1/4 h
0	100.0%	100.0%	100.0%	100.0%	100.0%	100.0%	100.0%
1	1.6%	0.6%	0.9%	0.9%	0.5%	1.1%	0.3%
2	-0.1%	0.6%	-0.2%	0.3%	0.1%	0.3%	0.5%
3	0.5%	0.0%	0.3%	-0.3%	0.8%	-0.7%	0.2%
4	-0.1%	0.1%	0.0%	-0.3%	0.0%	-0.8%	0.1%
5	0.1%	-0.1%	0.0%	0.3%	0.9%	-0.5%	-1.6%
6	-0.1%	0.0%	-0.1%	0.2%	-1.0%	0.2%	1.9%
7	0.1%	0.1%	-0.1%	0.2%	0.0%	-1.2%	-0.4%
8	-0.1%	0.0%	0.0%	0.8%	-0.3%	0.3%	0.4%
9	0.0%	0.1%	-0.1%	-0.2%	-0.4%	-1.5%	0.0%
10	0.1%	-0.3%	0.2%	-0.1%	-0.4%	-0.4%	0.0%

Here, the discrepancy between Tables 6 and 8 is blatant. The simulated process does not exhibit the slightest trace of long–memory of the squared returns or of persistence of the volatility. This is not surprising since we did not explicitly build this pattern in the simulated process. It simply proves that the long–memory of the squared returns or the persistence of the volatility was not a hidden consequence of the spectral structure of the process and/or of the form of the distribution of the returns.

As the undulations of the variations were linked to this long–memory pattern, one should logically expect that they are absent for our simulated process.

Let us therefore finally consider the three types of variations (QV, AV, SV) that we computed for the eleven periods of discretisations for the actual process. As we expected, the behaviours of the actual and simulated processes do not perfectly coincide, as the undulations actually disappeared. The two behaviours show however resemblances in other regards.

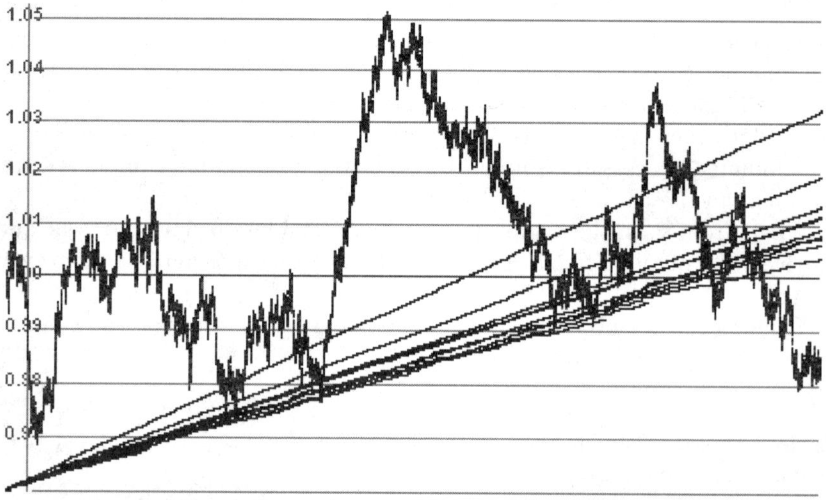

Fig. 13. Simulation: spot and realized quadratic variations

We compare figures 5–7 with figures 13–15 and make the following observations:

- The figures show that the variations are straight lines with different slopes. They are actually straighter than it was the case for the actual log euro–dollar. So we see at once that the "persistence of the volatility", which is a characteristic feature of the actual process, is not reproduced by the simulated process.
- The ranking of the slopes of the quadratic variations is correctly reproduced. The highest slope corresponds to the quadratic variation of lowest

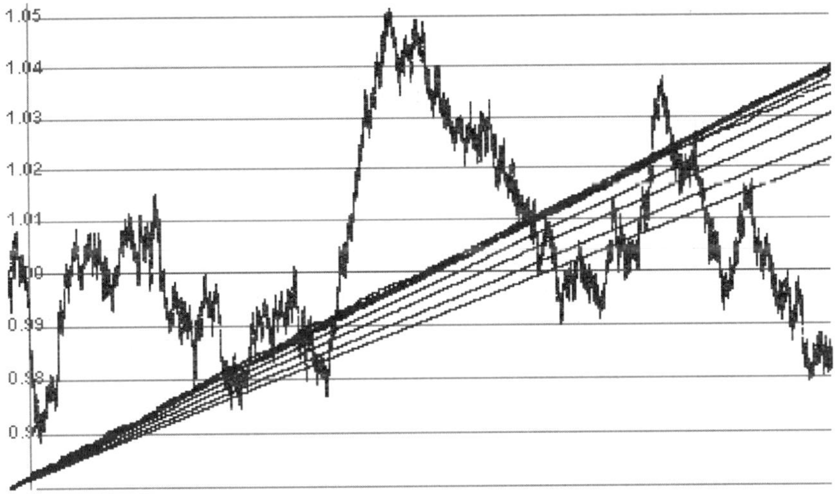

Fig. 14. Simulation: spot and realized absolute variations

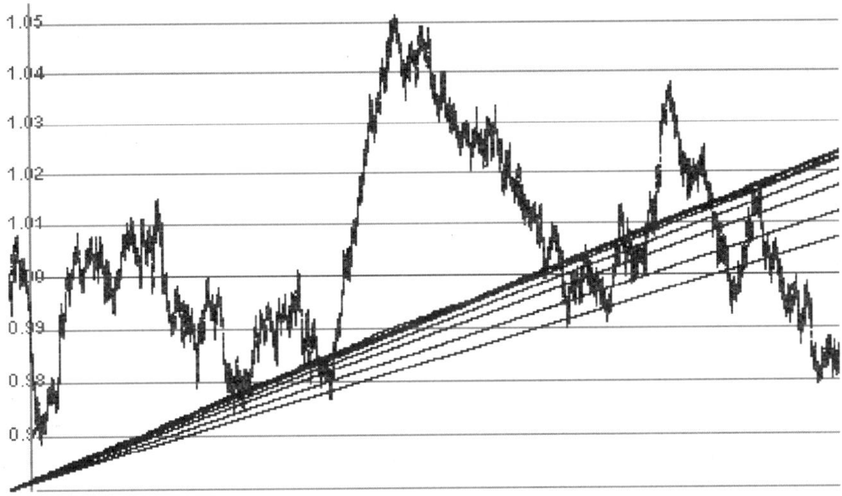

Fig. 15. Simulation: spot and realized squared root variations

discretisation period and the higher a slope, the lower the discretisation period. This common ranking was a feature of the actual log euro–dollar that we interpreted as a symptom of negative short–term correlations; we find it again in the simulated process.

- The reverse ranking (15–second, lower slope, 4–hour, higher slope) that we found for the squared root variation in the case of the actual process is equally present in the case of the simulated process. This reverse ranking is also conserved for the absolute variation, however here there is a difference:

The slopes are clearly more differentiated in the case of the simulated process than in the case of the actual process.

Finally, let us mention that the simulated process has a dimension which is as well fitted and as closed to 1.5 that was the case of the actual process and that would be the case of a Wiener process of comparable size.

The overall assessment seems to be that the simulated process reproduces well enough some statistical regularities and misses some other ones. The regularities that are missed are the ones that are linked to that empirical pattern showing through the undulations of the variations, through the long–memory of squared returns and through the persistence of the volatility.

Can we better that? This empirical pattern could be built in a simulated process by introducing a distortion of time. Instead of a process $Z(t)$, without long–memory of the squared returns, consider the process $Y(t) = Z(\theta(t))$, where $\theta(t)$ is increasing, but non–linear, function of t, $\theta(t)$ can be random, provided that it is independent of Z. The process Y will display some memory of squared returns, and the variations of Y will exhibit the required undulations. However, the shape of the distributions of the returns and/or the spectral structure of Z might be also affected by this transformation. The design of a process reproducing together the spectral structure, the distribution of returns *and* the long–memory of returns is probably not as simple as what we have done here. If it could be obtained with an acceptably simple mathematical construction, this would be a satisfactory description of the behaviour of the euro–dollar.

4 Conclusion

We have tried to categorise some empirical regularities of the euro–dollar log–price process and to disentangle some simple mathematical structure that would cause or explain each of those regularities. One of the empirical regularity on which we give a special attention was the spectral structure of euro–dollar log–price process. We were interested in the identification of a simple and recurrent shape for the spectrum of this process. To correctly measure this particular empirical regularity, the spectral structure, we have been led to examine some points regarding the quoted prices that constitute the data set. In particular, we identified a problem, namely the quick and random succession of quotes of heterogeneous qualities. This quick and random succession generates a high frequency noise that can severely perturb the measurement of the spectrum of the process.

To cope with this technical problem, we introduced a specific filtering algorithm, called "dynamical filtering", and we compare the output of this algorithm to traded (instead of quoted) data that are certainly free of this high–frequency noise. We actually identified a particular shape, form, or pattern, for the spectrum of the log–price of the euro–dollar. We also considered two other empirical regularities of this process, one being the shape of the short–term returns, the other being the long–memory of the squared returns. We did not find a tractable way to simulate a process that would reproduce the three empirical regularities, namely the spectrum, the distribution of short–term return and the long–memory. We shown that it was easy to simulate a process that would reproduce two out of those three empirical regularities, namely the spectrum and the distribution of short–term returns.

References

1. Bachelier, L. (1900). Théorie de la spéculation. *Annales de l'École Normale Supérieure*, **3,** 21–86.
2. Besicovitch, A.S. (1929). Linear set of points of fractional dimension, *Mathematische Annalen*, **101,** 161–198.
3. Black, F. and Scholes, M. (1973). The Pricing of Options and Corporate Liabilities. *Journal of Political Economy*, **81,** 637–654.
4. Garman, M.B. and Kohlhagen, S.W. (1983). Foreign Currency Option Values. *Journal of International Money and Finance*, **2,** 231–237.
5. Hausdorff, F. (1918). Dimension und äußeres Maß. *Mathematische Annalen*, **79,** 157–179.
6. Ito, K. and McKean, H.P. (1974). *Diffusion Processes and their Sample Paths.* Springer Verlag.
7. Mandelbrot, B.B. (1973a). Formes nouvelles du hasard dans les sciences. *Économie Appliquée*, **26,** 307–319.
8. Mandelbrot, B.B. (1973b). Le problème de la réalité de cycles lents et le "syndrome de Joseph". *Économie Appliquée*, **26,** 349–365.
9. Mandelbrot, B.B. (1973c). Le syndrome de la variance infinie et ses rapports avec la discontinuité des prix. *Économie Appliquée*, **26,** 321–348.
10. Mandelbrot, B.B. (1997). *Fractales, Hasard et Finance.* Flammarion.
11. Mandelbrot, B.B. (1999). *Multifractals and $1/f$ Noise.* Springer Verlag.
12. Müller U. A., Dacorogna, M.M. Davé, R.D. Pictet, O.V., Olsen, R.B., and Ward, J.R. (1993). Fractals and intrinsic time - a challenge to econometricians. *Olsen and Associates Research Group, Preprint.*
13. Press, W.H., Teukolsky, S.A., Vetterling, W.T. and Flannery, B.P. (1988). *Numrerical Recipes in C. The Art of Scientific Computing.* Cambridge University Press.

Hölderian Invariance Principles and Some Applications for Testing Epidemic Changes

Alfredas Račkauskas[1] and Charles Suquet[2]

[1] Vilnius University and Institute of Mathematics and Informatics, Department of Mathematics, Vilnius University, Naugarduko 24, Lt-2006 Vilnius, Lithuania `Alfredas.Rackauskas@maf.vu.lt`

[2] Université des Sciences et Technologies de Lille, Laboratoire de Mathématiques appliquées F.R.E. CNRS 2222. Bât. M2, U.F.R. de Mathématiques, F-59655 Villeneuve d'Ascq Cedex, France. `Charles.Suquet@univ-lille1.fr`

1 Introduction

The well known invariance principle of probability theory introduced by Donsker and Prokhorov received and continue to receive many compliments from both theoreticians and practitioners in very broad area of researches. Econometricians use it to tackle the "unit root" problems (see e.g. Stock (1994) and cited literature therein) to analyze structural changes (see e.g. Csörgő and Horváth (1997) and cited literature therein) or "bubbles" phenomenon (see e.g. Kirman and Teyssière (2002) and cited literature therein) etc. It is impossible to name all problems even in econometrics only, the solutions of which were obtained with help of invariance principle. In this review we present some of recent results (mainly obtained by the authors of this review) on invariance principles with respect to Hölder topologies. We emphasize a new construction of polygonal line process and applications to the structural change problems.

In more modern mathematical form, invariance principle can be formulated as follows. A sequence of stochastic processes $\{\xi_n(t), t \in T\}_{n\in\mathbb{N}}$ induces a sequence of measures $\{\mu_n, n \in \mathbb{N}\}$ on some suitable topological function space, say, E. One proves then that the measures converge weakly to a measure μ corresponding to a limiting process, say $\{\xi(t), t \in T\}$. Weak convergence of measures, in the case where E is a separable metric space, gives

$$f(\xi_n) \xrightarrow{\mathcal{D}} f(\xi), \tag{1}$$

for any f a real-valued function continuous μ-almost everywhere on E, where $\xrightarrow{\mathcal{D}}$ denotes convergence in distribution. Usually $\xi_n, n \in \mathbb{N}$, are polygonal line processes constructed from partial sums of random variables. More modern theory deals also with polygonal line processes based on data with values in a completely arbitrary, perhaps infinite-dimensional sample space. This general approach is important for statisticians either working with data such as "functions" or "pictures" (seismographs, electrocardiograms, high-dimensional biomedical data, high-frequency finance etc.) or dealing with "infinite-dimensional parameters" such as distribution function, density, characteristic function etc. For statistical inference it is usually necessary to rely on the asymptotic convergence result (1) for a certain class $\mathcal{F}$ of functions f. The class $\mathcal{F}$ aims to highlight appropriate geometry of paths of processes under consideration which on its turn, in the case of partial sums polygonal line process, reflects certain properties of the data under consideration. Examples of paths spaces customary include L_2, Skorohod's $D[0,1]$ and $\mathcal{C}[0,1]$ the space of continuous functions. Each of these spaces corresponds to a particular geometry of paths of processes. So, L_2 spaces describe mean square properties (variance, mean square errors, energy etc.) the space $D[0,1]$ is responsible for extreme values and $\mathcal{C}[0,1]$ deals with continuity (increments, modulus of continuity etc.) of processes.

This review bestow attention to Hölder spaces which particularly well highlight variation properties (p-variation, α-slopes) of processes. Variation of polygonal line process reflects various epidemic type changes in a structure of data (epidemic or bubble type changes in a mean, in distribution or other perhaps infinite-dimensional parameters). That's precisely why we advocate to use Hölder continuous functionals of polygonal line processes as test statistics of structural stability in data under various epidemic alternatives. Another motivations to bestow Hölder spaces is many-sided properties of Wiener process. Paths of Wiener process are not only square integrable, bounded or continuous. They have also certain Hölder continuity. This fact to our opinion is not yet exploited in full extent.

The paper is organized as follows. In Section 2 we introduce Hölder spaces of continuous functions. Sections 3 and 4 are devoted to invariance principles with respect to Hölder spaces. Firstly we consider classical polygonal line process constructed from partial sums of random variables (elements with values in a Banach space). Necessary and sufficient conditions for their convergence in a Hölder space are discussed. Next we advocate another construction of polygonal line process. Differently from the classical case, this new construction puts abscissas of vertices at random points. It is shown that so constructed polygonal line process in many aspects behaves better than the classical one.

Sections 5 and 6 present in a rather broad perspective some applications of Hölderian invariance principles to the problem of testing stability in the sample under epidemic alternatives.

2 Some Facts About Hölder Spaces

The introduction of Hölder spaces answers the purpose to quantify the global smoothness of functions by controlling their modulus of uniform continuity.

Let us precise this by an informal description of the most familiar case. For fixed $0 < \alpha < 1$, $\mathcal{H}^\alpha$ is the set of functions $x : [0,1] \to \mathbb{R}$ such that $|x(t) - x(s)| \le K|t-s|^\alpha$ for some constant K depending only on x and α. The best constant K in this uniform estimate defines a semi-norm on the vector space $\mathcal{H}^\alpha$. By adding $|x(0)|$ to this semi-norm we obtain a norm $\|x\|_\alpha$ which makes $\mathcal{H}^\alpha$ a non separable Banach space. Clearly if $0 < \alpha < \beta < 1$, $\mathcal{H}^\beta$ is topologically embedded in $\mathcal{H}^\alpha$ and all these Hölder spaces are topologically embedded in the classical Banach space $\mathcal{C}$ of continuous functions $[0,1] \to \mathbb{R}$.

To remedy the non separability drawback of $\mathcal{H}^\alpha$, one introduces its subspace $\mathcal{H}^{\alpha,o}$ of functions x such that $|x(t) - x(s)| = o(|t-s|^\alpha)$ uniformly. This subspace is *closed* (hence also a Banach space for the same norm $\|x\|_\alpha$) and *separable.*

One interesting feature of the spaces $\mathcal{H}^{\alpha,o}$ is the existence of a basis of triangular functions, see Ciesielski (1960). It is convenient to write this basis as a triangular array of functions, indexed by the dyadic numbers. Let us denote by D_j the set of dyadic numbers in $[0,1]$ of level j, i.e.

$$\mathrm{D}_0 = \{0,1\}, \qquad \mathrm{D}_j = \{(2l-1)2^{-j};\ 1 \le l \le 2^{j-1}\}, \quad j \ge 1.$$

Write for $r \in \mathrm{D}_j$, $j \ge 0$,

$$r^- := r - 2^{-j}, \quad r^+ := r + 2^{-j}.$$

For $r \in \mathrm{D}_j$, $j \ge 1$, the triangular Faber-Schauder functions Λ_r are continuous, piecewise affine with support $[r^-, r^+]$ and taking the value 1 at r:

$$\Lambda_r(t) = \begin{cases} 2^j(t - r^-) & \text{if } t \in (r^-, r]; \\ 2^j(r^+ - t) & \text{if } t \in (r, r^+]; \\ 0 & \text{else.} \end{cases}$$

When $j = 0$, we just take the restriction to $[0,1]$ in the above formula, so

$$\Lambda_0(t) = 1 - t, \quad \Lambda_1(t) = t, \quad t \in [0,1].$$

The sequence $\{\Lambda_r; r \in \mathrm{D}_j,\ j \ge 0\}$ is a Schauder basis of $\mathcal{C}$. Each $x \in \mathcal{C}$ has a unique expansion

$$x = \sum_{j=0}^{\infty} \sum_{r \in \mathrm{D}_j} \lambda_r(x) \Lambda_r, \tag{2}$$

with uniform convergence on $[0,1]$. The Schauder scalar coefficients $\lambda_r(x)$ are given by

$$\lambda_r(x) = x(r) - \frac{x(r^+) + x(r^-)}{2}, \quad r \in \mathrm{D}_j,\ j \ge 1, \tag{3}$$

and in the special case $j = 0$ by

$$\lambda_0(x) = x(0), \quad \lambda_1(x) = x(1). \tag{4}$$

The partial sum $\sum_{j=0}^{n}$ in the series (2) gives the linear interpolation of x by a polygonal line between the dyadic points of level at most n.

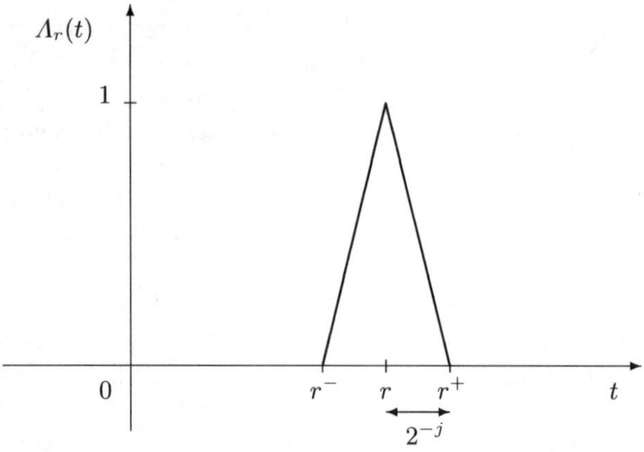

Fig. 1. The Faber-Schauder triangular function Λ_r

Ciesielski (1960) proved that $\{\Lambda_r; r \in \mathrm{D}_j,\ j \geq 0\}$ is also a Schauder basis of each space $\mathcal{H}^{\alpha,o}$ (hence the convergence (2) holds in the $\mathcal{H}^\alpha$ topology when $x \in \mathcal{H}^{\alpha,o}$) and that the norm $\|x\|_\alpha$ is equivalent to the following sequence norm :

$$\|x\|_\alpha^{\mathrm{seq}} := \sup_{j \geq 0} 2^{j\alpha} \max_{r \in \mathrm{D}_j} |\lambda_r(x)|.$$

This equivalence of norms provides a very convenient discretization procedure to deal with Hölder spaces and is extended in Račkauskas and Suquet (2001b) to the more general setting of Hölder spaces of Banach space valued functions x, with a modulus of continuity controlled by some weight function ρ.

Let $(\mathbb{B}, \|\ \|)$ be a separable Banach space. We write $\mathcal{C}_\mathbb{B}$ for the Banach space of continuous functions $x : [0,1] \to \mathbb{B}$ endowed with the supremum norm $\|x\|_\infty := \sup\{\|x(t)\|\,;\ t \in [0,1]\}$. Let ρ be a real valued non decreasing function on $[0,1]$, null and right continuous at 0. Put

$$\omega_\rho(x,\delta) := \sup_{\substack{s,t \in [0,1],\\ 0<t-s<\delta}} \frac{\|x(t)-x(s)\|}{\rho(t-s)}.$$

Denote by $\mathcal{H}_\mathbb{B}^\rho$ the set of continuous functions $x : [0,1] \to \mathbb{B}$ such that $\omega_\rho(x,1) < \infty$. The set $\mathcal{H}_\mathbb{B}^\rho$ is a Banach space when endowed with the norm

$$\|x\|_\rho := \|x(0)\| + \omega_\rho(x,1).$$

Define

$$\mathcal{H}_\mathbb{B}^{\rho,o} = \{x \in \mathcal{C}_\mathbb{B} :\ \lim_{\delta\to 0} \omega_\rho(x,\delta) = 0\}.$$

Then $\mathcal{H}_\mathbb{B}^{\rho,o}$ is a closed separable subspace of $\mathcal{H}_\mathbb{B}^\rho$. We shall abbreviate $\mathcal{C}_\mathbb{R}$, $\mathcal{H}_\mathbb{R}^\rho$ and $\mathcal{H}_\mathbb{R}^{\rho,o}$ in $\mathcal{C}$, $\mathcal{H}^\rho$ and $\mathcal{H}^{\rho,o}$ correspondingly. Our main examples of Hölder spaces use as weight function $\rho = \rho_{\alpha,\beta}$, $0 < \alpha < 1$, $\beta \in \mathbb{R}$ defined by:

$$\rho_{\alpha,\beta}(h) := h^\alpha \ln^\beta(c/h), \quad 0 < h \leq 1,$$

for a suitable constant c. For $\rho = \rho_{\alpha,\beta}$, we shall write $\mathcal{H}_{\mathbb{B}}^{\alpha,\beta}$ and $\mathcal{H}_{\mathbb{B}}^{\alpha,\beta,o}$ for $\mathcal{H}_{\mathbb{B}}^{\rho}$ and $\mathcal{H}_{\mathbb{B}}^{\rho,o}$ respectively and we abbreviate $\mathcal{H}_{\mathbb{B}}^{\alpha,0,o}$ in $\mathcal{H}_{\mathbb{B}}^{\alpha,o}$. As above, the subscript $\mathbb{B}$ will be omitted when $\mathbb{B} = \mathbb{R}$.

In what follows, we assume that the modulus of smoothness ρ satisfies the following technical conditions where c_1, c_2 and c_3 are positive constants:

$$\rho(0) = 0,\ \rho(\delta) > 0,\ 0 < \delta \leq 1; \tag{5}$$

$$\rho \text{ is non decreasing on } [0,1]; \tag{6}$$

$$\rho(2\delta) \leq c_1 \rho(\delta), \quad 0 \leq \delta \leq 1/2; \tag{7}$$

$$\int_0^\delta \frac{\rho(u)}{u}\,du \leq c_2 \rho(\delta), \quad 0 < \delta \leq 1; \tag{8}$$

$$\delta \int_\delta^1 \frac{\rho(u)}{u^2}\,du \leq c_3 \rho(\delta), \quad 0 < \delta \leq 1. \tag{9}$$

For instance, elementary computations show that the functions $\rho_{\alpha,\beta}$ satisfy Conditions (5) to (9), for a suitable choice of the constant c, namely $c \geq \exp(\beta/\alpha)$ if $\beta > 0$ and $c > \exp(-\beta/(1-\alpha))$ if $\beta < 0$. For any ρ satisfying (5) to (9), we have the equivalence of norms :

$$\|x\|_\rho \sim \|x\|_\rho^{\mathrm{seq}} := \sup_{j \geq 0} \frac{1}{\rho(2^{-j})} \max_{r \in \mathrm{D}_j} \|\lambda_r(x)\|,$$

where the $\mathbb{B}$-valued coefficients $\lambda_r(x)$ are still defined by (3) and (4).

The space $E = \mathcal{H}_{\mathbb{B}}^{\rho,o}$ may be used as the topological framework for functional central limit theorems. Among various continuous functionals f to which we can then apply the convergence (1), let us mention the norms $f_1(x) = \|x\|_\rho$ and $f_2(x) = \|x\|_\rho^{\mathrm{seq}}$, which are closely connected to the test statistics $\mathrm{UI}(n,\rho)$ and $\mathrm{DI}(n,\rho)$ proposed below for the detection of epidemic changes. Other examples of Hölder continuous functional and operators like p-variation, fractional derivatives are given in Hamadouche (2000).

To deal with convergences in distribution of stochastic processes considered as random elements in $\mathcal{H}_{\mathbb{B}}^{\rho,o}$, a key tool is the following tightness criterion established in Račkauskas and Suquet (2003b).

Theorem 1. *Suppose the Banach space $\mathbb{B}$ is separable. Then the sequence $(\xi_n)_{n\geq 1}$ of random elements in $\mathcal{H}_{\mathbb{B}}^{\rho,o}$ is tight if and only if it satisfies the two following conditions.*

(i) For each dyadic $t \in [0,1]$, *the sequence of $\mathbb{B}$-valued random variables $(\xi_n(t))_{n\geq 1}$ is tight on $\mathbb{B}$.*

(ii) For each positive ε,

$$\lim_{J\to\infty} \sup_{n\geq 1} \mathbf{P}\left\{ \sup_{j>J} \frac{1}{\rho(2^{-j})} \max_{r\in \mathrm{D}_j} \|\lambda_r(\xi_n)\| > \varepsilon \right\} = 0.$$

3 Hölderian Invariance Principles

Let $X_1, \dots, X_n, \dots$ be i.i.d. random elements in the separable Banach space $\mathbb{B}$. Set $S_0 = 0$,

$$S_k = X_1 + \cdots + X_k, \quad \text{for} \quad k = 1, 2, \dots$$

and consider the partial sums processes

$$\xi_n(t) = S_{[nt]} + (nt - [nt])X_{[nt]+1}, \quad t \in [0, 1].$$

In the familiar case where $\mathbb{B}$ is the real line $\mathbb{R}$, classical Donsker-Prohorov invariance principle states, that if $\mathbf{E}\, X_1 = 0$ and $\mathbf{E}\, X_1^2 = \sigma^2 < \infty$, then

$$n^{-1/2}\sigma^{-1}\xi_n \xrightarrow{\mathcal{D}} W, \tag{10}$$

in $C[0, 1]$, where $(W(t), t \in \mathbb{R})$ is a standard Wiener process. The finiteness of the second moment of X_1 is clearly necessary here, since (10) yields that $n^{-1/2}\xi_n(1)$ satisfies the central limit theorem.

Lamperti (1962) was the first who considered the convergence (10) with respect to some Hölderian topology. He proved that if $0 < \alpha < 1/2$ and $\mathbf{E}\, |X_1|^p < \infty$, where $p > p(\alpha) := 1/(1/2 - \alpha)$, then (10) takes place in $\mathcal{H}^{\alpha,o}$. This result was derived again by Kerkyacharian and Roynette (1991) by another method based on Ciesielski (1960) analysis of Hölder spaces by triangular functions. Further generalizations were given by Erickson (1981) (partial sums processes indexed by $[0, 1]^d$), Hamadouche (2000) (weakly dependent sequence (X_n)), Račkauskas and Suquet (2002) (Banach space valued X_i's and Hölder spaces built on the modulus $\rho(h) = h^\alpha \ln^\beta(1/h)$). The following result is proved in Račkauskas and Suquet (2002a).

Theorem 2. *Let $0 < \alpha < 1/2$ and $p(\alpha) = 1/(1/2 - \alpha)$. Then*

$$n^{-1/2}\sigma^{-1}\xi_n \xrightarrow[n\to\infty]{\mathcal{D}} W \quad \text{in the space } \mathcal{H}^{\alpha,o}$$

if and only if $\mathbf{E}\, X_1 = 0$ *and*

$$\lim_{t\to\infty} t^{p(\alpha)}\mathbf{P}\{|X_1| \ge t\} = 0. \tag{11}$$

Condition (11) yields the existence of moments $\mathbf{E}\, |X_1|^p$ for any $0 \le p < p(\alpha)$. If α approaches $1/2$ then $p(\alpha) \to \infty$. Hence, stronger invariance principle requires higher moments.

To describe more general results, some preparation is needed. Let $\mathbb{B}'$ be the topological dual of $\mathbb{B}$. For a random element X in $\mathbb{B}$ such that for every $f \in \mathbb{B}'$, $\mathbf{E}\, f(X) = 0$ and $\mathbf{E}\, f^2(X) < \infty$, the covariance operator $Q = Q(X)$ is the linear bounded operator from $\mathbb{B}'$ to $\mathbb{B}$ defined by $Qf = \mathbf{E}\, f(X)X$, $f \in \mathbb{B}'$. A random element $X \in \mathbb{B}$ (or covariance operator Q) is said to be pregaussian if there exists a mean zero Gaussian random element $Y \in \mathbb{B}$ with the same covariance operator as X, i.e. for all $f, g \in \mathbb{B}'$, $\mathbf{E} f(X)g(X) = \mathbf{E} f(Y)g(Y)$. Since the distribution of a centered Gaussian random element is defined by its covariance structure, we denote by Y_Q a zero mean Gaussian random element with covariance operator Q.

For any pregaussian covariance Q there exists a $\mathbb{B}$-valued Brownian motion W_Q with parameter Q, a centered Gaussian process indexed by $[0,1]$ with independent increments such that $W_Q(t) - W_Q(s)$ has the same distribution as $|t-s|^{1/2} Y_Q$.

We say that X_1 satisfies the *central limit theorem in* $\mathbb{B}$, which we denote by $X_1 \in \mathrm{CLT}(\mathbb{B})$, if $n^{-1/2} S_n$ converges in distribution in $\mathbb{B}$. This implies that $\mathbf{E} X_1 = 0$ and X_1 is pregaussian. It is well known (e.g. Ledoux and Talagrand, 1991) that the central limit theorem for X_1 cannot be characterized in general in terms of integrability of X_1 and involves the geometry of the Banach space $\mathbb{B}$.

We say that X_1 satisfies the *functional central limit theorem in* $\mathbb{B}$, which we denote by $X_1 \in \mathrm{FCLT}(\mathbb{B})$, if $n^{-1/2} \xi_n$ converges in distribution in $\mathcal{C}_{\mathbb{B}}$. Kuelbs (1973) extended the classical Donsker-Prohorov invariance principle to the case of $\mathbb{B}$-valued partial sums by proving that $n^{-1/2}\xi_n$ converges in distribution in $\mathcal{C}_{\mathbb{B}}$ to some Brownian motion W if and only if $X_1 \in \mathrm{CLT}(\mathbb{B})$ (in short $X_1 \in \mathrm{CLT}(\mathbb{B})$ if and only if $X_1 \in \mathrm{FCLT}(\mathbb{B})$). Of course in Kuelbs theorem, the parameter Q of W is the covariance operator of X_1.

The convergence in distribution of $n^{-1/2}\xi_n$ in $\mathcal{H}_{\mathbb{B}}^{\rho,o}$, which we denote by $X_1 \in \mathrm{FCLT}(\mathbb{B}, \rho)$, is clearly stronger than $X_1 \in \mathrm{FCLT}(\mathbb{B})$.

An obvious preliminary requirement for the FCLT in $\mathcal{H}_{\mathbb{B}}^{\rho,o}$ is that the $\mathbb{B}$-valued Brownian motion has a version in $\mathcal{H}_{\mathbb{B}}^{\rho,o}$. From this point of view, the critical ρ is $\rho_c(h) = \sqrt{h \ln(1/h)}$ due to Lévy's Theorem on the modulus of uniform continuity of the Brownian motion. So our interest will be restricted to functions ρ generating a weaker Hölder topology than ρ_c. More precisely, we consider the following class $\mathcal{R}$ of functions ρ.

Definition 1. *We denote by* $\mathcal{R}$ *the class of functions* ρ *satisfying*

i) for some $0 < \alpha \le 1/2$, *and some function* L *which is normalized slowly varying at infinity,*

$$\rho(h) = h^{\alpha} L(1/h), \quad 0 < h \le 1, \tag{12}$$

ii) $\theta(t) = t^{1/2}\rho(1/t)$ *is* C^1 *on* $[1, \infty)$,

iii) there is a $\beta > 1/2$ *and some* $a > 0$, *such that* $\theta(t)\ln^{-\beta}(t)$ *is non decreasing on* $[a, \infty)$.

The following result is proved in Račkauskas and Suquet (2002a).

Theorem 3. *Let* $\rho \in \mathcal{R}$. *Then* $X_1 \in \mathrm{FCLT}(\mathbb{B}, \rho)$ *if and only if* $X_1 \in \mathrm{CLT}(\mathbb{B})$ *and for every* $A > 0$,

$$\lim_{t \to \infty} t\mathbf{P}\{\|X_1\| \ge A t^{1/2}\rho(1/t)\} = 0. \tag{13}$$

If $\rho \in \mathcal{R}$ with $\alpha < 1/2$ in (12) then it suffices to check condition (13) for $A = 1$ only. Of course the special case $\mathbb{B} = \mathbb{R}$ and $\rho(h) = h^\alpha$ gives back Theorem 2.

In the case where $\rho(h) = h^{1/2} \ln^\beta(c/h)$ with $\beta > 1/2$, Condition (13) is equivalent to

$$\mathbf{E} \exp\Big(d \left\|X_1\right\|^{1/\beta}\Big) < \infty, \quad \text{for each } \ d > 0.$$

Let us note, that for the spaces $\mathbb{B} = L_p(0,1)$, $2 \le p < \infty$, as well as for each finite dimensional space, Condition (13) yields $X_1 \in \mathrm{CLT}(\mathbb{B})$. On the other hand it is well known that for some Banach spaces existence of moments of any order does not guarantee central limit theorem. It is also worth noticing that like in Kuelbs FCLT, all the influence of the geometry of the Banach space $\mathbb{B}$ is absorbed by the condition $X_1 \in \mathrm{CLT}(\mathbb{B})$.

It would be useful to extend the Hölderian FCLT to the case of dependent X_i's. This was done by Hamadouche (2000) in the special case where $\mathbb{B} = \mathbb{R}$ and under weak dependence (association and α-mixing). The result presented in Račkauskas and Suquet (2002a) provides a general approach for $\mathbb{B}$-valued X_i's and any dependence structure, subject to obtaining a good estimate of the partial sums. Laukaitis and Račkauskas (2002) obtained Hölderian FCLT for a polygonal line process based on residual partial sums of a stationary Hilbert space valued autoregression (ARH(1)) and applied it to the problem of testing stability of ARH(1) model under different types of alternatives.

4 Adaptive Self-Normalized Partial Sums Process

Various partial sums processes can be built from the sums $S_n = X_1 + \cdots + X_n$ of independent identically distributed mean zero random variables. In Račkauskas and Suquet (2001a) the so called *adaptive self-normalized* partial sums processes are defined. *Self-normalized* means that the classical normalization by $\sqrt{n}$ is replaced by

$$V_n = (X_1^2 + \cdots + X_n^2)^{1/2}.$$

Adaptive means that the vertices of the corresponding random polygonal line have their abscissas at the random points V_k^2/V_n^2 $(0 \le k \le n)$ instead of the deterministic equispaced points k/n. By this construction the slope of each line adapts itself to the value of the corresponding random variable.

By ζ_n (respect. ξ_n) we denote the random polygonal partial sums process defined on $[0,1]$ by linear interpolation between the vertices $(V_k^2/V_n^2, S_k)$, $k = 0, 1, \ldots, n$ (respect. $(k/n, S_k)$, $k = 0, 1, \ldots, n$), where

$$S_k = X_1 + \cdots + X_k, \quad V_k^2 = X_1^2 + \cdots + X_k^2.$$

For the special case $k = 0$, we put $S_0 = 0$, $V_0 = 0$. By convention the random functions $V_n^{-1}\xi_n$ and $V_n^{-1}\zeta_n$ are defined to be the null function on the event

$\{V_n = 0\}$. Figure 2 displays the polygonal lines $n^{-1/2}\xi_n$ and $V_n^{-1}\zeta_n$ built on a simulated sample of size $n = 1000$ of the symmetric distribution given by $\mathbf{P}(|X_1| > t) = 0.5\,t^{-2.3}\mathbf{1}_{[1,\infty)}(t)$.

From this picture one can observe how adaptive partition of the time interval improves slopes of polygonal line process.

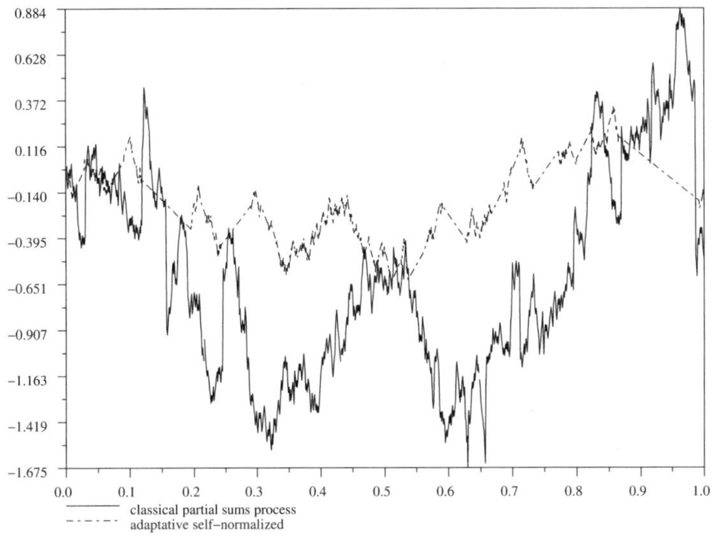

Fig. 2. Partial sums processes $n^{-1/2}\xi_n$ and $V_n^{-1}\zeta_n$

Recall that "X_1 belongs to the domain of attraction of the normal distribution" (denoted by $X_1 \in DAN$) means that there exists a sequence $b_n \uparrow \infty$ such that

$$b_n^{-1} S_n \xrightarrow{\mathcal{D}} N(0,1).$$

The following result is proved in Račkauskas and Suquet (2001a).

Theorem 4. *Assume that* ρ *satisfies Conditions* (5) *to* (9) *and*

$$\lim_{j\to\infty} \frac{2^j \rho^2(2^{-j})}{j} = \infty. \tag{14}$$

If X_1 *is symmetric then*

$$V_n^{-1}\zeta_n \xrightarrow{\mathcal{D}} W, \quad in \quad \mathcal{H}^{\rho,o}$$

if and only if $X_1 \in DAN$.

When tested with $\rho(h) = h^{1/2}\ln^{\beta}(c/h)$, (14) reduces to $j^{2\beta-1} \to \infty$. Due to the inclusions of Hölder spaces, this shows that Theorem 4 gives the best result possible in the scale of the separable Hölder spaces $\mathcal{H}^{\alpha,\beta,o}$. Moreover, no high order moments are needed except the condition $X_1 \in DAN$ which due to well known O'Briens result is equivalent to

$$V_n^{-1} \max_{1\le k\le n} |X_k| \xrightarrow{P} 0.$$

It seems worth noticing here, that without adaptive construction of the polygonal process, the existence of moments of order bigger than 2 is necessary for Hölder weak convergence. Indeed, if the process $V_n^{-1}\xi_n$ converges weakly to W in $\mathcal{H}^{\alpha,o}$ for some $\alpha > 0$, then its maximal slope $n^{-1/2}V_n^{-1}\max_{1\le k\le n}|X_k|$ converges to zero in probability. This on its turn yields $V_n^{-1}\max_{1\le k\le n}|X_k| \to 0$ almost surely, and according to Maller and Resnick (1984), $\mathbf{E}\,X_1^2 < \infty$. Hence $n^{-1}V_n^2$ converges almost surely to $\mathbf{E}\,X_1^2$ by the strong law of large numbers. Therefore $n^{-1/2}\xi_n$ converges weakly to W in $\mathcal{H}^{\alpha,o}$ and by Theorem 2 the moment restriction (11) is necessary.

Naturally it is very desirable to remove the symmetry assumption in Theorem 4. Although the problem remains open, we can propose the following partial result in this direction (for more on this problem see Račkauskas and Suquet (2001a)).

Theorem 5. *If for some $\varepsilon > 0$, $\mathbf{E}\,|X_1|^{2+\varepsilon} < \infty$, then for any $\beta > 1/2$, $V_n^{-1}\zeta_n$ converges weakly to W in the space $\mathcal{H}^{1/2,\beta,o}$.*

Some extensions of this result for the non i.i.d. case are given in Račkauskas and Suquet (2003a).

5 Hölder Norm Test Statistics

An important question in the large area of change point problems involves testing the null hypothesis of no parameter change in a sample versus the alternative that parameter changes do take place at an unknown time. For a survey we refer to the books by Brodsky and Darkhovsky (1993, 2000) or Csörgő and Horváth (1997). A widely accepted methodology to construct tests is to start with a functional limit theorems for processes build from partial sums $S(k), k = 0, \dots, n$. Functionals of paths of these processes can be used as tests statistics for null hypothesis under certain alternatives. In this way various test statistics are proposed in the literature. To name a few, the $KPSS$ statistic by Kwiatkowski *et al.* (1992) and the V/S statistic by Giraitis *et al.* (2003) are based on the L_2 space continuous functionals of partial sums process whereas Hurst (1951) R/S exploits $D[0,1]$ continuous functionals.

It is clear that a larger choice of admissible functionals on paths of polygonal line process corresponds to a bigger class of possible alternatives. Therefore it is important to have functional limit theorems established in a sufficiently rich functional framework. Working in the stronger topological framework of a space of Hölder functions allows to consider functionals based on increments $(\widehat{S}(k) - \widehat{S}(j),\ 0 \le k < j \le n)$ as well as on their ϑ-slopes $(|\widehat{S}(k) - \widehat{S}(j)|/\vartheta(k-j), 0 \le k < j \le n)$, where ϑ is certain function.

Using this methodology in Račkauskas and Suquet (2004) we propose some new statistics for testing change in the mean under the so called epidemic alternative. More precisely, given a sample $X_1, X_2, \dots, X_n$ of random variables, we want to test the standard null hypothesis of constant mean

(H_0): $X_1, \dots, X_n$ *all have the same mean denoted by* μ_0,

against the epidemic alternative

(H_A): *there are integers* $1 < k^* < m^* < n$ *and a constant* $\mu_1 \neq \mu_0$ *such that* $\mathbf{E}\, X_i = \mu_0 + (\mu_1 - \mu_0)\mathbf{1}_{\{k^* < i \le m^*\}}$, $i = 1, 2, \dots, n$.

Let us note, that other type of epidemics can be considered as well. If either $k^* = 1$ or $m^* = n$, then we have one abrupt change in the mean. This is a quite well studied case in the literature. If eiher $k^*/n \ge c_0 > 0$ or $m^*/n \le c_1 < 1$ then the problem can be considered with existing tests based on weighted partial sums processes. The situation when one naturally comes to Hölder topology concerns the cases, where either $l^*/n \to 0$ or $l^*/n \to 1$. Here $l^* := m^* - k^*$ denotes the length of the epidemic.

To motivate the definition of some test statistics, let us assume just for a moment that the changes times k^* and m^* are known. Suppose moreover under (H_0) that the $(X_i,\ 1 \le i \le n)$ are i.i.d. with finite variance σ^2. Under (H_A), suppose that the $(X_i,\ i \in \mathbb{I}_n)$ and the $(X_i,\ i \in \mathbb{I}_n^c)$ are separately i.i.d. with the same finite variance σ^2, where

$$\mathbb{I}_n := \{i \in \{1, \dots, n\};\ k^* < i \le m^*\}, \quad \mathbb{I}_n^c := \{1, \dots, n\} \setminus \mathbb{I}_n.$$

Then we simply have a two sample problem with known variances. It is then natural to accept (H_0) for small values of the statistics $|Q|$ and to reject it for large ones, where

$$Q := \frac{S(\mathbb{I}_n) - l^* S(n)/n}{\sqrt{l^*}} - \frac{S(\mathbb{I}_n^c) - (n - l^*)S(n)/n}{\sqrt{n - l^*}}.$$

After some algebra, Q may be recast as

$$Q = \frac{\sqrt{n}}{\sqrt{l^*(n-l^*)}}\Big(S(\mathbb{I}_n) - \frac{l^*}{n}S(n)\Big)\Big[\sqrt{1 - l^*/n} + \sqrt{l^*/n}\Big]$$

As the last factor into square brackets ranges between 1 and $\sqrt{2}$, we may drop it and introducing the notation

$$t_k = t_{n,k} := \frac{k}{n}, \quad 0 \le k \le n,$$

replace $|Q|$ by the statistics

$$R = n^{-1/2} \frac{\big|S(m^*) - S(k^*) - S(n)(t_{m^*} - t_{k^*})\big|}{\big[(t_{m^*} - t_{k^*})\big(1 - (t_{m^*} - t_{k^*})\big)\big]^{1/2}}.$$

Now in the more realistic situation where k^* and m^* are unknown it is reasonable to replace R by taking the maximum over all possible indexes for k^* and m^*. This leads to consider

$$\mathrm{UI}(n, 1/2) := \max_{1 \le i < j \le n} \frac{\big|S(j) - S(i) - S(n)(t_j - t_i)\big|}{\big[(t_j - t_i)\big(1 - (t_j - t_i)\big)\big]^{1/2}}.$$

It is worth to note that the same statistics arises from likelihood arguments in the special case where the observations X_i are Gaussian, see Yao (1993). The asymptotic distribution of $\mathrm{UI}(n, 1/2)$ is unknown, due to difficulties caused by the denominator (for historical remarks see Csörgő and Horváth p.183).

In our setting, the X_i's are not supposed to be Gaussian. Moreover it seems fair to pay something in terms of normalization when passing from R to $n^{-1/2}\mathrm{UI}(n, 1/2)$. Intuitively the cost should depend on the moment assumptions made about the X_i's. To discuss this, consider the polygonal partial sums process ξ_n. Then $\mathrm{UI}(n, 1/2)$ appears as the discretization through the grid $(t_k,\ 0 \le k \le n)$ of the functional $T_{1/2}(\xi_n)$ where

$$T_{1/2}(x) := \sup_{0<s<t<1} \frac{\big|x(t) - x(s) - \big(x(1) - x(0)\big)(t - s)\big|}{\big[(t - s)\big(1 - (t - s)\big)\big]^{1/2}}. \tag{15}$$

This functional is continuous in the Hölder space $\mathcal{H}^{1/2,o}$. Obviously finite dimensional distributions of $n^{-1/2}\sigma^{-1}\xi_n$ converge to those of a standard Brownian motion W. However Lvy's theorem on the modulus of uniform continuity of W implies that $\mathcal{H}^{1/2,o}$ has too strong a topology to support a version of W. So $n^{-1/2}\xi_n$ cannot converge in distribution to σW *in the space* $\mathcal{H}^{1/2,o}$. This forbid us to obtain limiting distribution for $T_{1/2}(n^{-1/2}\xi_n)$ by invariance principle in $\mathcal{H}^{1/2,o}$ via continuous mapping. As discussed in the previous section, Hölderian invariance principles do exist for, roughly speaking, all the scale of Hölder spaces $\mathcal{H}^{\rho,o}$, provided that the weight function ρ satisfies $\lim_{h\downarrow 0} \rho(h)(h \ln|h|)^{-1/2} = \infty$. This leads us to replace $T_{1/2}(n^{-1/2}\xi_n)$ by $T_\alpha(n^{-1/2}\xi_n)$ obtained substituting the denominator in (15) by $(t-s)^\alpha\big(1 - (t - s)\big)^\alpha$. Going back to the discretization we finally suggest the class UI (uniform increments) of statistics which includes particularly

$$\mathrm{UI}(n, \alpha) := \max_{1 \le i < j \le n} \frac{\big|S(j) - S(i) - S(n)(j/n - i/n)\big|}{\big[(j/n - i/n)\big(1 - (j - i)/n\big)\big]^\alpha}.$$

and similar ones $\mathrm{UI}(n, \rho)$ built with a general weight $\rho(h)$ instead of h^α. Together with UI we consider the class of DI (dyadic increments) statistics, which includes particularly

$$\mathrm{DI}(n,\alpha) = \max_{1\le j\le \log n} 2^{j\alpha} \max_{r\in \mathrm{D}_j} \left| S(nr) - \frac{1}{2}\big(S(nr+n2^{-j}) + S(nr-n2^{-j})\big)\right|,$$

where D_j is the set of dyadic numbers of the level j. We use here log to denote the logarithm *with basis* 2, while ln denotes the natural logarithm ($\log(2^t) = t = \ln(\mathrm{e}^t)$). $\mathrm{DI}(n,\alpha)$ and $\mathrm{UI}(n,\alpha)$ have similar asymptotic behaviors. Moreover, dyadic increments statistics are of particular interest since their limiting distributions are completely specified (see Theorem 11 below).

To simplify notation put

$$\varrho(h) := \rho\big(h(1-h)\big), \quad 0 \le h \le 1.$$

For $\rho \in \mathcal{R}$, define (recalling that $t_k = k/n$, $0 \le k \le n$),

$$\mathrm{UI}(n,\rho) = \max_{1\le i<j\le n} \frac{\big|S(j) - S(i) - S(n)(t_j - t_i)\big|}{\varrho(t_j - t_i)}$$

and

$$\mathrm{DI}(n,\rho) = \max_{1\le j\le \log n} \frac{1}{\rho(2^{-j})} \max_{r\in \mathrm{D}_j}\left| S(nr) - \frac{1}{2}\big(S(nr^+) + S(nr^-)\big)\right|.$$

Let $W = (W(t), t \in [0,1])$ be a standard Wiener process and $B = (B(t), t \in [0,1])$ the corresponding Brownian bridge $B(t) = W(t) - tW(1)$, $t \in [0,1]$. Consider for ρ in $\mathcal{R}$, the following random variables

$$\mathrm{UI}(\rho) := \sup_{0<t-s<1} \frac{|B(t) - B(s)|}{\varrho(t-s)} \tag{16}$$

and

$$\mathrm{DI}(\rho) = \sup_{j\ge 1} \frac{1}{\rho(2^{-j})} \max_{r\in \mathrm{D}_j}\left| W(r) - \frac{1}{2}W(r^+) - \frac{1}{2}W(r^-)\right| = \|B\|_\rho^{\mathrm{seq}}. \tag{17}$$

These variables serve as limiting for uniform increment (UI) and dyadic increment (DI) statistics respectively. No analytical form seems to be known for the distribution function of $\mathrm{UI}(\rho)$, whereas the distribution of $\mathrm{DI}(\rho)$ is completely specified below.

To obtain limiting distribution for statistics $\mathrm{UI}(n,\rho)$ and $\mathrm{DI}(n,\rho)$ we consider a bit stronger null hypothesis, namely

(H_0'): $X_1, \dots, X_n$ *are i.i.d. random variables.*

Theorem 6. *Under* (H_0'), *assume that* $\rho \in \mathcal{R}$ *and for every* $A > 0$,

$$\lim_{t\to\infty} t\,\mathbf{P}\big\{|X_1| > At^{1/2}\rho(1/t)\big\} = 0.$$

Then

$$\sigma^{-1}n^{-1/2}\mathrm{UI}(n,\rho) \xrightarrow[n\to\infty]{\mathcal{D}} \mathrm{UI}(\rho) \quad \textit{and} \quad \sigma^{-1}n^{-1/2}\mathrm{DI}(n,\rho) \xrightarrow[n\to\infty]{\mathcal{D}} \mathrm{DI}(\rho),$$

where $\sigma^2 = \operatorname{Var} X_1$ *and* $\mathrm{UI}(\rho)$, $\mathrm{DI}(\rho)$ *are defined by (16) and (17).*

Remark 1. In the case where the variance σ^2 is unknown the results remain valid if σ^2 is substituted by its standard estimator $\widehat{\sigma}^2$.

Remark 2. Under (H_0), the statistics $n^{-1/2}\mathrm{UI}(n,\rho)$ keeps the same value when each X_i is substituted by $X_i' := X_i - \mathbf{E}\,X_i$. This property is only asymptotically true for $n^{-1/2}\mathrm{DI}(n,\rho)$. For practical use of $\mathrm{DI}(n,\rho)$, it is preferable to replace X_i by X_i' if μ_0 is known or else by $X_i - \overline{X}$ where $\overline{X} = n^{-1}(X_1 + \cdots + X_n)$. This will avoid a bias term which may be of the order of $|\mathbf{E}\,X_1| \ln^{-\beta} n$ in the worst cases.

To see a consistency of tests to reject null hypothesis versus epidemic alternative H_A for large values of $n^{-1/2}\mathrm{UI}(n,\rho)$, we naturally assume that the numbers of observations k^*, $m^* - k^*$, $n - m^*$ before, during and after the epidemic go to infinity with n. Write $l^* := m^* - k^*$ for the length of the epidemic.

Theorem 7. *Let $\rho \in \mathcal{R}$. Assume under (H_A) that the X_i's are independent and $\sigma_0^2 := \sup_{k\geq 1} \operatorname{Var} X_k$ is finite. If*

$$\lim_{n\to\infty} n^{1/2}\frac{h_n}{\rho(h_n)} = \infty, \quad \textit{where} \quad h_n := \frac{l^*}{n}\Big(1 - \frac{l^*}{n}\Big), \tag{18}$$

then

$$n^{-1/2}\mathrm{UI}(n,\rho) \xrightarrow[n\to\infty]{\mathrm{P}} \infty, \quad \textit{and} \quad n^{-1/2}\mathrm{DI}(n,\rho) \xrightarrow[n\to\infty]{\mathrm{P}} \infty.$$

When $\rho(h) = h^\alpha$, (18) allows us to detect *short epidemics* such that $l^* = o(n)$ and $l^* n^{-\delta} \to \infty$, where $\delta = (1-2\alpha)(2-2\alpha)^{-1}$. Symmetrically one can detect *long epidemics* such that $n - l^* = o(n)$ and $(n - l^*)n^{-\delta} \to \infty$.

When $\rho(h) = h^{1/2}\ln^\beta(c/h)$ with $\beta > 1/2$, (18) is satisfied provided that $h_n = n^{-1}\ln^\gamma n$, with $\gamma > 2\beta$. This leads to detection of short epidemics such that $l^* = o(n)$ and $l^* \ln^{-\gamma} n \to \infty$ as well as of long ones verifying $n - l^* = o(n)$ and $(n - l^*)\ln^{-\gamma} n \to \infty$.

One of the disadvantages of DI and UI statistics is the moment restriction for their convergence. The introduction of adaptive self-normalized analogues for these statistics resolves this problem.

For any index set $A \subset \{1, \dots, n\}$, define

$$V^2(A) := \sum_{i\in A} X_i^2.$$

Then $V_k^2 = V^2(\{1, \dots, k\})$, $V_0^2 = 0$. To simplify notation we write v_k for the random points of $[0,1]$

$$v_k := \frac{V_k^2}{V_n^2}, \quad k = 0, 1, \dots, n.$$

For $\rho \in \mathcal{R}$ define

$$\mathrm{SUI}(n,\rho) := \max_{1\le i<j\le n} \frac{\left|S(j)-S(i)-S(n)(v_j-v_i)\right|}{\varrho(v_j-v_i)}.$$

Introduce for any $t\in[0,1]$,

$$\tau_n(t) := \max\{i\le n;\ V_i^2\le tV_n^2\}$$

and

$$J_n := \log(V_n^2/X_{n:n}^2) \quad \text{where} \quad X_{n:n} := \max_{1\le k\le n}|X_k|.$$

Now define

$$\mathrm{SDI}(n,\rho) := \max_{1\le j\le J_n} \frac{1}{\rho(2^{-j})} \max_{r\in \mathrm{D}_j} \left| S\big(\tau_n(r)\big) - \frac{1}{2}\Big[S\big(\tau_n(r^+)\big)+S\big(\tau_n(r^-)\big)\Big]\right|.$$

Theorem 8. *Assume that under* (H_0'), $\mu_0 = 0$ *and the random variables* $X_1,\dots,X_n$ *are either symmetric and belong to the domain of attraction of normal law or* $\mathbf{E}\,|X_1|^{2+\varepsilon}<\infty$ *for some* $\varepsilon>0$. *Then for every* $\rho\in\mathcal{R}$,

$$V_n^{-1}\mathrm{SUI}(n,\rho) \xrightarrow[n\to\infty]{\mathcal{D}} \mathrm{UI}(\rho) \quad \text{and} \quad V_n^{-1}\mathrm{SDI}(n,\rho) \xrightarrow[n\to\infty]{\mathcal{D}} \mathrm{DI}(\rho).$$

We restricted the null hypothesis assuming that $\mu_0 = 0$. Practically the reduction to this case by centering requires the knowledge of μ_0. Although this seems a quite reasonable assumption, it should be pointed out that this knowledge was not required for the UI and DI statistics.

Theorem 9. *Let* $\rho\in\mathcal{R}$. *Under* (H_A), *assume that the* X_i*'s satisfy*

$$S(\mathbb{I}_n) = l^*\mu_1 + O_p\big(\sqrt{l^*}\big), \quad S(\mathbb{I}_n^c) = O_p\big(\sqrt{n-l^*}\big), \tag{19}$$

and

$$\frac{V^2(\mathbb{I}_n)}{l^*} \xrightarrow[n\to\infty]{\mathrm{P}} b_1, \quad \frac{V^2(\mathbb{I}_n^c)}{n-l^*} \xrightarrow[n\to\infty]{\mathrm{P}} b_0, \tag{20}$$

for some finite constants b_0 *and* b_1. *Assume that* l^*/n *converges to a limit* $c\in[0,1]$ *and when* $c=0$ *or* 1, *assume moreover that*

$$\lim_{n\to\infty} n^{1/2}\frac{h_n}{\rho(h_n)} = \infty, \quad \text{where} \quad h_n := \frac{l^*}{n}\Big(1-\frac{l^*}{n}\Big). \tag{21}$$

Then

$$V_n^{-1}\mathrm{SUI}(n,\rho) \xrightarrow[n\to\infty]{\mathrm{P}} \infty.$$

Note that in Theorem 9, no assumption is made about the dependence structure of the X_i's. It is easy to verify the general hypotheses (19) and (20) in various situations of weak dependence like mixing or association. Under independence of the X_i's (without assuming identical distributions inside each block $\mathbb{I}_n$ and $\mathbb{I}_n^c$), it is enough to have $\sup_{k\ge1}\mathbf{E}\,|X_k|^{2+\varepsilon}<\infty$ and $\mathbf{E}\,V^2(\mathbb{I}_n)/l^*\to b_1$, $\mathbf{E}\,V^2(\mathbb{I}_n^c)/(n-l^*)\to b_0$. This follows easily from Lindebergh's condition for the central limit theorem (giving (19)) and of Theorem 5 p.261 in Petrov (1975) giving the weak law of large numbers required for the X_i's.

Theorem 10. *Let $\rho \in \mathcal{R}$ and set $X_i' := X_i - \mathbf{E}(X_i)$. Under (H_A) with $\mu_0 = 0$, assume that the random variables X_i' are independent identically distributed and that $\mathbf{E}|X_1'|^{2+\varepsilon}$ is finite for some $\varepsilon > 0$. Then under (21),*

$$V_n^{-1}\mathrm{SDI}(n,\rho) \xrightarrow[n\to\infty]{\mathrm{P}} \infty.$$

The distribution function of $\mathrm{DI}(\rho)$ may be conveniently expressed in terms of the *error function*:

$$\operatorname{erf} x = \frac{2}{\sqrt{\pi}} \int_0^x \exp(-s^2)\,\mathrm{d}s.$$

Theorem 11. *Let $c = \limsup_{j\to\infty} j^{1/2}/\theta(2^j)$, where $\theta(t) = t^{1/2}\rho(1/t)$.*

i) If $c = \infty$ then $\mathrm{DI}(\rho) = \infty$ almost surely.
ii) If $0 \le c < \infty$, then $\mathrm{DI}(\rho)$ is almost surely finite and its distribution function is given by

$$\mathbf{P}\bigl(\mathrm{DI}(\rho) \le x\bigr) = \prod_{j=1}^{\infty} \bigl\{\operatorname{erf}\bigl(\theta(2^j)x\bigr)\bigr\}^{2^{j-1}}, \quad x > 0.$$

The distribution function of $\mathrm{DI}(\rho)$ is continuous with support $\bigl[c\sqrt{\ln 2}, \infty\bigr)$.

Figure 1 below displays a table of the distribution function of $\mathrm{DI}(\rho)$ with $\rho(h) = h^{0.4}$.

Theorems 6 to 11 are proved in Račkauskas and Suquet (2003a).

6 Testing Changes of Infinite Dimensional Parameters

As mentioned in the introduction, polygonal line processes based on sample taking values in an infinite dimensional sample space gives tools for statisticians working with functional data or using nonparametric approach to the testing problems. In this section we just provide one example showing how general approach to the simple problem of testing stability in the mean allows to obtain prototypical Cramér-von Mises test statistic for change in distribution problem. In Račkauskas and Suquet (2003c) more examples are available including the prototypical Kolmogorov-Smirnov test.

Let $X_1, \dots, X_n$ be the real valued random variables with distribution functions $F_1, \dots, F_n$ respectively. Consider the null hypothesis

(H_0): $F_1 = \dots = F_n = F$

and the following epidemic alternative:

(H_A): *there are integers $1 < k^* < m^* < n$ such that*
$F_1 = F_2 = \dots = F_{k^*} = F_{m^*+1} = \dots = F_n$,
$F_{k^*+1} = \dots = F_{m^*}$ *and* $F_{k^*} \ne F_{k^*+1}$.

Table 1. Distribution function $F(x) = \mathbf{P}(\mathrm{DI}(h^{0.4}) \le x)$

x	F(x)	x	F(x)	x	F(x)	x	F(x)
1.06	0.0000075	1.40	0.5110985	1.74	0.9424161	2.08	0.9940030
1.08	0.0000573	1.42	0.5615323	1.76	0.9495661	2.10	0.9947547
1.10	0.0003004	1.44	0.6084394	1.78	0.9558321	2.12	0.995413
1.12	0.0011653	1.46	0.6515890	1.80	0.9613221	2.14	0.9959895
1.14	0.0035455	1.48	0.6909226	1.82	0.9661314	2.16	0.9964943
1.16	0.0088640	1.50	0.7265088	1.84	0.9703438	2.18	0.9969363
1.18	0.0188969	1.52	0.7585053	1.86	0.9740332	2.20	0.9973232
1.20	0.0353857	1.54	0.7871269	1.88	0.9772645	2.22	0.9976618
1.22	0.0596012	1.56	0.8126221	1.90	0.9800944	2.24	0.9979582
1.24	0.0920395	1.58	0.8352537	1.92	0.9825729	2.26	0.9982175
1.26	0.1323403	1.60	0.8552866	1.94	0.9847436	2.28	0.9984444
1.28	0.1794090	1.62	0.8729782	1.96	0.9866448	2.30	0.9986428
1.30	0.2316604	1.64	0.8885729	1.98	0.9883099	2.32	0.9988163
1.32	0.2872905	1.66	0.9022985	2.00	0.9897684	2.34	0.9989680
1.34	0.3445122	1.68	0.9143643	2.02	0.9910458	2.36	0.9991006
1.36	0.4017209	1.70	0.9249609	2.04	0.9921647	2.38	0.9992165
1.38	0.4575890	1.72	0.9342602	2.06	0.9931446	2.40	0.9993176

For simplicity let us consider only the case of continuous function F. Then we can restrict to uniform empirical process.

Let $U_1, \dots, U_n$ be independent uniformly on $[0,1]$ distributed random variables. Define $u_0 = 0$,

$$u_v(t) = \sum_{j \le v} \left(\mathbf{1}_{\{U_j \le t\}} - t \right), \quad 1 \le v \le n; \quad t \in [0,1].$$

For $r \in \mathrm{D}$ set

$$\mathcal{D}^2 u_{nr} = u_{nr} - \frac{1}{2}(u_{nr^+} + u_{nr^-})$$

and

$$T^2(n,r) = \int_0^1 \left| \mathcal{D}^2 u_{nr}(t) \right|^2 \mathrm{d}t.$$

The Cramér-von Mises type dyadic increments statistic is defined by

$$\mathcal{T}^2(n,\rho) = \max_{1 \le j \le \log n} \frac{1}{\rho^2(2^{-j})} \max_{r \in \mathrm{D}_j} T^2(n,r).$$

Its limiting distribution is completely defined by the limiting distribution of the classical Cramr-von Mises statistic, namely, by the distribution function

$$L_2(x) = \mathbf{P}\left\{ \int_0^1 B^2(t)\,\mathrm{d}t \le x \right\}, \quad x \ge 0,$$

where $B(t), t \in [0,1]$ is a standard Brownian bridge. The distribution function $L_2(x)$ is well known and several its representations are available (see Shorack and Wellner, 1986).

Theorem 12. *If $\rho \in \mathcal{R}$ then*

$$\lim_{n\to\infty} \mathbf{P}(n^{-1}\mathcal{T}^2(n,\rho) \le x) = L_{2,\rho}(x),$$

for each $x > 0$, where

$$L_{2,\rho}(x) = \prod_{j=1}^{\infty} \Big[L_2\big(2\theta(2^j)x\big)\Big]^{2^{j-1}}.$$

The result is an application of functional central limit Theorem 3 to the random elements $Y_1, \dots, Y_n$ with values in the space $L_2(0,1)$ defined by

$$X_k(t) = \mathbf{1}_{\{U_k \le t\}} - t, \quad t \in [0,1], \quad k = 1, \dots, n.$$

If $\sum_j j\theta^{-2}(2^j) < \infty$, for practical uses the following estimate of the tails of $L_{2,\rho}$ can be helpful. Define for $x > 0$

$$L_{2,\rho}^{(J)}(x) = \prod_{j=1}^{J} \Big[L_2\big(2\theta(2^j)x\big)\Big]^{2^{j-1}}.$$

Proposition 1. *Put for $J = 0, 1, 2, \dots$*

$$c_J := \inf_{\gamma > 0} \Big\{ \gamma + 8 \sum_{j=J+1}^{\infty} \frac{2^j}{\theta^2(2^j)} \exp\Big(\frac{-\gamma\theta^2(2^j)}{4}\Big)\Big\}.$$

i) If $x > 0$ then

$$1 - L_{2,\rho}(x) \le 4 \exp\Big\{ -\frac{3x}{4c_0}\Big\}.$$

ii) For each $x > 0$ and $J \ge 1$

$$\Big(1 - 4\exp\Big\{-\frac{3x}{8c_J}\Big\}\Big) L_{2,\rho}^{(J)}(x) \le L_{2,\rho}(x) \le L_{2,\rho}^{(J)}(x).$$

The null hypothesis (H_0) is rejected for big values of the statistic $\mathcal{T}^2(n,\rho)$ calculated with $U_i = F(X_i)$, $i = 1, \dots, n$.

References

1. Brodsky, B. E.; Darkhovsky, B. S. (2000). *Non-parametric statistical diagnosis. Problems and methods.* Mathematics and its Applications, 509. Kluwer Academic Publishers, Dordrecht.
2. Brodsky, B., E. and Darkhovsky, B., S. (1993). *Non parametric methods in change point problems.* Kluwer Academic Publishers, Dordrecht.
3. Csörgő, M. and Horváth, L. (1997). *Limit Theorems in Change-Point Analysis.* John Wiley and Sons, New York.

4. Ciesielski, Z. (1960). On the isomorphisms of the spaces H_α and m. *Bull. Acad. Pol. Sci. Ser. Sci. Math. Phys.*, **8,** 217–222.
5. Erickson, R. V. (1981). Lipschitz smoothness and convergence with applications to the central limit theorem for summation processes. *Annals of Probabability*, **9,** 831–851.
6. Giraitis, L., Kokoszka, P.S., Leipus, R. and Teyssière, G. (2003). Rescaled variance and related test for long memory in volatility and levels. *Journal of Econometrics*, **112,** 265–294.
7. Hamadouche, D. (2000). Invariance principles in Hölder spaces. *Portugaliae Mathematica*, **57,** 127–151.
8. Hurst, H. (1951). Long term storage capacity of reservoirs. *Transactions of the American Society of Civil Engineers*, **116,** 770–799.
9. Kerkyacharian, G. and Roynette, B. (1991). Une démonstration simple des théorèmes de Kolmogorov, Donsker et Ito-Nisio. *Comptes Rendus de l' Académie des Sciences Paris, Série I* **312,** 877–882.
10. Kirman. A, Teyssière, G. (2002). Bubbles and Long Range Dependence in Asset Prices Volatilities, In: *Equilibrium, Markets and Dynamics.* Cars Hommes, Roald Ramer and Cees Withagen editors, p. 307–327. Springer Verlag.
11. Kuelbs, J. (1973). The invariance principle for Banach space valued random variables. *Journal of Multivariate Analysis,* **3,** 161–172.
12. Kwiatkowski, D., Philips, P.C.B., Schmidt, P. and Shin, Y. (1992). Testing the null hypothesis of stationarity against the alternative of a unit root: how sure are we that economic time series have a unit root? *Journal of Econometrics*, **54,** 159–178.
13. Lamperti, J. (1962). On convergence of stochastic processes. *Transactions of the American Mathematical Society.* **104,** 430–435.
14. Laukaitis A. and Račkauskas A. (2002). Testing changes in Hilbert space autoregressive models. *Lietuvos Matematikos Rinkinys*, **42,** 434–447.
15. Ledoux, M. and Talagrand, M. (1991). *Probability in Banach Spaces.* Springer-Verlag, Berlin, Heidelberg.
16. Maller, R. A. and Resnick, S.I. (1984). Limiting behaviour of sums and the term of maximum modulus. *Proceedings of London Mathematical Society,* **49,** 385–422.
17. Petrov, V. V. (1975) *Sums of independent random variables*, Springer.
18. Račkauskas, A. and Suquet, Ch. (1999a). Central limit theorem in Hölder spaces. *Probability and Mathematical Statistics*, **19,** 155–174.
19. Račkauskas, A. and Suquet, Ch. (1999b). Random fields and central limit theorem in some generalized Hölder spaces. *Prob. Theory and Math. Statist., Proceedings of the 7th Vilnius Conference (1998)*, B. Grigelionis *et al.* (Eds), 599–616, TEV Vilnius and VSP Utrecht.
20. Račkauskas, A. and Suquet, Ch. (2000). Convergence of self-normalized partial sums processes in $C[0,1]$ and $D[0,1]$. *Pub. IRMA Lille*, **53-VI**. *Preprint.*
21. Račkauskas, A. and Suquet, Ch. (2001a). Invariance principles for adaptive self-normalized partial sums processes. *Stochastic Processes and their Applications*, **95,** 63–81.
22. Račkauskas, A. and Suquet, Ch. (2001b) Hölder versions of Banach spaces valued random fields. *Georgian Mathematical Journal*, **8,** No 2, 347–362.
23. Račkauskas, A. and Suquet, Ch. (2002a). Necessary and sufficient condition for the Hölderian functional central limit theorem. *Pub. IRMA Lille*, **57-I**. *Preprint.*

24. Račkauskas, A. and Suquet, Ch. (2002b). On the Hölderian functional central limit theorem for i.i.d. random elements in Banach space. In *Limit Theorems in Probability and Statistics*, Balatonlelle 1999 (I. Berkes, E. Csáki, M. Csörgő, eds.) János Bolyai Mathematical Society, Budapest, **2,** 485–498.
25. Račkauskas, A. and Suquet, Ch. (2003a). Invariance principle under self-normalization for nonidentically distributed random variables. *Acta Applicandae Mathematicae*, **79,** 83–103.
26. Račkauskas, A. and Suquet, Ch. (2003b) Central limit theorems in Hölder topologies for Banach space valued random fields. To appear in *Theory of Probability and Applications.*
27. Račkauskas, A. and Suquet, Ch. (2003c). Testing epidemic change of infinite dimensional parameters. *Pub. IRMA Lille* **60-VIII**. *Preprint.*
28. Račkauskas, A. and Suquet, Ch. (2004) Hölder norm test statistics for epidemic change. *Journal of Statistical Planning and Inference*, **126,** 495–520.
29. Shorack G.R. and Wellner J.A. (1986). *Empirical processes with applications to statistics.* Wiley.
30. Stock James H. (1994). Unit root, structural breaks and trends, In *Handbook of Econometrics,* p. 2740–2843. North Holland, Amsterdam.
31. Yao, Q. (1993). Tests for change-points with epidemic alternatives. *Biometrika*, **80,** 179–191.

Adaptive Detection of Multiple Change–Points in Asset Price Volatility

Marc Lavielle[1] and Gilles Teyssière[2]

[1] Université René Descartes and Université Paris–Sud, Laboratoire de Mathématiques. Marc.Lavielle@math.u-psud.fr
[2] Laboratoire de Statistique Appliquée et MOdélisation Stochastique (SAMOS), Université Paris 1. stats@gillesteyssiere.net

Summary. This chapter considers the multiple change–point problem for time series, including strongly dependent processes, with an unknown number of change–points. We propose an adaptive method for finding the segmentation, i.e., the sequence of change–points $\boldsymbol{\tau}$ with the optimal level of resolution. This optimal segmentation $\hat{\boldsymbol{\tau}}$ is obtained by minimizing a penalized contrast function $J(\boldsymbol{\tau}, \boldsymbol{y}) + \beta \text{pen}(\boldsymbol{\tau})$. For a given contrast function $J(\boldsymbol{\tau}, \boldsymbol{y})$ and a given penalty function $\text{pen}(\boldsymbol{\tau})$, the adaptive procedure for automatically choosing the penalization parameter β is such that the segmentation $\hat{\boldsymbol{\tau}}$ does not strongly depend on β. This algorithm is applied to the problem of detection of change–points in the volatility of financial time series, and compared with Vostrikova's (1981) binary segmentation procedure.

1 Introduction

The change–point analysis of volatility processes is a recent and important research topic in financial econometrics. Volatility processes, i.e., absolute and squared returns on asset prices, are characterized by a hyperbolic decay of their autocorrelation function (ACF), and then have been first considered as the realization of a strongly dependent, or long–range dependent or long–memory process. Most of the applied research works in this field resorted to the class of long–range dependent volatility processes introduced by Robinson (1991), developed by Granger and Ding (1995) and other authors, defined as

$$Y_t = \sigma_t \varepsilon_t, \quad \varepsilon_t \sim \text{iid}, \quad E\varepsilon_0 = 0, \quad \text{Var}\,\varepsilon_0 = 1, \quad \sigma_t^2 = \omega + \sum_{j=1}^{\infty} \alpha_j Y_{t-j}^2, \tag{1}$$

with hyperbolically decaying positive weights $\alpha_j \asymp j^{-(1+\vartheta/2)}$, $\vartheta \in (0,1)$, $\sum_j \alpha_j \leqslant 1$. Volatility processes were then implicitly viewed as the realization of a homogeneous process defined by the single scaling parameter ϑ. The estimated intensity of long–range dependence $\hat{\vartheta}$ of asset price volatility is usually strong, near the stationarity limit.

Alternatively, the returns series are modeled by the Integrated ARCH(∞) volatility process, defined by equation (1) with exponentially decaying coefficients α_j and $\sum_j \alpha_j = 1$, i.e., a process with infinite variance since $\omega > 0$. This IARCH representation is incompatible with the conclusion on the presence of long–memory in volatility grounded on the hyperbolic decay of the sample ACF of power transformations of this returns process, since this sample ACF is not properly defined. This contradiction should have questioned the relevance of the hypothesis of a homogeneous volatility process as it was already well known in the statistical literature that change–point processes, nonstationary processes and long–range dependent processes might be confused; see e.g., Bhattacharya *et al.* (1983).

Pioneering works by Mikosch and Stărică (1999, 2003, 2004) advocated the change–point alternative for the analysis of volatility processes, and claimed that the empirical long–range dependence of volatility process was the consequence of nonstationarities. The standard short–range dependent volatility models still provide an accurate representation of the volatility process, provided that we estimate them on intervals of homogeneity. This idea of approximating nonstationary processes with locally stationary processes has been considered by Dalhaus (1997). The statistical theory for volatility processes with a change–point was developed very recently; see, besides the references mentioned above, Chu (1995), Kokoszka and Leipus (1999, 2000), Horváth, Kokoszka and Teyssière (2001), Kokoszka and Teyssière (2002), Berkes, Horváth and Kokoszka (2004), Berkes, Gombay, Horváth and Kokoszka (2004). Interested readers are referred to the chapter on GARCH volatility models by Giraitis, Leipus and Surgailis (2005) in this volume.

Furthermore, if we stick to the parametric framework of long–range dependent volatility processes, the estimates of the long memory parameter on the whole sample and on different subsamples significantly differ. Statistical tests for change in the memory parameter by Horváth (2001), Horváth and Shao (1999) and Beran and Terrin (1999) would reject the null hypothesis of constant long–memory parameter. Thus, even in this framework, the hypothesis of a homogeneous process might be too strong.

We consider here a semiparametric approach, i.e., without reference to a parametric volatility model. The time series cannot be modeled as a stationary process but rather as a piecewise stationary process. Some abrupt changes affect the variance of the time-series at random times, but the distribution of the data does not vary between two successive sudden changes. In what follows, we propose a method which allows to systematically detect these sudden changes and to locate their positions. This method also allows the estimation of the distribution of the data between the abrupt changes. This semiparametric approach is also of interest in a parametric framework, as it might suggest a partition of the series in intervals of homogeneity were stationary volatility models can be estimated; see Aggarwal *et al.* (1999).

The issue of multiple change–points detection has been first viewed as an extension of the single change–point problem by using Vostrikova's (1981) binary segmentation (BS) procedure, which consists in iteratively applying the single change–point detection procedure, i.e., apply first the test for change–point on the whole sample of observations, and if such a point is found, use the same testing procedure on the two resulting sub–segments and on subsequent partitions, until no further change–point is found. This method has been extended by Whitcher *et al* (2002) to the case of long–range dependent processes by applying it to the discrete wavelet transform of the long–memory process with changes in variance. The BS method was also used by Berkes, Horváth, Kokoszka and Shao (2003) for adjudicating between multiple change–points and long–range dependence in levels.

The BS method is very simple, but has a serious drawback: the number of change–points might be overestimated and their location might be wrong, as one transforms the global problem of change–point detection in a sequence of local change–point detections. The resulting segmentation is not optimal.

We shall adopt here a global approach, where all the change–points are simultaneously detected by minimizing a penalized contrast function of the form

$$J(\boldsymbol{\tau}, \boldsymbol{y}) + \beta \text{pen}(\boldsymbol{\tau}),$$

see Braun *et al.* (2000), Lavielle (1999), Lavielle and Ludeña (2000) and Yao (1988). Here, $J(\boldsymbol{\tau}, \boldsymbol{y})$ measures the fit of $\boldsymbol{\tau}$ with $\boldsymbol{y}$, with $\boldsymbol{y} = Y_1, \ldots, Y_n$. Its role is to locate the change–points as accurately as possible. The penalty term $\text{pen}(\boldsymbol{\tau})$ only depends on the dimension $K(\boldsymbol{\tau})$ of the model $\boldsymbol{\tau}$ and increases with $K(\boldsymbol{\tau})$. Thus, it is used for determinating the number of change–points. The penalization parameter β adjusts the trade-off between the minimization of $J(\boldsymbol{\tau}, \boldsymbol{y})$ (obtained with a high dimension of $\boldsymbol{\tau}$), and the minimization of $\text{pen}(\boldsymbol{\tau})$ (obtained with a small dimension of $\boldsymbol{\tau}$). Lavielle (1999) applied this method, with an arbitrary choice for β, to the series of French CAC 40 index and uncovered changes in the distribution of returns.

Asymptotic results concerning penalized least–squares estimates have been obtained in theoretical general contexts in Lavielle (1999) and Lavielle and Ludeña (2000), extending the previous results by Yao (1988). We shall show that this kind of contrast can also be useful in practice. The main problem is the optimal choice for a penalty function and a coefficient β. In the Gaussian case, Yao (1988) suggested the Schwarz criterion. A complete discussion of the most popular criteria (AIC, Mallow's C_p, BIC), and many other references can be found in Birgé and Massart (2001). In a more general context, we can use a contrast other than the least-squares criterion, since the variables are not necessarily Gaussian and independent. We propose an adaptive procedure for automatically choosing the penalty parameter β in section **2**. We present in section **3** the binary segmentation procedure. An application to financial time series, daily returns on the FTSE 100 index and on 30–minutes spaced returns on FX rates, and simulated returns from artificial financial markets based on microeconomic models with interactive agents, is considered in section **4**.

2 A Penalized Contrast Estimate for the Change–Point Problem

2.1 The Contrast Function

We assume that the process $\{Y_t\}$ is abruptly changing and is characterized by a parameter $\theta \in \Theta$ that remains constant between two changes. We will strongly use this assumption to define our contrast function $J(\boldsymbol{\tau}, \boldsymbol{y})$.

Let K be some integer and let $\boldsymbol{\tau} = \{\tau_1, \tau_2, \ldots, \tau_{K-1}\}$ be an ordered sequence of integers satisfying $0 < \tau_1 < \tau_2 < \ldots < \tau_{K-1} < n$. For any $1 \leqslant k \leqslant K$, let $U(Y_{\tau_{k-1}+1}, \ldots, Y_{\tau_k}; \theta)$ be a contrast function useful for estimating the unknown true value of the parameter in the segment k. In other words, the minimum contrast estimate $\hat{\theta}(Y_{\tau_{k-1}+1}, \ldots, Y_{\tau_k})$, computed on segment k of $\boldsymbol{\tau}$, is defined as a solution to the following minimization problem:

$$U\left(Y_{\tau_{k-1}+1}, \ldots, Y_{\tau_k}; \hat{\theta}(Y_{\tau_{k-1}+1}, \ldots, Y_{\tau_k})\right) \leqslant U(Y_{\tau_{k-1}+1}, \ldots, Y_{\tau_k}; \theta) \ , \ \forall \theta \in \Theta. \tag{2}$$

For any $1 \leqslant k \leqslant K$, let G be defined as

$$G(Y_{\tau_{k-1}+1}, \ldots, Y_{\tau_k}) = U\left(Y_{\tau_{k-1}+1}, \ldots, Y_{\tau_k}; \hat{\theta}(Y_{\tau_{k-1}+1}, \ldots, Y_{\tau_k})\right). \tag{3}$$

Then, define the contrast function $J(\boldsymbol{\tau}, \boldsymbol{y})$ as

$$J(\boldsymbol{\tau}, \boldsymbol{y}) = \frac{1}{n} \sum_{k=1}^{K} G(Y_{\tau_{k-1}+1}, \ldots, Y_{\tau_k}), \tag{4}$$

where $\tau_0 = 0$ and $\tau_K = n$.

For the detection of changes in the variance of a sequence of random variables, the following contrast function, based on a Gaussian log–likelihood function, can be used:

$$J_n(\boldsymbol{\tau}, \boldsymbol{y}) = \frac{1}{n} \sum_{k=1}^{K} n_k \log(\hat{\sigma}_k^2), \tag{5}$$

where $n_k = \tau_k - \tau_{k-1}$ is the length of segment k, $\hat{\sigma}_k^2$ is the empirical variance computed on that segment k, $\hat{\sigma}_k^2 = n_k^{-1} \sum_{i=\tau_{k-1}+1}^{\tau_k} (Y_i - \bar{Y})^2$, and $\bar{Y}$ is the empirical mean of $Y_1, \ldots, Y_n$.

When the true number $K^\star$ of segments is known, the sequence $\hat{\boldsymbol{\tau}}_n$ of change–point instants that minimizes this kind of contrast function has the property, that under extremely general conditions, for any $1 \leqslant k \leqslant K^\star - 1$,

$$\mathrm{P}(|\hat{\tau}_{n,k} - \tau_k^\star| > \delta) \to 0, \quad \text{when } \delta \to \infty \text{ and } n \to \infty, \tag{6}$$

see Lavielle (1999), Lavielle and Ludeña (2000). In particular, this result holds for weakly and strongly dependent processes.

2.2 Penalty Function for the Change–Point Problem

When the number of change–points is unknown, we estimate it by minimizing a penalized version of the function $J(\boldsymbol{\tau}, \boldsymbol{y})$. For any sequence of change–point instants $\boldsymbol{\tau}$, let $\text{pen}(\boldsymbol{\tau})$ be a function of $\boldsymbol{\tau}$ that increases with the number $K(\boldsymbol{\tau})$ of segments of $\boldsymbol{\tau}$. Then, let $\{\hat{\boldsymbol{\tau}}_n\}$ be the sequence of change–point instants that minimizes

$$U(\boldsymbol{\tau}) = J(\boldsymbol{\tau}, \boldsymbol{y}) + \beta \text{pen}(\boldsymbol{\tau}). \tag{7}$$

The procedure is intuitively simple: the adjustment criteria must be compensated for in a way such that the over-segmentation would be penalized. However, the compensation must not be very important as a too large penalty function yields an underestimation of the number of segments.

If β is a function of n that goes to 0 at an appropriate rate as n goes to infinity, the estimated number of segments $K(\hat{\boldsymbol{\tau}}_n)$ converges in probability to $K^\star$ and condition (6) still holds; see Lavielle (1999), Lavielle and Ludeña (2000) for more details.

In practice, asymptotic results are not very useful for selecting the penalty term $\beta \text{pen}(\boldsymbol{\tau})$. Indeed, given a real observed signal with a fixed and finite length n, the parameter β must be fixed to some arbitrary value. When the parameter β is chosen to be very large, only the more significant abrupt changes are detected. However, a small value of β produces a high number of the estimated changes. Therefore, a trade-off must be made, i.e., we have to select a value of β which yields a reasonable level of resolution in the segmentation.

Various authors suggest different penalty functions according to the model they consider. For example, the Schwarz criterion is used by Braun *et al.* (2000) for detecting changes in a DNA sequence.

Consider first the penalty function $\text{pen}(\boldsymbol{\tau})$. By definition, $\text{pen}(\boldsymbol{\tau})$ should increase with the number of segments $K(\boldsymbol{\tau})$. Following the most popular information criteria such the AIC and the Schwarz criteria, we suggest to use in practice the simplest penalty function $\text{pen}(\boldsymbol{\tau}) = K(\boldsymbol{\tau})$.

Remark 1. We can argue this specific choice for the penalty function with theoretical considerations. Indeed, precise results have been recently obtained by Birgé and Massart (2001) in the following model:

$$Y_i = s^\star(i) + \sigma \varepsilon_i, \quad 1 \leqslant i \leqslant n, \tag{8}$$

where $s^\star(i) = \sum_{k=1}^{K^\star} m_k \mathbf{1}_{\{\tau_{k-1}^\star + 1 \leqslant i \leqslant \tau_k^\star\}}$ is a piecewise constant function. The sequence $\{\varepsilon_i\}$ is a sequence of Gaussian white noise, with variance 1. A penalized least-squares estimate is obtained by minimizing

$$J(\boldsymbol{\tau}, \boldsymbol{y}) = \frac{1}{n} \sum_{k=1}^{K(\boldsymbol{\tau})} \sum_{i=\tau_{k-1}+1}^{\tau_k} (Y_i - \bar{Y}_k)^2 + \beta \text{pen}(\boldsymbol{\tau}). \tag{9}$$

In a non asymptotic context, Birgé and Massart (2001) have shown that a penalty function of the form

$$\text{pen}(\boldsymbol{\tau}) = K(\boldsymbol{\tau})\left(1 + c\log\frac{n}{K(\boldsymbol{\tau})}\right), \quad \beta = \frac{2\sigma^2}{n}, \tag{10}$$

is optimal for minimizing $\mathbb{E}\left(\|\hat{s}_{\boldsymbol{\tau}} - s^{\star}\|^2\right)$, where the estimated sequence of means $\{\hat{s}_{\boldsymbol{\tau}}(i)\}$ is defined as $\hat{s}_{\boldsymbol{\tau}}(i) = \sum_{k=1}^{K(\boldsymbol{\tau})} \bar{Y}_k \mathbf{1}_{\{\tau_{k-1}+1 \leqslant i \leqslant \tau_k\}}$. Based on some numerical experiments, the authors suggest to use $c = 2.5$. Note that when the number $K^{\star}$ of segments is small in comparison with the length n of the series, this optimal penalty function is an almost linear function of K. Furthermore, Yao (1988) has proved the consistency of the Schwarz criterion for this model, with $\text{pen}(\boldsymbol{\tau}) = K(\boldsymbol{\tau})$ and $\beta = 2\sigma^2(\log n)/n$.

2.3 An Adaptive Choice for the Penalization Parameter

For a given contrast function J and a given penalty function $\text{pen}(\boldsymbol{\tau})$, the problem now reduces to the choice for the parameter β.

Let K_{MAX} be an upper bound on the dimension of $\boldsymbol{\tau}$. For any $1 \leqslant K \leqslant K_{MAX}$, let $\mathcal{T}_K$ be the set of all the models of dimension K:

$$\mathcal{T}_K = \{\boldsymbol{\tau} = (\tau_0, \ldots, \tau_K) \in \mathbb{N}^{K+1}, \; \tau_0 = 0 < \tau_1 < \tau_2 < \ldots \tau_{K-1} < \tau_K = n\}.$$

By definition the best model $\hat{\boldsymbol{\tau}}_K$ of dimension K minimizes the contrast function J:

$$\hat{\boldsymbol{\tau}}_K = \arg\min_{\boldsymbol{\tau} \in \mathcal{T}_K} J(\boldsymbol{\tau}, \boldsymbol{y}). \tag{11}$$

Note that the sequence $\{\hat{\boldsymbol{\tau}}_K, 1 \leqslant K \leqslant K_{MAX}\}$ can easily be computed. Indeed, let $\mathcal{G}$ be the upper triangular matrix of dimension $n \times n$ such that the element (i, j), for $j \geqslant i$ is $\mathcal{G}_{i,j} = G(Y_i, Y_{i+1}, \ldots Y_j)$, where $G(Y_i, \ldots Y_j)$ is the contrast function computed with $(Y_i, Y_{i+1}, \ldots Y_j)$. Thus, for any $1 \leqslant K \leqslant K_{MAX}$, we have to find a path $\tau_0 = 0 < \tau_1 < \tau_2 < \ldots, < \tau_{K-1} < \tau_K = n$ that minimizes the total cost

$$J(\boldsymbol{\tau}, \boldsymbol{y}) = \frac{1}{n}\sum_{k=1}^{K} \mathcal{G}_{\tau_{k-1}, \tau_k}. \tag{12}$$

A dynamic programming algorithm can recursively compute the optimal paths $(\hat{\boldsymbol{\tau}}_K, 1 \leqslant K_{MAX})$, see Kay (1998). This algorithm requires $\mathcal{O}(n^2)$ operations. Then, let

$$J_K = J(\hat{\boldsymbol{\tau}}_K, \boldsymbol{y}), \tag{13}$$
$$p_K = \text{pen}(\boldsymbol{\tau}), \quad \forall \boldsymbol{\tau} \in \mathcal{T}_K. \tag{14}$$

As mentioned above, we suggest to use $p_K = K$.

Thus, for any penalization parameter $\beta > 0$, the solution $\hat{\boldsymbol{\tau}}(\beta)$ minimizes the penalized contrast:

$$\hat{\boldsymbol{\tau}}(\beta) = \arg\min(J(\boldsymbol{\tau}, \boldsymbol{y}) + \beta\text{pen}(\boldsymbol{\tau})) \tag{15}$$
$$= \hat{\boldsymbol{\tau}}_{\hat{K}(\beta)} \tag{16}$$

where

$$\hat{K}(\beta) = \arg\min_{K\geqslant 1}\{J_K + \beta p_K\}. \tag{17}$$

The way how the solution $\hat{K}(\beta)$ varies with the penalization parameter β is given by the following proposition:

Proposition 1. *There exists a sequence* $\{K_1 = 1 < K_2 < \ldots\}$, *and a sequence* $\{\beta_0 = \infty > \beta_1 > \ldots\}$, *with*

$$\beta_i = \frac{J_{K_i} - J_{K_{i+1}}}{p_{K_{i+1}} - p_{K_i}}\,, \quad i \geqslant 1, \tag{18}$$

such that $\hat{K}(\beta) = K_i$, $\forall \beta \in [\beta_i, \beta_{i-1})$.
The subset $\{(p_{K_i}, J_{K_i}), i \geqslant 1\}$ *is the convex hull of the set* $\{(p_K, J_K), K \geqslant 1\}$.

Proof. For any $K \geqslant 1$, let $\hat{K}(\beta) = K$. Then

$$J_K + \beta p_K < \min_{L>K}(J_L + \beta p_L), \tag{19}$$

$$J_K + \beta p_K < \min_{L<K}(J_L + \beta p_L). \tag{20}$$

Thus, β must satisfy

$$\max_{L>K} \frac{J_K - J_L}{p_L - p_K} < \beta < \min_{L<K} \frac{J_L - J_K}{p_K - p_L}. \tag{21}$$

■

The estimated sequence $\hat{\boldsymbol{\tau}}$ should not strongly depend on the choice for the penalization coefficient β. In other words, a small change of β should not lead to a radically different solution $\hat{\boldsymbol{\tau}}$. This stability of the solution with respect to the choice for β will be ensured if we only retain the largest intervals $[\beta_i, \beta_{i-1}), i \geqslant 1$.

In summary, we propose the following procedure:

1. For $K = 1, 2, \ldots, K_{MAX}$, compute $\hat{\boldsymbol{\tau}}_K$, $J_K = J(\hat{\boldsymbol{\tau}}_K, \boldsymbol{y})$ and $p_K = \text{pen}(\hat{\boldsymbol{\tau}}_K)$,
2. compute the sequences $\{K_i\}$ and $\{\beta_i\}$, and the lengths $\{l_{K_i}\}$ of the intervals $[\beta_i, \beta_{i-1})$,
3. retain the greatest value(s) of K_i such that $l_{K_i} \gg l_{K_j}$, for $j > i$.

Remark 2. Choosing the largest interval usually underestimates the number of changes. Indeed, this interval usually corresponds to a very small number of change–points and we only detect the most drastic changes with such a penalty function. This explains why we should better look for the highest dimension K_i such that $l_{K_i} \gg l_{K_j}$, for any $j > i$, to recover the smallest details.

Instead of computing only one configuration of change–points, this method allows us to put forward different solutions with different dimensions. Indeed, it would be an illusion to believe that a completely blind method can give the "best" solution in any situation. If two dimensions K_i and K_j satisfy the criteria suggested in step 3, it is more suitable to propose these two solutions to the user, instead of removing one of them with an arbitrary criterion.

Remark 3. A classical and natural graphical method for selecting the dimension K can be summarized as follows:

i) examine how the contrast J_K decreases when K (that is, p_K) increases,
ii) select the dimension K for which J_K ceases to decrease significatively.

In other words, this heuristic approach looks for the maximum curvature in the plot $(p_K\ ,\ J_K)$. Proposition 1 states that the second derivative of this curve is directly related to the length of the intervals $([\beta_i, \beta_{i-1}), i \geqslant 1)$. Indeed, if we represent the points $(p_K\ ,\ J_K)$, for $1 \leqslant K \leqslant K_{MAX}$, β_i is the slope between the points (p_{K_i}, J_{K_i}) and $(p_{K_{i+1}}, J_{K_{i+1}})$. Thus, looking for where J_K ceases to decrease means looking for a break in the slope of this curve. Now, the variation of the slope at the point (p_K, J_K) is precisely the length l_{K_i} of the interval $[\beta_i, \beta_{i-1})$.

2.4 An Automatic Procedure for Estimating *K*

Without any changes in the variance, the joint distribution of the sequence $\{J_K\}$ is very difficult to compute in a closed form, but some Monte-Carlo experiments show that this sequence decreases as $c_1 K + c_2 K \log(K)$.

A numerical example is displayed Figure 1. We have simulated ten sequences of i.i.d. Gaussian variables and computed the series (J_K) for each of them. The fit with a function $c_1 K + c_2 K \log(K)$ is always almost perfect $(r^2 > 0.999)$. Nevertheless, the coefficients c_1 and c_2 are different for each of these series. Thus, we propose the following algorithm:

Algorithm 1
For $i = 1, 2, \ldots,$

1. *Fit the model*
$$J_K = c_1 K + c_2 K \log(K) + e_K,$$
to the sequence $\{J_K, K \geqslant K_i\}$, *assuming that* $\{e_K\}$ *is a sequence of iid centered Gaussian random variables,*
2. *Evaluate the probability that* $J_{K_i - 1}$ *follows also this model, i.e., estimate the probability*
$$\mathcal{P}_{K_i} = P\left(e_{K_i-1} \geqslant J_{K_i-1} - \hat{c}_1(K_i - 1) + \hat{c}_2(K_i - 1)\log(K_i - 1)\right), \quad (22)$$
under this estimated model.

Then, the estimated number of segments will be the largest value of K_i *such that the* P*–value* $\mathcal{P}_{K_i}$ *is smaller than a given threshold* α. *We set* $\alpha = 10^{-5}$ *in the numerical examples.*

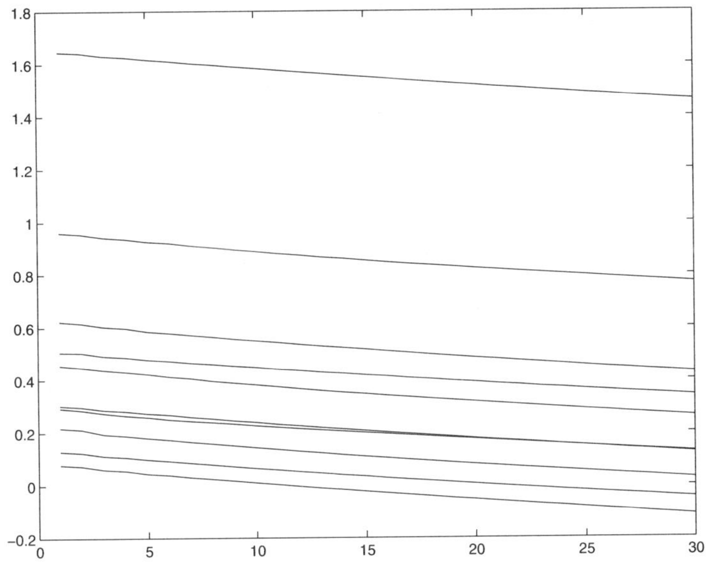

Fig. 1. Ten sequences of contrast functions (J_K) computed from ten sequences of i.i.d. Gaussian variables

3 An Alternative Method: The Binary Segmentation Procedure

We present here the local approach for finding multiple change–points, i.e., finding the configuration $\boldsymbol{\tau} = (\tau_1, \ldots, \tau_{K-1})$ with break dates $\{0 < \tau_1 < \ldots < \tau_{K-1} < n\}$, which rely on single change–points tests. Since financial time series are very large, i.e., over several thousands of observations, single change–point tests are of limited practical interest.

The binary segmentation procedure, studied by Vostrikova (1981), is the standard method for detecting multiple change–points by using a test for single change–point: we split the series at the point detected by the single change–point test, i.e., the point where the test statistic reaches its maximum over the critical value, and repeat the detection procedure on the new segments until no further change–point is found. However, the problem of optimal resolution in the segmentation $\boldsymbol{\tau}$ is not solved as no penalized objective function is considered.

Remark 4. Most applied econometrics research papers supposedly using the multiple change–point tests by Lavielle (1999) and Lavielle and Moulines (2000) do in fact resort to the binary segmentation algorithm, which is the less we can say missleading.

3.1 Weakly Dependent Processes

We present here the tests for single change–point in the variance of time series by Inclán and Tiao (1994) and Kokoszka and Leipus (1999). The test by Inclán and Tiao (1994) for change in the variance of a weakly dependent process $\{Y_t\}$ is based on the process $\{D_n(h), h \in [0,1]\}$ defined as

$$D_n(h) := \frac{\sum_{j=1}^{[nh]} Y_j^2}{\sum_{j=1}^{n} Y_j^2} - \frac{[nh]}{n}, \quad h \in [0,1]. \tag{23}$$

Under the null hypothesis of constant unconditional variance, the process $\{D_n(h), h \in [0,1]\}$ converges to a Brownian bridge on $[0,1]$. A test for constancy of the unconditional variance is based on the following functional of the process $\{D_n(h)\}$, which under this null hypothesis of constant unconditional variance converges in distribution to the supremum of a Brownian bridge on $[0,1]$

$$\sqrt{n/2} \sup_{0 \leqslant h \leq 1} |D_n(h)| \xrightarrow{d} \sup_{0 \leqslant h \leqslant 1} \left|W^0(h)\right|. \tag{24}$$

where $W^0(h)$ is the Brownian bridge on the unit interval $[0,1]$ defined as $W^0(h) = W(h) - hW(1)$, $W(h)$ is the Wiener process.

Kokoszka and Leipus (1999) made the assumption that the process $\{Y_t\}$ follows an ARCH(∞) process defined as

$$\begin{aligned} Y_t &= \sigma_t \varepsilon_t, \quad \varepsilon_t \sim \text{iid}, \quad E\varepsilon_0 = 0, \quad \operatorname{Var}\varepsilon_0 = 1, \\ \sigma_t^2 &= \omega + \sum_{j=1}^{\infty} \alpha_j Y_{t-j}^2, \quad t = 1, \ldots, t_0, \\ \sigma_t^2 &= \omega^\star + \sum_{j=1}^{\infty} \alpha_j^\star Y_{t-j}^2, \quad t = t_0 + 1, \ldots, n, \end{aligned} \tag{25}$$

with the assumption that the unconditional variance of the process changes at an unknown time t_0, i.e.,

$$\Delta(n) = \frac{\omega}{1 - \sum_{j=1}^{\infty} \alpha_j} - \frac{\omega^\star}{1 - \sum_{j=1}^{\infty} \alpha_j^\star} \neq 0. \tag{26}$$

The null hypothesis is $H_0 : \omega = \omega^\star$, $\alpha_j = \alpha_j^\star$ for all j, while under the alternative hypothesis $H_A : \omega \neq \omega^\star$ or $\alpha_j \neq \alpha_j^\star$ for some j. The change–point test is based on the process $\{U_n(h), h \in [0,1]\}$ defined as

$$U_n(h) := \sqrt{n} \frac{[nh](n - [nh])}{n^2} \left(\frac{1}{[nh]} \sum_{j=1}^{[nh]} Y_j^2 - \frac{1}{n - [nh]} \sum_{j=[nh]+1}^{n} Y_j^2 \right), \tag{27}$$

which under H_0 converges to the process $\left\{\sigma W^0(h), h \in [0,1]\right\}$, i.e.,

$$U_n(h) \stackrel{\mathcal{D}[0,1]}{\longrightarrow} \sigma W^0(h), \tag{28}$$

where $\stackrel{\mathcal{D}[0,1]}{\longrightarrow}$ means weak convergence in the space $\mathcal{D}[0,1]$ endowed with the Skorokhod topology. We consider here as test statistic the functional based on the process $\{U_n(h), h \in [0,1]\}$

$$\sup_{0 \leqslant h \leqslant 1} |U_n(h)| / \sigma \stackrel{d}{\longrightarrow} \sup_{0 \leqslant h \leqslant 1} \left| W^0(h) \right|, \tag{29}$$

where the long–run variance σ^2 is usually estimated by nonparametric kernel methods. We use here the heteroskedastic and autocorrelation consistent (HAC) estimator by Newey and West (1987) with the truncations order $q = 0, 2, 5, 10, 15$.

The location of the change–point $\hat{\tau}$ is detected by the CUSUM–type estimator based on the same process $\{U_n(h), h \in [0,1]\}$, and defined by

$$\hat{\tau} = [n\hat{h}], \quad \hat{h} = \min \left\{ h : |U_n(h)| = \max_{0 < h \leqslant 1} |U_n(h)| \right\}. \tag{30}$$

This estimator is consistent if $\Delta(n) \to 0$ as $n \to \infty$ but at a slower rate than $n^{1/2}$ as

$$|\Delta(n)| n^{1/2} \to \infty, \quad n \to \infty,$$

see Kokoszka and Leipus (2000) for further details.

3.2 Strongly Dependent Processes

In the previous section, the process $\{Y_t\}$ was assumed weakly dependent. Whitcher *et al.* (2002) proposed to deal with long–range dependent processes with an unknown number of change–points in the unconditional variance, by applying the BS procedure to the discrete wavelet transform of the long–memory process $\{Y_t\}$.

4 Detecting Change–Points in the Volatility of Financial Time Series

We consider two series, the FTSE 100 index and the US dollar–Japanese yen intra–day foreign exchange (FX) rate.

4.1 Application to The FTSE 100 Index

The FTSE 100 index, or Footsie, consists of 100 blue chip stocks that trade on the London Stock Exchange. This series of 4381 observations has been observed between January 1984 and November 2002. Figure 2 displays the series of indices in levels:

Fig. 2. The series of FTSE 100 indices

Table 1 below displays the sequence of change–points K_i, lengths l_{K_i} and P–values given by Algorithm **1**, see also equation (22).

Table 1. Sequences of number of change–points K_i, lengths l_{K_i} and corresponding P–values $\mathcal{P}_{K_i}$ given by Algorithm **1**

K_i	l_{K_i}	$\mathcal{P}_{K_i}$
1	∞	5.0000e-05
3	152.9601	9.7200e-07
4	68.9379	6.8018e-04
6	50.9085	9.1889e-07
7	32.4764	6.5439e-06
8	7.4296	2.8738e-01
11	5.6321	2.6108e-01
13	5.6107	1.3535e-03
15	5.4152	3.7485e-02

Figure 3 below shows that Algorithm **1** is able to pick the main changes in the unconditional variance of the series of returns on FTSE 100 .

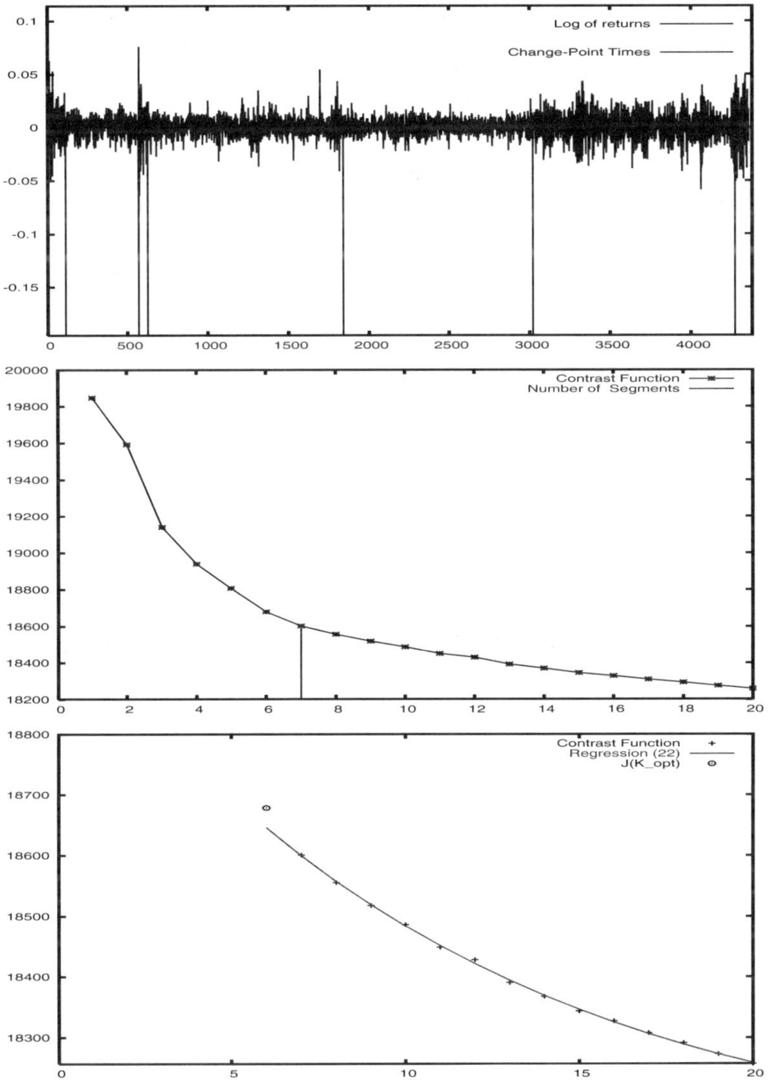

Fig. 3. Above: the series with the estimated change–points represented by vertical lines; Middle: The sequence of contrasts $(J_K, 1 \leqslant K \leqslant K_{MAX})$, the vertical line indicates the estimated number of segments $(\hat{K}, J_{\hat{K}})$; Below: the sequence of contrasts $\{J_K, \hat{K} \leqslant K \leqslant K_{MAX}\}$ are indicated with +, the fitted function $\hat{c}_1(K)+\hat{c}_2 K\log(K)$ is in solid line and $J_{\hat{K}}$ is represented with a circle

We obtain the segmentation $\hat{\boldsymbol{\tau}} = \{112, 568, 624, 1840, 3020, 4272\}$. The point $\boldsymbol{\tau}_1 = 112$ matches a change in the sampling frequency, as we have weekly data before January 1986, and daily observations after that date. Thus, the procedure detects this heterogeneity in the process. The point $\tau_2 = 568$ is simply the 14th October 1987, i.e., the stock market crash, while the increase of volatility after $\tau_5 = 3020$ (June 26, 1997) indicates the conjunction of two opposite phenomena: the Footsie has broken the psychological 5,000 barrier in August 1997, as a consequence of a series of positive earnings for the companies composing the index. On the other side, the Asian crisis of Summer 1997 increased the uncertainty, and then the volatility, as the extent of the consequences of this crisis on economic activity were unpredictable.

Figure 4 below displays the sample autocorrelation function (ACF) for the whole series of absolute returns on the FTSE index:

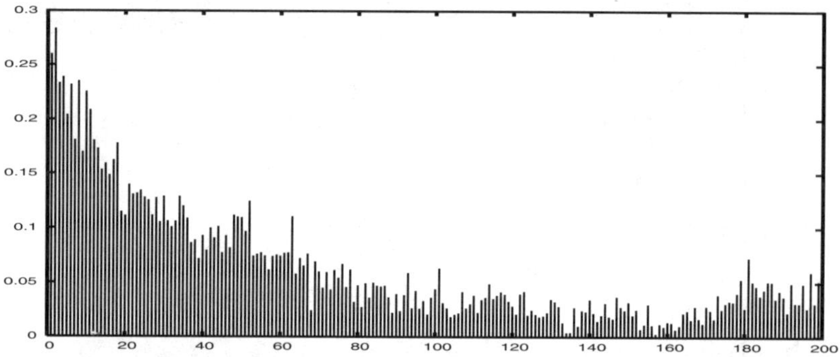

Fig. 4. Sample ACF of the absolute value of returns $|r_t|$ on FTSE

This sample ACF has a hyperbolic decay which is similar to the one of a strongly dependent process: the ACF are always positive, with a plateau for the larger orders of autocorrelation. However, when displaying the sample ACF for the sub–intervals defined by Algorithm **1**, we get the following pictures, the shape of which are very different from Figure 4:

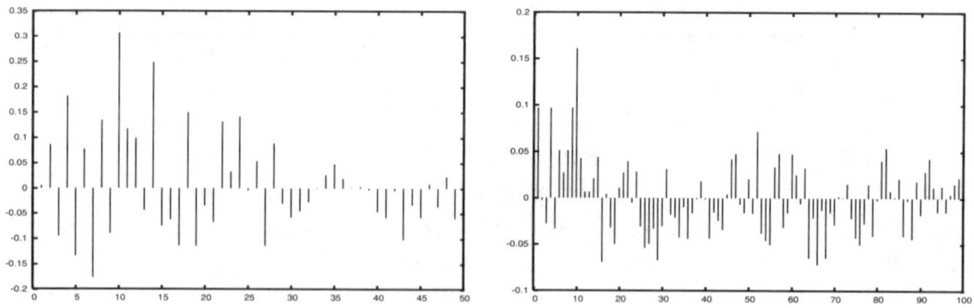

Fig. 5. Sample ACFs of the absolute value of returns $|r_t|$ on FTSE for the time interval $[1, 112]$ (left), and for the time interval $[113, 568]$ (right)

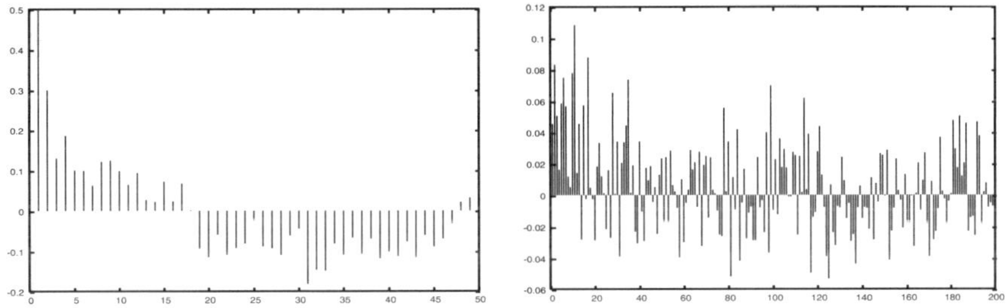

Fig. 6. Sample ACF of the absolute value of returns $|r_t|$ on FTSE for the time interval [569, 624] (left), and for the time interval [625, 1840] (right)

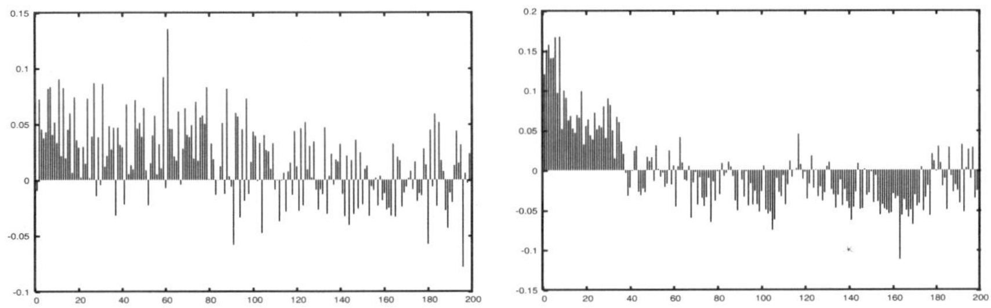

Fig. 7. Sample ACF of the absolute value of returns $|r_t|$ on FTSE for the time interval [1841, 3020] (left), and for the time interval [3021, 4272] (right)

The sample ACF of the absolute value of returns $|r_t|$ on FTSE for the time interval [4273, 4380] displays a pattern similar to Figure 6 (right) and is not displayed here.

For all the sub–samples, the sample ACF displayed in figures 5–7 do not indicate the same degree of persistence as the one observed for the whole sample in Figure 4: some autocorrelations are negative, and these ACF do not display the "plateau effect" for the higher orders. Thus, as mentioned in the introduction of this chapter, the hypothesis of homogeneity and stationarity of the returns process is inappropriate, and the global procedure for finding the optimal resolution for the process is able to pick the nonstationarities of the process out.

Choosing the level of resolution just below, i.e., with 6 segments, would have given $\hat{\boldsymbol{\tau}} = \{112, 568, 624, 3048, 4272\}$. However, considering the period between $t = 624$ and $t = 3048$, i.e., between the 5th of January 1988 and the 4th of August 1997 as homogeneous is rather unlikely. Figure 8 below displays the sample ACF for the absolute returns for this time interval.

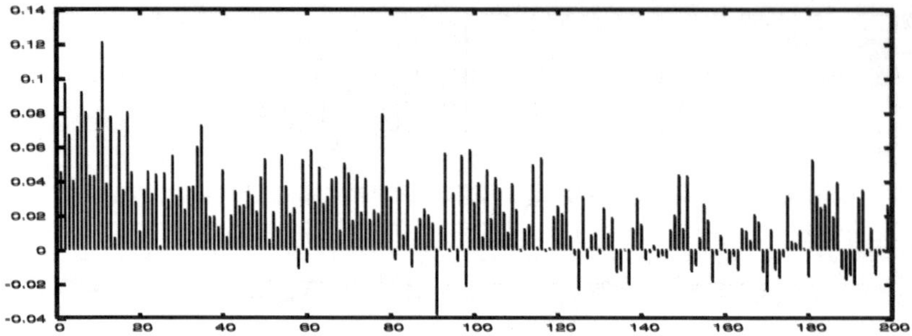

Fig. 8. Sample ACF of the absolute value of returns $|r_t|$ on FTSE for the time interval [624, 3047]

The sample ACF resembles the one of a long–range dependent process, with a mild degree of persistence. However, we have seen before that the sample ACF for the two sub–periods [624, 1847] and [1848, 3019] have a different shape, which indicates that the mild persistence for the interval [624, 3047] is a statistical artefact. Thus, the resolution with 7 segments looks preferable.

We compare the selected segmentation with the one obtained with the binary segmentation procedure. Table 2 below reports the segmentation yield by the BS method, using both statistics given by equations (24) and (29). We observe that the two statistics give a quite similar segmentation, which however has a higher resolution than the one yield by Algorithm **1**. The segmentation given by the BS method includes the points found by Algorithm **1**, or points close to those of $\hat{\boldsymbol{\tau}}$.

Table 2. Segmentation found by the BS method, Inclán and Tiao (1994), henceforth IT, and Kokoszka and Leipus (1999), henceforth KL

KL statistic	Change–point date	IT statistic	Change–point date
9.0585	110	5.5376	110
12.2929	570	6.1560	570
13.2107	648	10.0611	648
1.8179	1062	1.6352	1062
1.3997	1113	1.9091	1273
2.6116	1273	2.0503	1324
2.5349	1324	2.3089	1703
2.6729	1703	2.6606	1838
2.8486	1838	1.5992	1943
2.3037	2117	2.3103	2117
2.1048	2458	2.2611	2458
4.2924	3000	4.3451	3000
1.6788	3173	1.3736	3173
11.2412	3284	7.4494	3284
4.8933	3418	2.4585	3418
3.2717	3654	1.9844	3654
2.6650	3761	1.7056	3761
5.0372	3955	2.5301	3955
3.2697	4128	1.7171	4128
8.6904	4270	4.3607	4270

If we try to refine the segmentation by choosing a number of change–points similar to the BS method, we get the following picture

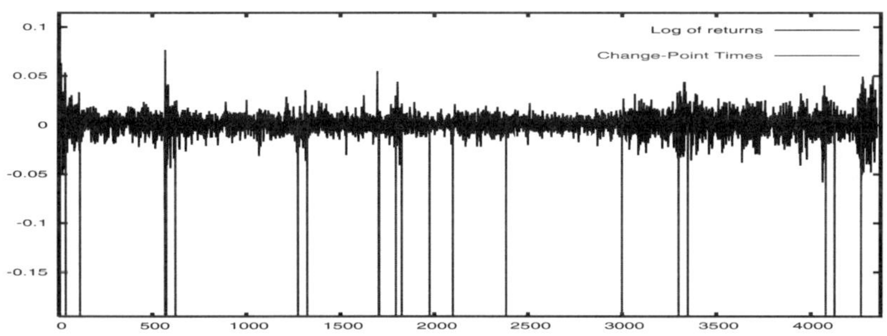

Fig. 9. The series with the 20 estimated change–points represented by vertical lines

We capture more and more details of the variations of the process, but the gain is rather marginal, as the main variations are captured with 7 segments.

4.2 Application to the US Dollar–Japanese Yen FX Rate

We consider here a sample of 30 minute–spaced observations observed in the year 1996. These data, provided by Olsen & Associates are in ϑ time, i.e., all intra–day seasonal components have been removed.

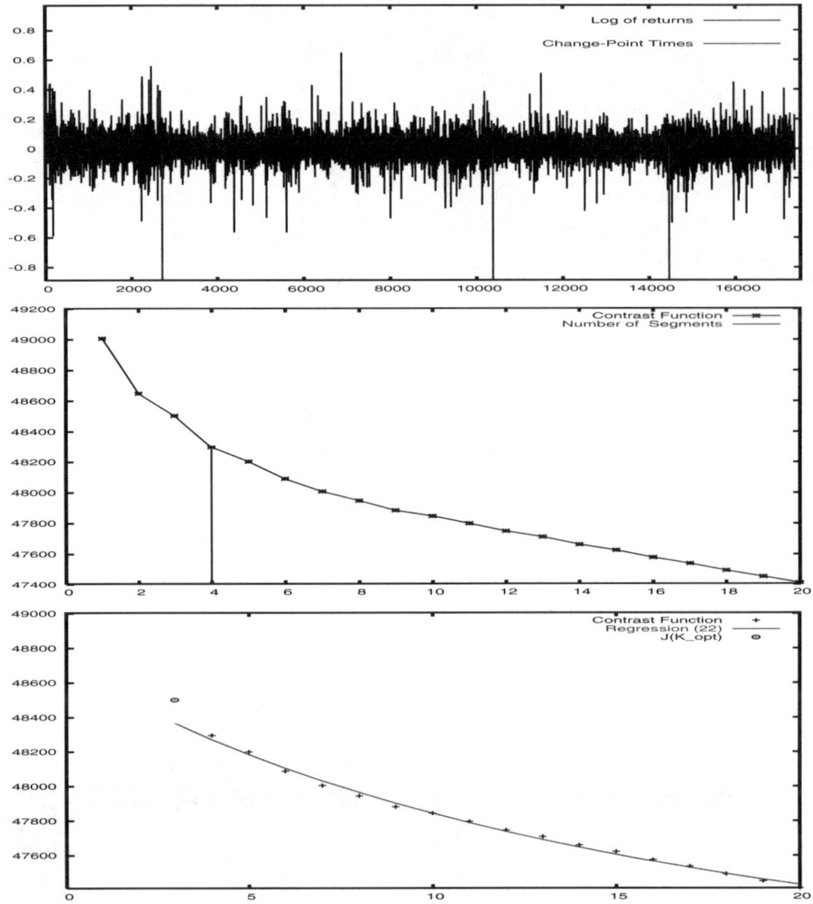

Fig. 10. Above: the series with the estimated change–points represented by vertical lines; Middle: The sequence of contrasts $(J_K, 1 \leqslant K \leqslant K_{MAX})$, the vertical line indicates the estimated number of segments $(\hat{K}, J_{\hat{K}})$; Below: the sequence of contrasts $\{J_K, \hat{K} \leqslant K \leqslant K_{MAX}\}$ are indicated with $+$, the fitted function $\hat{c}_1(K)+\hat{c}_2 K \log(K)$ is in solid line and $J_{\hat{K}}$ is represented with a circle

Figure 10 displays the series with the estimated change–points, the contrast function J_K, and the fitted function $\hat{c}_1(K) + \hat{c}_2 K \log(K)$.

The detected number of change points is rather low for the sample size considered, but we have to keep in mind that this series represents only a year of observations, so that structural changes are rather rare, even for data sampled with a high frequency.

Table 2 below displays the sequence of the number of change–points K_i found by Algorithm **1**.

Table 3. Sequences of number of change–points K_i, lengths l_{K_i} and corresponding P–values $\mathcal{P}_{K_i}$ given by Algorithm **1**

K_i	l_{K_i}	$\mathcal{P}_{K_i}$
1	∞	5.0000e-05
2	183.5677	6.8982e-05
4	72.2762	3.1535e-05
7	21.4666	2.4397e-04
9	16.3938	9.4880e-03
12	1.7163	1.7154e-01

The chapter by Teyssière and Abry (2005) in this volume considers the wavelet analysis of this series: they compare the wavelet estimator of the degree of persistence for the absolute returns of this series with the local Whittle and log–periodogram estimators. The discrepancy between the estimation results obtained with the wavelet based estimator and the ones obtained from the spectral based estimators is interpreted as the consequence of nonstationarities in the returns process.

The BS procedure finds a far larger number of change–points, i.e., 103 for the KL statistic and 95 for the IT statistic. The graphical representation of the segmentation yield by algorithm **1** looks however more sensible.

In fact, the segmentations yield by competing statistical methods are very different. Mikosch and Stărică (1999) and Granger and Hyung (2004) studied the series of S&P 500, using respectively parametric and semiparametric tests. While Mikosch and Stărică (1999) found a rather parsimonious segmentation, the number of change–points found by Granger and Hyung (2004) is huge.

However, this parsimonious segmentation looks relevant when comparing the sample ACF of the absolute returns of the whole series and of the sub–intervals defined by Algorithm **1**, see figures 11–13: while the sample ACF, computed on the whole sample, Figure 11, is similar to the one of a strongly dependent process, the patterns of the sample ACFs for the sub–intervals defined by Algorithm **1** show that the persistence on these sub–intervals is far smaller than the one for the whole sample.

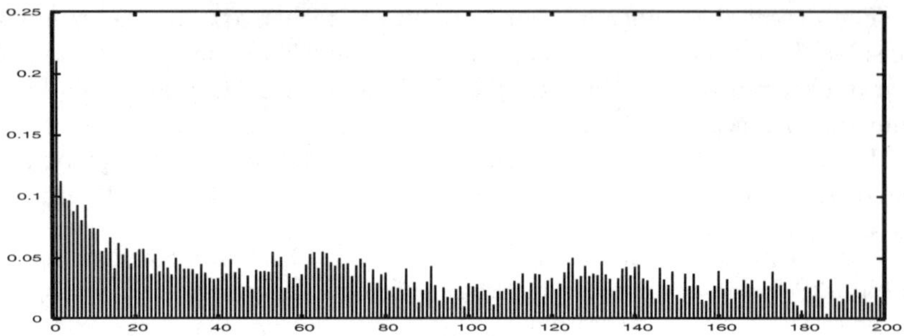

Fig. 11. Sample ACF of the absolute value of returns $|r_t|$ on USD–JPY FX rate

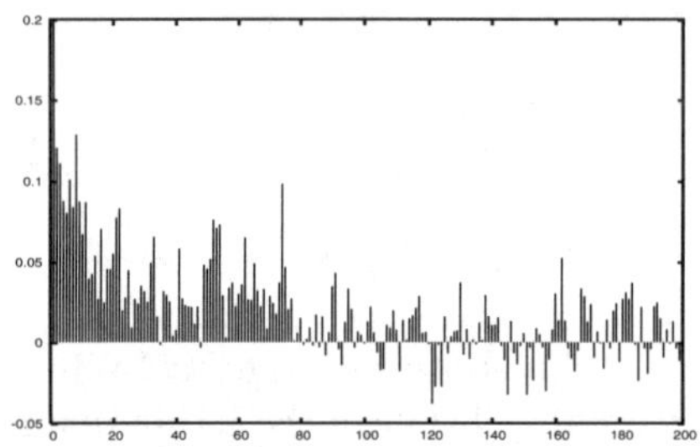

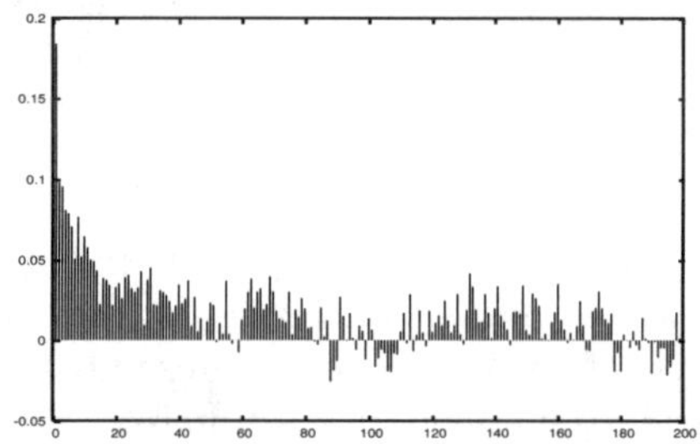

Fig. 12. Sample ACF of the absolute value of returns $|r_t|$ on USD–JPY FX rate for the time interval $[1, 2736]$ (left), and for the time interval $[2737, 10386]$ (right)

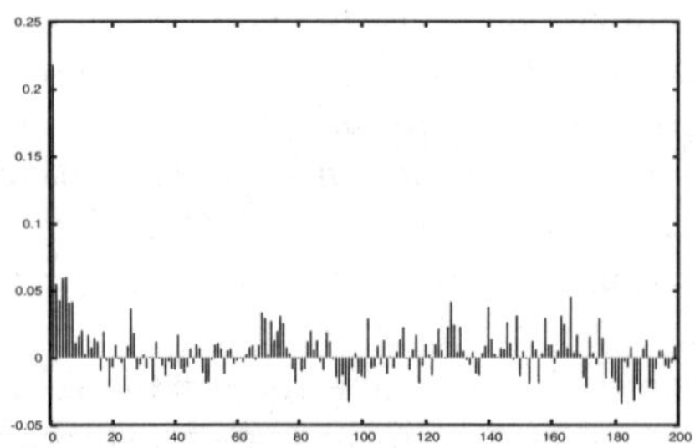

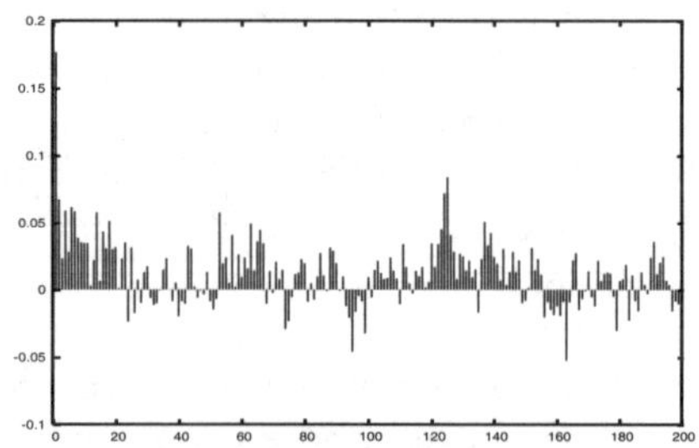

Fig. 13. Sample ACF of the absolute value of returns $|r_t|$ on USD–JPY FX rate for the time interval $[10387, 14454]$ (left), and for the time interval $[14455, 17508]$ (right)

4.3 Application to Micro–Simulated Data

We consider here simulated series from an artificial financial market, i.e., a dynamic system which models financial markets with interacting agents. Although these models do not resort to statistical distributions leading to the generation of long–range dependent processes, the volatility series generated by these models display the same dependence properties as the ones of the volatility of asset prices.

We consider that agents i on financial markets differ by their forecasting function $E^i(P_{t+1}|I_t)$ of the future price as a function of the information set I_t. Chartists extrapolate the exchange rate P_{t+1} by using is a linear function of the previous prices, i.e.,

$$E^c\left(P_{t+1}|I_t\right) = \sum_{j=0}^{M^c} h_j P_{t-j}, \tag{31}$$

where h_j, $j = 0, \ldots, M^c$ are constants, M^c is the 'memory' of the chartists, while fundamentalists forecast this next price as:

$$E^f\left(P_{t+1}|I_t\right) = \bar{P}_t + \sum_{j=1}^{M^f} \nu_j(P_{t-j+1} - \bar{P}_{t-j}), \tag{32}$$

where ν_j, $j = 1, \ldots, M^f$ are positive constants, representing the degree of reversion to the fundamentals, M^f is the 'memory' of the fundamentalists. We assume that the series of 'fundamentals' $\bar{P}_t$, which can be thought as the price if it were only to be explained by a set of relevant variables, follows a random walk:

$$\bar{P}_t = \bar{P}_{t-1} + \varepsilon_t, \quad \varepsilon_t \sim N(0, \sigma_\varepsilon^2). \tag{33}$$

Agents have the possibility of investing at home in a risk free asset or investing abroad in a risky asset. We denote by ρ_t the foreign interest rate, by d_t^i the demand by the ith agent for foreign currency, and by r the domestic interest rate, with $\rho_t > r$. The exchange rate P_t and the foreign interest rate ρ_t are considered by agents as independent random variables, with $\rho_t \sim N(\rho, \sigma_\rho^2)$. The cumulated wealth of individual i at time $t+1$, W_{t+1}^i is given by:

$$W_{t+1}^i = (1 + \rho_{t+1})P_{t+1}d_t^i + (W_t^i - P_t d_t^i)(1 + r). \tag{34}$$

Agents i have a standard mean–variance utility function:

$$U(W_{t+1}^i) = E(W_{t+1}^i) - \lambda \mathrm{Var}(W_{t+1}^i), \tag{35}$$

where λ denotes the risk aversion coefficient, and

$$E(W_{t+1}^i|I_t) = (1 + \rho)E^i(P_{t+1}|I_t)d_t^i + (W_t^i - P_t d_t^i)(1 + r), \tag{36}$$

$$\mathrm{Var}(W_{t+1}^i|I_t) = (d_t^i)^2 \zeta_t, \quad \zeta_t = \mathrm{Var}\left(P_{t+1}(1 + \rho_{t+1})\right). \tag{37}$$

Demand d_t^i is found by maximizing utility. First order condition gives

$$(1+\rho)E^i(P_{t+1}|I_t) - (1+r)P_t - 2\zeta_t\lambda d_t^i = 0, \tag{38}$$

where $E^i(.|I_t)$ denotes the forecast of an agent of type i. Let k_t be the proportion of fundamentalists at time t, the market demand is:

$$d_t = \frac{(1+\rho)\left(k_t E^f(P_{t+1}|I_t) + (1-k_t)E^c(P_{t+1}|I_t)\right) - (1+r)P_t}{2\zeta_t\lambda}. \tag{39}$$

Now consider the exogenous supply of foreign exchange X_t, then the market is in equilibrium if aggregate supply is equal to aggregate demand, i.e., $X_t = d_t$, which gives

$$P_t = \frac{1+\rho}{1+r}\left(k_t E^f(P_{t+1}|I_t) + (1-k_t)E^c(P_{t+1}|I_t)\right) - \frac{2\zeta_t\lambda X_t}{1+r}. \tag{40}$$

From equation (40), the dynamics of the price process $\{P_t\}$ depends on the evolution of the process $\{k_t\}$, i.e., the proportion of fundamentalists, which governs the transition between the two forecast functions $E^f(P_{t+1}|I_t)$ and $E^c(P_{t+1}|I_t)$. Several mechanisms for the evolution of the opinion process $\{k_t\}$ have been proposed in the literature, which are either based on epidemiologic phenomenon, or on a preference given to the most performing forecasting function, or on the accumulated wealth gained with each forecast function, etc. Interested readers are referred to Teyssière (2003), and the chapter by Gaunersdorfer and Hommes (2005) in this volume.

We consider a multivariate extension of this model, i.e., a bivariate process $(P_{1,t}, P_{2,t})$ of foreign exchange rates. This is motivated by the fact that structural changes do not affect singles markets, i.e., the same swing in opinions from chartists to fundamentalists affects linked markets. It has been suggested in the 1999 version of the work by Granger and Hyung that these common breaks might explain the common persistence of asset prices volatility. Indeed, the bivariate common break process by Teyssière (2003) used here, generates the same type of dependence as the one observed in multivariate financial time series.

We then consider that the opinion process $\{k_t\}$ is the same for both markets. This bivariate foreign exchange rate process depends on a pair of foreign interest rates (ρ_1, ρ_2). We assume that $2\zeta_{i,t}\lambda X_{i,t}/(1+\rho_i) = \gamma_i \bar{P}_{i,t}$ for $i = 1, 2$, and $M^f = M^c = 1$, the equilibrium price for the bivariate model is given by

$$\begin{pmatrix} P_{1,t} \\ P_{2,t} \end{pmatrix} = \begin{pmatrix} \frac{k_t-\gamma}{A_1}\bar{P}_{1,t} - \frac{k_t\nu_{1,1}}{A_1}\bar{P}_{1,t-1} + \frac{(1-k_t)h_{1,1}}{A_1}P_{1,t-1} \\ \frac{k_t-\gamma}{A_2}\bar{P}_{2,t} - \frac{k_t\nu_{2,1}}{A_2}\bar{P}_{2,t-1} + \frac{(1-k_t)h_{2,1}}{A_2}P_{2,t-1} \end{pmatrix}, \tag{41}$$

with

$$A_i = \frac{1+r}{1+\rho_i} - (1-k_t)h_{i,0} - k_t\nu_{i,1}. \tag{42}$$

We assume that the bivariate process of fundamentals $(\bar{P}_{1,t}, \bar{P}_{2,t})$ is positively correlated as follows:

$$\begin{pmatrix} \bar{P}_{1,t} \\ \bar{P}_{2,t} \end{pmatrix} = \begin{pmatrix} \bar{P}_{1,t-1} \\ \bar{P}_{2,t-1} \end{pmatrix} + \begin{pmatrix} \varepsilon_{1,t} \\ \varepsilon_{2,t} \end{pmatrix}, \quad \begin{pmatrix} \varepsilon_{1,t} \\ \varepsilon_{2,t} \end{pmatrix} \sim N\left[\begin{pmatrix} 0 \\ 0 \end{pmatrix}, \begin{pmatrix} \sigma_{1,1}^2 & \sigma_{1,2} \\ \sigma_{1,2} & \sigma_{2,2}^2 \end{pmatrix} \right], \tag{43}$$

with $\sigma_{1,2} > 0$. In the example considered here, we set $\sigma_{1,2}$ so that the coefficient of correlation between the two processes $\varepsilon_{1,t}$ and $\varepsilon_{2,t}$ is equal to 0.75, a choice motivated by empirical results; see Teyssière (1997, 2003).

We generate here a bivariate series of returns. Figure 14 below displays the two generated series of returns $r_{1,t}, r_{2,t}$, with $r_{1,t} = \ln(P_{1,t}/P_{1,t-1})$ and $r_{2,t} = \ln(P_{2,t}/P_{2,t-1})$, and the detected changes in their unconditional variance. We can see that the detected changes for the series $r_{2,t}$ are very close to some of the detected changes for the series $r_{1,t}$, i.e., $\hat{\tau} = \{676, 868, 1360, 1584, 2408, 4032, 4148\}$ for the series $r_{1,t}$, while $\hat{\tau} = \{1360, 1580, 4144\}$ for the series $r_{2,t}$, which is not very surprising as the opinion process $\{k_t\}$ is common for both processes $r_{1,t}$ and $r_{2,t}$. The joint detection of change–points in multivariate time series is considered in a subsequent paper; see Lavielle and Teyssière (2005).

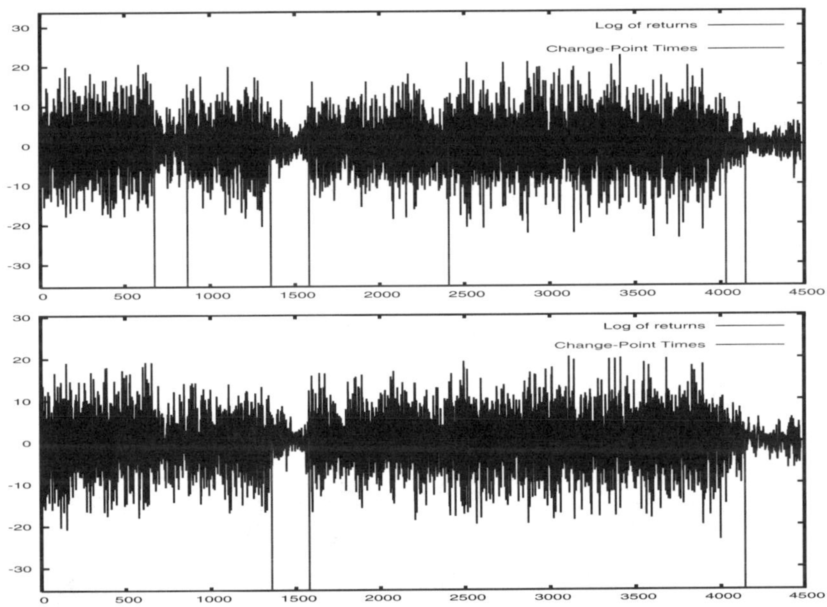

Fig. 14. The two jointly simulated series, $r_{1,t}$ above and $r_{2,t}$ below, with the estimated change–points represented by vertical lines

We focus on the second returns series $r_{2,t}$, with the lowest resolution level: its sample ACF, see Figure 15, is similar to the one of a strongly dependent process. However, the sample ACF for the sub–intervals defined by Algorithm **1**, see figures 16 and 17, display a different pattern with both positive and negative autocorrelations, a property similar to what has been observed with the two previous examples.

Thus, Algorithm **1** is able to detect the nonstationarities of the returns process generated by the artificial financial market.

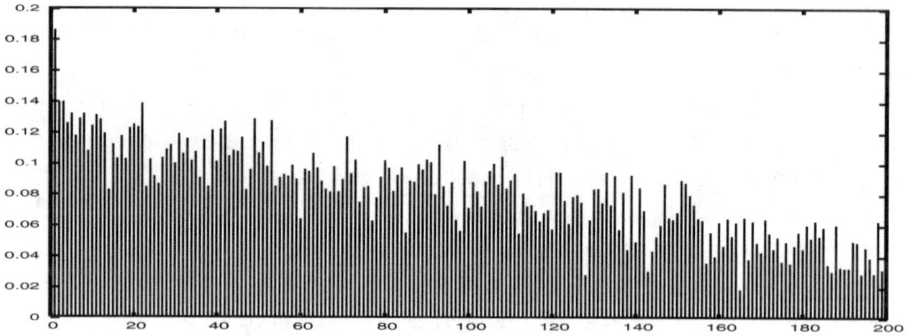

Fig. 15. Sample ACF of the absolute value of simulated returns $|r_{2,t}|$

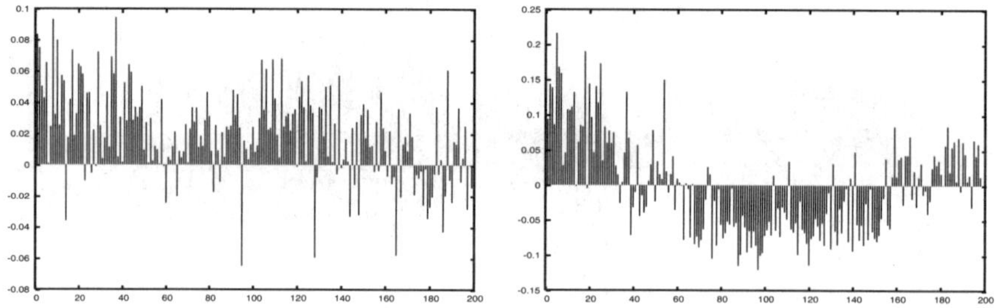

Fig. 16. Sample ACF of the absolute value of simulated returns $|r_{2,t}|$ for the time interval [1, 1360] (left), and for the time interval [1361, 1580] (right)

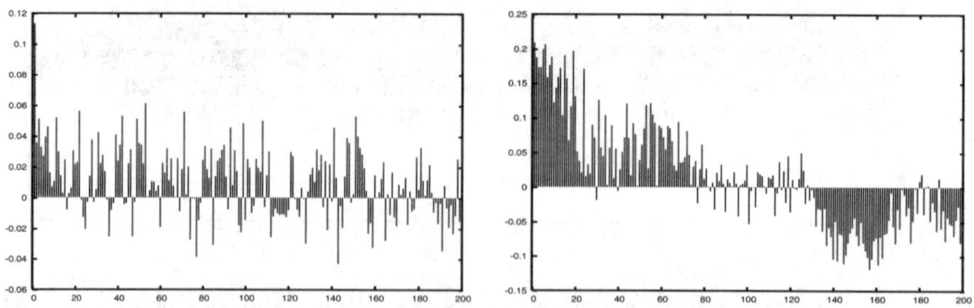

Fig. 17. Sample ACF of the absolute value of simulated returns $|r_{2,t}|$ for the time interval [1581, 4144] (left), and for the time interval [4145, 4500] (right)

5 Conclusion: Detecting Break in the Variance of Returns or in the Mean of Absolute Returns?

Since we checked the adequacy of the resolution by looking at the ACF of the sequence of absolute returns, one might think that changes in the volatility are uncovered by detecting changes in the mean of the absolute returns series, i.e., instead of the contrast function given by equation (5) one might consider the following contrast function based again on a Gaussian likelihood function

$$J_n(\boldsymbol{\tau}, \boldsymbol{y}) = \frac{1}{n} \sum_{k=1}^{K} \|Y_{\tau_k} - \bar{Y}_{\tau_k}\|. \tag{44}$$

Applying again Algorithm **1** to the series of absolute returns $|r_t|$ on the FTSE 100 index, we select different values for the number of change–points K_i from 9 to 12, and obtained the following segmentations:

- $\hat{\tau} = \{112, 568, 576, 624, 3300, 3348, 4080, 4092, 4272\}$,
- $\hat{\tau} = \{112, 568, 576, 624, 3300, 3348, 4080, 4092, 4284, 4304\}$,
- $\hat{\tau} = \{112, 568, 576, 624, 3300, 3340, 3348, 4080, 4092, 4284, 4304\}$,
- $\hat{\tau} = \{112, 568, 576, 624, 1856, 3004, 3312, 3348, 4080, 4092, 4284, 4304\}$,

i.e., we have to consider a large number of segments, 13, for splitting the interval $[624, 3300]$. Thus, considering the series of absolute returns is not suitable for finding the optimal resolution of the volatility series.

Alternatively, one might detect both changes in the mean and the variance for the series of absolute returns by considering the following contrast function:

$$J_n(\boldsymbol{\tau}, \boldsymbol{y}) = \frac{1}{n} \sum_{k=1}^{K} \frac{\|Y_{\tau_k} - \bar{Y}_{\tau_k}\|}{\hat{\sigma}_k^2} + n_k \log(\hat{\sigma}_k^2). \tag{45}$$

In that case, algorithm **1** selected the following partition with 12 change–points:

$$\hat{\tau} = \{112, 568, 576, 624, 1796, 1828, 3020, 3304, 3348, 4080, 4128, 4272\},$$

which has a level of resolution far higher than the one obtained when detecting changes in the unconditional variance of the returns process. We obtain similar results for the series of returns on US dollar–Japanese Yen FX rate, and the series of results generated by the artificial microeconomic model. Thus, the straightforward and natural way for detecting changes in the volatility is to consider the contrast function defined by equation (5).

References

1. Aggarwal, R., Inclán, C. and Leal, R. (1999). Volatility in emerging stock markets. *Journal of Financial and Quantitative Analysis*, **34,** 1–17.

2. Basseville, M. and Nikiforov, N. (1993). *The Detection of Abrupt Changes - Theory and Applications.* Prentice-Hall: Information and System Sciences Series.
3. Battacharya, P.K. (1987). Maximum Likelihood estimation of a change–point in the distribution of independent random variables: general multiparameter case. *Journal of Multivariate Analysis*, **32,** 183–208.
4. Beran, J. and Terrin, N. (1999). Correction: Testing for a change of the long–memory parameter. *Biometrika*, **86,** 233.
5. Beran, J. and Terrin, N. (1996). Testing for a change of the long–memory parameter. *Biometrika*, **83,** 627–638.
6. Berkes, I., Horváth, L. and Kokoszka, P.S. (2004). Testing for parameter constancy in GARCH(p, q) models. *Statistics and Probability Letters*, **70,** 263–273.
7. Berkes, I., Gombay, E., Horváth, L. and Kokoszka, P.S. (2004). Sequential change–point detection in GARCH(p, q) models. *Econometric Theory*, **20,** 1140–1167.
8. Berkes, I., Horváth, L., Kokoszka, P.S. and Shao, Q-M. (2003). On discriminating between long–range dependence and changes in mean. *The Annals of Statistics*, forthcoming.
9. Bhattacharya, R.N., Gupta, V.K. and Waymire, E. (1983). The Hurst effect under trends. *Journal of Applied Probability*, **20,** 649–662.
10. Birgé, L. and Massart, P. (2001). A generalized C_p criterion for Gaussian model selection. *Preprint Université Paris-Sud.*
11. Braun, J.V., Braun, R.K. and Muller, H.G. (2000). Multiple changepoint fitting via quasilikehood, with application to DNA sequence segmentation. *Biometrika*, **87,** 301–314.
12. Brodsky. B. and Darkhovsky, B. (1993). *Nonparametric Methods in Change–Point Problems.* Kluwer Academic Publishers, The Netherlands.
13. Chu, C.-S.J. (1995). Detecting parameter shift in GARCH models. *Econometric Reviews*, **14,** 241–266.
14. Csörgö, M. and Horváth, L. (1997). *Limit Theorems in Change–Point Analysis.* Wiley.
15. Dalhaus, R. (1997). Fitting time series models to nonstationary processes. *The Annals of Statistics*, **25,** 1–37.
16. Gaunersdorfer, A. and Hommes, C.H. (2005). A nonlinear structural model for volatility clustering. In: Teyssière, G. and Kirman, A., (Eds.), *Long–Memory in Economics.* Springer Verlag, Berlin. Appears in this volume.
17. Gijbels, I., Hall, P. and Kneip, A. (1999). On the estimation of jump points in smooth curves. *The Annals of the Institute of Statistical Mathematics*, **51,** 231–251.
18. Giraitis, L., Leipus, R. and Surgailis, D. (2005). Recent advances in ARCH modelling. In: Teyssière, G. and Kirman, A., (Eds.), *Long–Memory in Economics.* Springer Verlag, Berlin. Appears in this volume.
19. Giraitis, L. and Leipus, R. (1992). Testing and estimating in the change–point problem of the spectral function. *Lithuanian Mathematical Journal*, **32,** 20–38.
20. Giraitis, L. and Leipus, R. (1990). Functional CLT for nonparametric estimates of the spectrum and change–point problem for a spectral function. *Lithuanian Mathematical Journal*, **30,** 674–679.
21. Giraitis, L., Leipus, R. and Surgailis, D. (1996). The change–point problem for dependent observations. *Journal of Statistical Planning and Inference*, **53,** 297–310.

22. Granger, C.W.J. (2002). Long memory, volatility, risk and distribution. *Preprint.*
23. Granger, C.W.J. (2000). Current perspectives on long memory processes. *Academia Economic Papers*, **28,** 1–16.
24. Granger, C.W.J. and Hyung, N. (2004). Occasional structural breaks and long memory with an application to the S&P 500 absolute stock returns. *Journal of Empirical Finance*, **11,** 399–421.
25. Granger, C.W.J. and Hyung, N. (1999). Occasional structural breaks and long–memory. *Preprint.*
26. Granger, C.W.J. and Ding, Z. (1995). Some properties of absolute returns, an alternative measure of risk. *Annales d'Économie et de Statistique*, **40,** 67–91.
27. Hawkins, D.L. (1977). Testing a sequence of observations for a shift in location. *Journal of the American Statistical Association*, **72,** 180–186.
28. Horváth, L. (2001). Change–point detection in long–memory processes. *Journal of Multivariate Analysis*, **78,** 218–234.
29. Horváth, L., Kokoszka, P.S. and Teyssière, G. (2001). Empirical process of the squared residuals of an ARCH sequence. *The Annals of Statistics*, **29,** 445–469.
30. Horváth, L. and Shao, Q-M. (2001). Limit theorems for quadratic forms with application to the Whittle's estimate. *Annals of Applied Probability,* **9,** 146–187.
31. Inclán, C. and Tiao, G.C. (1994). Use of cumulative sums of squares for retrospective detection of changes in variance. *Journal of the American Statistical Association*, **89,** 913–923.
32. Kay, S.M. (1998). *Fundamentals of Statistical Signal Processing: Detection Theory.* Prentice Hall: Signal Processing Series.
33. Kim, H.J. (1994). Likelihood ratio and cumulative sum tests for a change–point in linear regression. *Journal of Multivariate Analysis*, **51,** 54–70.
34. Kokoszka, P.S. and Leipus, R. (1999). Testing for parameter changes in ARCH models. *Lithuanian Mathematical Journal*, **39,** 231–247.
35. Kokoszka, P.S. and Leipus, R. (2000). Change–point estimation in ARCH models. *Bernoulli*, **6,** 513–539.
36. Kokoszka, P.S. and Leipus, R. (2003). Detection and estimation of changes in regime. In: Doukhan, P.,Oppenheim, G. and Taqqu, M.S. (Eds.), *Long-Range Dependence: Theory and Applications*, 325–337. Birkhauser.
37. Kokoszka, P.S. and Teyssière, G. (2002). Change–point detection in GARCH models: Asymptotic and bootstrap tests. *Preprint.*
38. Lavielle, M. (1999). Detection of multiple changes in a sequence of dependent variables. *Stochastic Processes and their Applications*, **83,** 79–102.
39. Lavielle, M. (1998). Optimal segmentation of random processes. *IEEE Transactions on Signal Processing*, **46,** 1365–1373.
40. Lavielle, M. (1993). Detection of changes in the spectrum of a multidimensional process. *IEEE Transactions on Signal Processing*, **41,** 742–749.
41. Lavielle, M. and Teyssière, G. (2005). Detection of multiple change–points in multivariate time series. *Preprint.*
42. Lavielle, M. and Lebarbier, E. (2001). An application of MCMC methods to the multiple change–points problem. *Signal Processing*, **81,** 39–53.
43. Lavielle, M. and Moulines, E. (2000). Least–squares estimation of a unknown number of shifts in a time series. *Journal of Time Series Analysis*, **21,** 33–59.
44. Lavielle, M. and Ludeña, C. (2000). The multiple change–points problem for the spectral distribution. *Bernoulli*, **6,** 845–869.
45. Lebarbier, E. (2001). Adaptive estimation of a piecewise continuous signal in Gaussian noise. *Preprint Université Paris-Sud.*

46. Lombard, F. (1987). Rank tests for change point problems. *Biometrika*, **74,** 615–624.
47. Mia, B.Q. and Zhao, L.C. (1988). Detection of change points using rank methods. *Communications in Statistics - Theory and Methods*, **17,** 3207–3217.
48. Mikosch, T. and Stărică, C. (2004). Change of structure in financial time series and the GARCH model. *REVSTAT Statistical Journal*, **2,** 41–73.
49. Mikosch, T. and Stărică, C. (2003) Long–range dependence effects and ARCH modeling. In: Doukhan, P., Oppenheim, G. and Taqqu, M.S. (Eds.), *Long–Range Dependence: Theory and Applications.* Birkhauser, Boston, pp. 439–459.
50. Mikosch, T. and Stărică, C. (1999). Change of structure in financial time series, long range dependence and the GARCH model. *Preprint.*
51. Newey, W.K. and West, K.D. (1987). A simple positive definite, heteroskedasticity and autocorrelation consistent covariance matrix. *Econometrica*, **55,** 703–708.
52. Robinson, P.M. (1991). Testing for strong serial correlation and dynamic conditional heteroskedasticity in multiple regression. *Journal of Econometrics*, **47,** 67–84.
53. Schechtman, E. and Wolfe, D.A. (1985). Multiple change points problem - nonparametric procedures for estimation of the points of change. *Communications in Statistics - Simulation and Computation*, **14,** 615–631.
54. Sen, A. and Srivastava, M.S. (1975). On tests for detecting change in the mean. *The Annals of Statistics*, **3,** 96–103.
55. Stărică, C. and Granger, C.W.J. (2005). Nonstationarities in stock returns. *Review of Economics and Statistics*, **87,** 503–522
56. Teyssière, G. and Abry, P.(2005). Wavelet analysis of nonlinear long–range dependent processes. Applications to financial time series. In: Teyssière, G. and Kirman, A., (Eds.), *Long–Memory in Economics.* Springer Verlag, Berlin. Appears in this volume.
57. Teyssière, G. (2003). Interaction models for common long–range dependence in asset price volatility. In: Rangarajan, G., Ding, M. (Eds.), *Processes with Long Range Correlations: Theory and Applications,* Lecture Notes in Physics, Vol. 621. Springer Verlag, Berlin, pp. 251–269.
58. Teyssière, G. (1997). Modelling exchange rates volatility with multivariate long–memory ARCH processes. *Preprint.*
59. Vostrikova, L.Ju. (1981). Detection of "disorder" in multidimensional random processes. *Soviet Mathematics Doklady*, **24,** 55–59.
60. Whitcher, B., Byers, S.D., Guttorp, P. and Percival D.B. (2002). Testing for homogeneity of variance in time series: long memory, wavelets and the nile river. *Water Resources Research*, **38,** 10.1029/2001WR000509.
61. Yao, Y.C. (1988). Estimating the number of change–points via Schwarz criterion. *Statistics & Probability Letters*, **6,** 181–189.

Bandwidth Choice, Optimal Rates and Adaptivity in Semiparametric Estimation of Long Memory

Marc Henry[1]

Department of Economics, Columbia University. marc.henry@columbia.edu

Summary. Semiparametric estimation of long memory refers to periodogram based estimation of the shape of the spectral density $f(\lambda)$ at low frequencies, where all but the lowest harmonics of the periodogram are discarded, so as to forego specification of the short range dynamic structure of the time series, and avoid bias incurred when the latter is misspecified. Such a procedure entails an order of magnitude loss of efficiency with respect to parametric estimation, but may be warranted when long series (earth scientific or financial) can be obtained. This paper presents strategies proposed for the choice of bandwidth, i.e. the number of periodogram harmonics used in estimation, with the aim of minimizing this loss of efficiency. Such strategies are assessed with respect to minimax rates of convergence, that depend on the smoothness of $|\lambda|^{-2d}f(\lambda)$ (where d is the long memory parameter) in the neighbourhood of frequency zero. The plug-in strategy is discussed in the case where the degree of local smoothness is known a priori, and adaptive estimation of d is discussed for the case where the degree of local smoothness is unknown.

1 Local Spectral Estimation of Long Memory

Throughout this chapter, we shall consider n observations $X_1, \ldots, X_n$ from a covariance stationary time series $\{X_t\}_{t=-\infty}^{\infty}$ with mean μ and spectral density f satisfying

$$f(\lambda) = c|\lambda|^{-2d}\, h(\lambda), \;\lambda \in [-\pi, \pi],\; d \in (-\frac{1}{2}, \frac{1}{2}), \tag{1}$$

where $c \in (0, \infty)$, and $h(\lambda)$ converges to 1 when λ converges to 0, and the condition on d is necessary for stationarity and invertibility of the process. Such a process may be derived through fractional filtering

$$X_t = (1 - L)^{-d}\, Y_t, \tag{2}$$

of a short memory process Y_t with spectral density $h(\lambda)\; |2\sin(\lambda/2)/\lambda|^{-2d}$, where L is the lag operator, and

$$(1-L)^{-d} = \sum_{-\infty}^{\infty} \frac{\Gamma(d+j)}{\Gamma(d)\Gamma(j+1)} L^j, \tag{3}$$

with $\Gamma(.)$ denoting the Gamma function.

The process $\{X_t\}$ is said to exhibit long memory when $0 < d < 1/2$, short memory when $d = 0$, and "intermittent memory" or "antipersistence" when $-1/2 < d < 0$. The latter case is empirically relevant insofar as it may characterize first-differences of series that were incorrectly believed to hold a unit root.

Estimation of parametric versions of (1) is discussed in Giraitis and Robinson (2003b). However, when the parameter of interest is d, and the sample size is deemed large enough (as in most financial applications), local methods are preferable, and we shall concentrate on the latter.

We call "local," estimation methods which rely on versions of (1), where $h(\lambda)$ is left unspecified outside a neighbourhood of zero frequency to avoid bias in the estimation of d under certain misspecifications of the short-range dynamics (i.e. the shape of $h(\lambda)$) outside a neighbourhood of frequency zero). One thereby incurs an order of magnitude loss of efficiency, and much of the focus of the present chapter will be on strategies to minimize this loss of efficiency through optimal choice of smoothing parameter (called bandwidth by analogy with kernel estimation).

The minimum order of magnitude loss of efficiency is identified by minimax lower bounds for the quadratic risk of any estimator $\hat{d}$ of d under local-to-zero smoothness assumptions on $h(\lambda)$. Bearing in mind that the parametric rate of convergence is $\sqrt{n}$ (see for instance Giraitis and Robinson, 2003b), the best attainable rate for local estimators is $n^{s/(1+2s)}$, where the degree of local smoothness s of $h(\lambda)$ is defined as follows: for $0 < s \leq 1$, $h(\lambda)$ has smoothness s if it satisfies a Lipschitz condition of degree s around $\lambda = 0$; and for $s > 1$, $h(\lambda)$ has degree of smoothness s if it is $[s]$ times differentiable in a neighbourhood of $\lambda = 0$ and its $[s]$th derivative satisfies a Lipschitz condition of degree $s - [s]$ around $\lambda = 0$.

Several "semiparametric" estimators of d have been proposed (they are "semiparametric" in the sense that they rely only on extensions of the local specification (1)), two of which, the "local Whittle" (proposed by Künsch (1987), hereafter LW), and the "log-periodogram" (proposed by Geweke and Porter–Hudak (1983), hereafter LP) have been shown by Robinson (1995a, 1995b) to be asymptotically normal with asymptotic variances that are free of unknown parameters. Both LW and LP are based on the periodogram of the observations

$$I(\lambda) = |\phi(\lambda)|^2 \tag{4}$$

where $\phi(\lambda)$ is the Discrete Fourier Transform (hereafter DFT) of the data, defined as

$$\phi(\lambda) = (2\pi n)^{-1/2} \sum_{t=1}^{n} X_t \exp{(it\lambda)}. \tag{5}$$

Both estimators rely only on the first m harmonics of the periodogram, i.e. $I_j = I(\lambda_j)$, $j = 1, \ldots, m$, with $\lambda_j = 2\pi j/n$, and the thus defined "bandwidth" satisfies at least

$$\frac{1}{m} + \frac{m}{n} \to 0, \text{ when } n \to 0, \tag{6}$$

so as to conform to local-to-zero specifications.

Finally, both LW and LP can be construed as special cases of a general class of m-estimators $\hat{d}$ proposed by Robinson and Henry (2003), which solve[1]

$$\sum_{j=1}^{m} w_j \psi(I_j/\lambda_j^{-2\hat{d}}) = 0, \tag{7}$$

where w_j are suitably defined weights and $\psi(z)$ is a real valued monotonic function. The weights are chosen to satisfy at least

$$\sum_{j=1}^{m} w_j = 0 \tag{8}$$

to avoid the presence of higher-order cumulants in the asymptotic variances of such estimators (under A2 below).

When

$$w_j = \nu_j \stackrel{\text{def}}{=} \log j - \frac{1}{m}\sum_{l=1}^{m} \log \lambda_l, \tag{9}$$

the LW obtains with

$$\psi(z) = z - 1, \tag{10}$$

and the LP obtains with

$$\psi(z) = \log z. \tag{11}$$

The LP has the computational advantage of being defined in closed form as the least squares estimator for d in the simple regression

$$\log I_j = \alpha + d(-2\log \lambda_j) + \epsilon_j, \tag{12}$$

for $j = 1, \ldots, m$, with α the "intercept" and ϵ_j the "error term."

The LW is not defined in closed form, so that a consistency proof is required prior to a proof of asymptotic normality based on a mean value argument and the Bartlett decomposition:

[1] Note that the omission of $\lambda_0 = 0$ permits X_t to have unknown mean μ since I_j is invariant to location shift for $1 \leq j < n/2$.

$$I(\lambda) = 2\pi f(\lambda) I^{\epsilon}(\lambda) + Q(\lambda), \tag{13}$$

where I^{ϵ} denotes the periodogram of the innovations driving the generalized linear process X_t (i.e. the ϵ_t's in assumption A2 below), and $Q(\lambda)$ is a remainder term. The principle of the proof is then to prove a central limit theorem for the relevant function (i.e. the function G from (27) in the case of the LW) of the "pseudo-periodogram" $2\pi f(\lambda) I^{\epsilon}(\lambda)$ and show that the relevant function of the remainder term is asymptotically negligible. Given the method of proof, however, the linearity of ψ in the LW case allows derivation of asymptotic results under much weaker assumptions on the process (A2 with weak conditions on the conditional moments of the martingale difference innovations).

Since the central limit theorem is applied to averages of rescaled periodogram ordinates, a short discussion of their properties is in order. When the data has long memory, Hurvich and Beltrao (1993) showed that the periodogram fails to be asymptotically unbiased for the spectral density, and that periodogram ordinates at distinct frequencies generally fail to be asymptotically uncorrelated. However, the maintained use of periodogram ordinates in this framework is vindicated by the analogous, if weaker, results under long memory proved in Robinson (1995b) and summarized below: for all n, and $1 \leq k \leq [(n-1)/2]$, and $k < j \leq [(n-1)/2]$, and putting $f_k = f(\lambda_k)$,

$$|E(I_k/f_k) - 1| \leq \alpha_k, \qquad \lim_k \alpha_k = 0 \tag{14}$$

$$|\mathrm{Cov}(I_k/f_k, I_j/f_j)| \leq \alpha_{k,j},$$
$$\sum_{1\leq k<j<[(n-1)/2]} \alpha_{k,j} = O(\log^r n),\ r > 0, \tag{15}$$

$$\left[\frac{I_{k_1}}{f_{k_1}}, \ldots, \frac{I_{k_l}}{f_{k_l}}\right] \to_d \text{Vector of } \mathcal{X}_2^2/2, \tag{16}$$

for a given l-uple $k_1, \ldots, k_l$.

The properties above underlie asymptotic results for semiparametric estimators of long memory under a variety of assumptions on the process including the following:

A1: $\{X_t\}$ is a Gaussian process.
A2: $X_t = \mu + \sum_{j=0}^{\infty} \alpha_j \epsilon_{t-j}$, where $\sum_{j=0}^{\infty} \alpha_j^2 < \infty$ and ϵ_j are martingale differences with finite fourth moments.
A3: A2 is satisfied with i.i.d. ϵ_j.

The linearity of LW allows for asymptotic normality under A2, as proved in Robinson (1995a), and Robinson and Henry (1999) for an extension to time dependent conditional variances in the innovations ϵ_j. Robinson (1995b) shows asymptotic normality for LP under A3, and Robinson and Henry (2003) base formal expansions in the general m-estimation case on assumption A1. Asymptotic variances are all proportional to $1/m$, so that restrictions needed on the rate of divergence of m for asymptotic normality to hold are restrictions on efficiency. Higher-order local-to-zero smoothness in $h(\lambda)$ (cases with $s \geq 2$) can be exploited to implement bias reduction techniques and achieve in all cases asymptotic normality under

$$\frac{1}{m} + \frac{m^s}{n^{1+s}} \to 0, \quad \text{as } n \to \infty, \tag{17}$$

which results in $n^{s/(1+2s)}$ rates of convergence, which can be arbitrarily close to $\sqrt{n}$ when $h(\lambda)$ is infinitely differentiable in the neighbourhood of zero. This rate was shown by Giraitis, Robinson and Samarov (1997) to be the minimax rate, assuming a priori knowledge of the local smoothness. The next two sections consider bandwidth choice in the implementation of rate optimal estimators when s is assumed known a priori, whereas the last section considers estimators that are adaptive to an unknown degree of smoothness s.

2 Optimal Rates of Convergence

Before assessing the quality of local estimation techniques, one needs to identify the best achievable rates of convergence by any measurable function of the data consistent with predefined compact sets of spectral densities $f(\lambda)$.

For ease of exposition, all such results will be expressed in terms of quadratic risk

$$E_f\left[\hat{d} - d\right]^2 \tag{18}$$

where the expectation is taken under the law of a process with spectral density f, and $\hat{d}$ is a measurable function of $(X_1, \ldots, X_n)$.

Using notation from Andrews and Guggenberger (2003), let s, δ, and the elements of $K = (K_1, K_2)'$ and $a = (a_0, a_{00}, a_1, \ldots, a_{[s/2]})'$ be positive finite constants, and define the following class of integrable spectral densities:

$$\mathcal{F}(s, a, \delta, K) = \left\{ f : f(\lambda) = c|\lambda|^{-2d} h(\lambda), \int_{-\pi}^{\pi} f \leq K_1, -\frac{1}{2} < d < \frac{1}{2} \right\} \tag{19}$$

where $h(\lambda)$ is an even function on $[-\pi, \pi]$ satisfying

$$a_0 < c < a_{00} \tag{20}$$

$$h(\lambda) = 1 + \sum_{k=1}^{[s/2]} \frac{h_k \lambda^{2k}}{(2k)!} I\!I_{\{s \geq 2\}} + R(\lambda) \tag{21}$$

where the h_k's are constants satisfying

$$|h_k| \leq a_k,\ k = 1, \ldots, [s/2], \tag{22}$$

and

$$|R(\lambda)| \leq K_2 |\lambda|^s, \text{ for all } |\lambda| < \delta. \tag{23}$$

The following theorem is proved in Giraitis, Robinson and Samarov (1997), with the extension to $s > 2$ given in Andrews and Guggenberger (2003).

Theorem 1.

$$\liminf_n \inf_{\hat{d}} \sup_{f \in \mathcal{F}(s,a,\delta,K)} n^{-\frac{2s}{1+2s}} E_f(\hat{d} - d)^2 > 0. \tag{24}$$

This theorem shows how smoothness restricts the rate of convergence of local estimators. Upper bounds presented below will show that these rates are attained.

Note 1. Note that the condition $s \geq 1$ and (iii) of Andrews and Guggenberger (2003) are not necessary for the lower bound result of Theorem 1.

Note 2. Implicit in the bound on the integral of the spectral density $\int f \leq K_1$, is the fact that d needs to be bounded away from 1/2.

Note 3. The lower bound on d need not be -1/2, but since this chapter is concerned with stationary and invertible processes, we impose it here for ease of notation.

Note 4. We have substituted a local condition on $R(\lambda)$ to the global condition in Giraitis, Robinson and Samarov (1997) which runs contrary to the spirit of local estimators.

Note 5. Condition $K_1 \geq 2\pi a_{00}$ would be necessary to insure that the family of spectral densities considered is not empty.

3 Semiparametric M-estimation of Long Memory

We now turn to estimation strategies that can be expected to produce estimators of d which attain the optimal rates of convergence.

3.1 Plug-in Method for the LW

Let us first consider the case where the spectral density $f(\lambda)$ of X_t satisfies the following extension of (1):

$$f(\lambda) = c|\lambda|^{-2d}(1 + E_\beta \lambda^\beta + o(\lambda^\beta)) \text{ when } \lambda \to 0, \tag{25}$$

with $0 < E_\beta < \infty$ and $\beta > 0$. Note that (25) is satisfied for spectral densities that belong to any $\mathcal{F}(s, a, \delta, K)$, $s \leq \beta$ and either $\beta \leq 2$ or $a_j = 0$, all $j \geq 1$ (the latter being a very specific configuration which is relaxed later in this section).

First, we briefly recall the plug-in method first proposed by Henry and Robinson (1996) to select bandwidth optimally for the LW estimator, denoted $\hat{d}_{LW}$. Recall from (7), (9) and (10) that $\hat{d}_{LW}$ solves

$$G(\hat{d}_{LW}) = 0, \tag{26}$$

with

$$G(d) = \sum_{j=1}^{m} \nu_j \left(\frac{I_j}{\lambda_j^{-2d}} - 1 \right). \tag{27}$$

The heuristics of the Henry and Robinson (1996) approach can be summarized as follows: the mean value theorem yields

$$\hat{d}_{LW} - d = -\frac{G(d)}{G'(\tilde{d})}, \quad |\tilde{d} - d| \leq |\hat{d}_{LW} - d|. \tag{28}$$

Robinson (1995b) shows that

$$\frac{1}{m} G'(\tilde{d}) \to_p 4. \tag{29}$$

The expectation of $G(d)$ can be approximated by

$$\sum_{j=1}^{m} \nu_j \left(\frac{f_j}{\lambda_j^{-2d}} - 1 \right) \tag{30}$$

which has first order term, from (25), equal to

$$E_\beta \sum_{j=1}^{m} \nu_j \lambda_j^\beta \tag{31}$$

so that the first order component of the asymptotic bias AB_{LW} for the LW was suggested to be

$$AB_{LW} = -\frac{\beta}{2(1+\beta)^2} E_\beta \lambda_m^\beta \stackrel{\text{def}}{=} -\theta \left(\frac{m}{n} \right)^\beta. \tag{32}$$

Similarly, the variance of $G(d)$ was approximated, using (15), by

$$\sum_{j=1}^{m} \nu_j E \left[\frac{I_j}{\lambda_j^{-2d}} - \frac{EI_j}{\lambda_j^{-2d}} \right] \sim \sum_{j=1}^{m} \nu_j^2 \sim 1. \tag{33}$$

Finally, balancing squared asymptotic bias and asymptotic variance yielded the "optimal bandwidth"

$$m_{\text{opt}} = \left[\frac{(1+\beta)^4}{2\beta^3 E_\beta^2 (2\pi)^{2\beta}} \right]^{\frac{1}{1+2\beta}} n^{\frac{2\beta}{1+2\beta}} \tag{34}$$

allowing $\sqrt{m}$-consistent LW to attain the minimax rate of convergence (i.e. rate $n^{\beta/(1+2\beta)}$).

The bias approximation above is vindicated by result (2.15) of Giraitis and Robinson (2003a), which states that, under assumption A1, (25), and

$$m \sim K n^{\frac{2\beta}{1+2\beta}}, \quad 0 < K < \infty, \tag{35}$$

the following expansion holds:

$$\sup_{y \in \mathbb{R}} \left| P\left\{ \sqrt{m}(\hat{d}_{LW} - d) \leq y \right\} - \Phi(y + \sqrt{m}\theta) \right| = o(1), \tag{36}$$

where Φ is the standard normal cumulative distribution function, and θ is defined in (32).

Even when the degree of smoothness β is known, the constant E_β needs to be estimated. Henry and Robinson (1996) proposed $\hat{E}_\beta$ equal to the ratio of the second and first coefficients in the least squares regression of the periodogram against

$$|\lambda|^{-2\hat{d}_{LW}(m)} \left[1 \ \lambda_j^\beta \right]', \ j = 1 \ldots, m', \tag{37}$$

where $\hat{d}_{LW}(m)$ denotes an initial LW estimate computed with bandwidth m. Giraitis and Robinson (2003b) show that when m' diverges faster than m, more precisely when

$$n m^{-\frac{1}{2\beta}+\epsilon} \leq m' \leq n^{1-\epsilon}, \text{ some } \epsilon > 0, \tag{38}$$

then $\hat{\theta}$ obtained as θ in (32), where E_β is substituted with $\hat{E}_\beta$, converges almost surely to θ, so that

$$\frac{\hat{m}_{\text{opt}}}{m_{\text{opt}}} \rightarrow_{\text{a.s.}} 1, \tag{39}$$

where $\hat{m}_{\text{opt}}$ is the feasible version of m_{opt} with E_β replaced by $\hat{E}_\beta$ in (34). This feasible version is shown in Henry and Robinson (1996) and Henry (2001) to perform well in small samples on a wide variety of models, including conditionally heteroscedastic ones.

This approach has two main drawbacks: the first is that selecting bandwidth satisfying (39) does not guarantee that we are as close as possible to minimum quadratic risk

$$\min_m E(\hat{d}_{LW}(m) - d)^2; \tag{40}$$

and the second is that it assumes a known value of the smoothness parameter β, and, as we shall see in the following example, this greatly limits the empirical relevance of this method at least when the degree of smoothness is believed to be strictly smaller than 2

3.2 Long Memory in Stochastic Volatility

The example of long memory in stochastic volatility we develop below shows the relevance of considering local smoothness $\beta < 2$, and shows the incompatibility of a priori knowledge of β in the estimation of the long memory parameter d, since β turns out to be a function of d.

A large amount of research has recently been focused on the degree of dependence in squared, log-squared and other nonlinear transformations of financial returns, themselves generally uncorrelated, as part of a general investigation of long memory in financial volatilities (see for instance Henry and Zaffaroni (2003) and references therein). Such estimation often relies on a mixture of distributions hypothesis on the market microstructure (first proposed by Clark, 1973), and the resulting stochastic volatility model, where the observables are martingale differences of the form

$$Y_t = \epsilon_t \sigma_t, \tag{41}$$

where the innovations ϵ_t are i.i.d. and independent of the volatilities σ_t, and the transformation of interest, say

$$X_t = |Y_t|^\alpha, \ \alpha > 0, \tag{42}$$

has spectral density

$$f(\lambda) = \tilde{f}(\lambda) + \frac{\tau}{2\pi}, \ \lambda \in [-\pi, \pi], \tag{43}$$

where $\tilde{f}$ is the spectral density of $|\epsilon_t \sigma_t|^\alpha$ and

$$\tau^2 = V(|\epsilon_t|^\alpha)\, E|\sigma_t|^{2\alpha}. \tag{44}$$

So, if we let $\tilde{f}(\lambda) = c|\lambda|^{-2d}\tilde{h}(\lambda)$, no matter how smooth $\tilde{h}$, the short memory part of the spectrum of X_t, i.e. $h(\lambda) = f(\lambda)|\lambda|^{2d}/c$ satisfies a Lipschitz condition of order at most $2d < 1$, thus ruling out differentiability (as initially noted by Deo and Hurvich, 2001). This also rules out rate optimal estimation of d under the assumption that the local degree of smoothness of $h(\lambda)$ is known a priori.

3.3 Smooth Models

In view of the preceeding example, the correct way to proceed in the choice of bandwidth, is via a procedure that is adaptive to unknown degrees of smoothness in $h(\lambda)$.

However, certain models are known to satisfy (1), where $h(\lambda)$ is smooth of arbitrary order (i.e. infinitely continuously differentiable). One example is the popular ARFIMA(p, d, q), defined as the model for a process X_t satisfying

$$X_t = (I - L)^{-d} b^{-1}(L) a(L) \epsilon_t \tag{45}$$

where $a(z)$ and $b(z)$ are polynomials of order q and p respectively, with zeros outside the unit circle in the complex plane.

Such a process does not, in general, satisfy (25) with β arbitrarily large. It does, however, satisfy an expansion of (1) of the form

$$f(\lambda) = c|\lambda|^{-2d} \left\{ 1 + \sum_{k=1}^{q} \frac{h_k \lambda^{2k}}{(2k)!} + o(\lambda^{2q}) \right\}, \ \lambda \to 0, \tag{46}$$

for q and arbitrary integer. Note that h_k is the $(2k)$th derivative of $h(\lambda)$ at $\lambda = 0$, and that a spectral density satisfying (46) belongs to $\mathcal{F}(s, a, \delta, K)$ for all $s \leq q$.

In cases where h is infinitely differentiable at 0, bias reduction techniques described below can be implemented, which increase the rate at which m is allowed to diverge at the cost of an increased constant term in the asymptotic variance. So, theoretically, rates can be achieved that are arbitrarily close to the parametric rate. However, the constant term in the variance of higher-order bias reducing estimators may be very large, so that, even with very large samples, only limited bias reduction may be envisaged, and in practice, the hypothesized degree of smoothness becomes, for all intents and purposes, a user chosen integer, so that it makes sense in those cases to rely on a priori "knowledge" of q for optimal estimation of d (when both rate and constant are considered).

It is apparent in expansion (46) that the extra smoothness can be exploited to control bias terms of increasing order and thus attain the rate lower bound. Two related strategies to do so have been proposed.

The first strategy, proposed by Robinson and Henry (2003) relies on basing the weights w_j in the m-estimates solving (7) on higher-order kernels in the following way: let $k_q(u)$, $0 \leq u \leq 1$, called a q-th order kernel, satisfy

$$\int_0^1 k_q(u) du = 1, \tag{47}$$

$$\int_0^1 (1 + \log u) u^{2i} k_q(u) du \stackrel{\text{def}}{=} U_{iq} = 0, \ 0 \leq i < q, \tag{48}$$

$$\neq 0, \ i = q. \tag{49}$$

Let the higher-order kernel m-estimating weights be defined by

$$w_j = k_q\left(\frac{j}{m}\right)\left\{\log\lambda_j - \frac{\sum_{j=1}^{m} k_q\left(\frac{j}{m}\right)\log\lambda_j}{\sum_{j=1}^{m} k_q\left(\frac{j}{m}\right)}\right\}, \tag{50}$$

and call $\hat{d}_{q\psi}(m)$ a solution to (7) under (50).

Assuming $k_q(u)$ is boundedly differentiable except perhaps at finitely many points, and that $\psi(z)$ is infinitely differentiable and such that there is some $0 < K < \infty$ such that for all $k \geq 1$,

$$E\left|(cZ)^k\psi^{(k)}(cZ)\right| = e_{k\psi} \leq K. \tag{51}$$

where $\psi^{(k)}(z)$ denotes the kth derivative of $\psi(z)$, and Z is a $\mathcal{X}_2^2/2$ random variable, the same heuristics can be applied here as in section 2.1 to show how the bias terms of increasing order are eliminated by the kernel weights to yield asymptotic bias of order $(m/n)^{2q}$, so that the optimal rate is attained when square bias and variance are balanced by $m = Kn^{4q/(1+4q)}$.

More precisely, if we define

$$V_q = \int_0^1 (1+\log u)^2 k_q(u)du, \tag{52}$$

$$W_q = \int_0^1 (1+\log u)^2 k_q^2(u)du, \tag{53}$$

and

$$Q(\psi) = \mathrm{Var}(\psi(cZ)), \tag{54}$$

$$P(\psi) = (2\pi)^{2q}\sum_{u=1}^{q}\frac{e_{u\psi}}{u!}\sum_{r_1+\ldots+r_u=q}\left(\prod_{l=1}^{u}\frac{h_{r_l}}{(2r_l)!}\right),$$

Robinson and Henry (2003) derive expressions for the asymptotic bias AB:

$$AB = -\frac{U_{qq}P(\psi)}{2e_{1\psi}V_q}\left(\frac{m}{n}\right)^q, \tag{55}$$

and variance AV:

$$AV = \frac{Q(\psi)W_q}{(e_{1\psi}V_q)^2}\frac{1}{m}, \tag{56}$$

so that the asymptotic mean squared error is minimized by

$$m_{\mathrm{opt}} = \left(\frac{e_{1\psi}W_q}{4qU_{qq}^2P^2(\psi)}\right)^{\frac{1}{1+4q}} n^{\frac{4q}{1+4q}}. \tag{57}$$

Andrews and Sun (2004) also achieve the optimal rate under (46) by a transformation of the LW that allows for joint estimation of the memory parameter d and the shape of $h(\lambda)$ in the neighbourhood of frequency zero. $\log h(\lambda)$ is fitted with a polynomial of order q in λ^2, with coefficients θ_k, $k = 1, \ldots, q$, so that the loss function that defines the LW becomes

$$L(d, c, \theta) = \frac{1}{m} \sum_{j=1}^{m} \left\{ \frac{I_j}{c\lambda_j^{-2d} \exp \sum_{k=1}^{q} \theta_k \lambda_j^{2k}} + \log \left[c\lambda_j^{-2d} \exp \sum_{k=1}^{q} \theta_k \lambda_j^{2k} \right] \right\} \quad (58)$$

where $\theta = (\theta_1, \ldots, \theta_q)'$, and the estimator minimizes $L(d, c, \theta)$.

3.4 Faster Rate or Smaller Constant?

The asymptotic variance of the local polynomial Whittle estimator defined above, and that of the higher-order kernel m-estimator, shrink at the faster rate $1/m = n^{-2q/1+2q}$ at the cost of a larger constant term. In the case of the local polynomial Whittle, the constant inflation term is

$$c_q = (1 - \mu_q' \Gamma_q^{-1} \mu_q)^{-1}, \quad (59)$$

with

$$\mu_q = \left(\frac{2k}{(1+2k)^2} \right)_{k=1}^{q} \quad (60)$$

and

$$\Gamma_q = \left(\frac{4lk}{(1+2l+2k)(1+2l)(1+2k)} \right)_{l,k=1}^{q} . \quad (61)$$

For instance, for smoothness $s \in (2, 4]$, $q = 2$ and $c_2 = 3.52$. In the same way, $c_3 = 4.79$ and $c_4 = 6.06$ are the inflating constants in the higher smoothness cases.

In the special case of higher-order kernel m-estimation of long memory corresponding to higher-order kernel LW estimation, since $Q(\psi)/e_{1\psi}$ in (56) equals 1 under (10), the inflating constant is

$$\left(\frac{W_q}{V_q^2} \right)_{q=2}^{4} = (5.82 \;\; 10799 \;\; 13.4)' \quad (62)$$

where V_q and W_q, defined in (52) and (53 respectively) are computed in the case

$$k_q(u) = \sum_{j=0}^{q} a_j u^{2j}, \quad (63)$$

and the a_j are such that (47) and (48) are satisfied. Of course, it is doubtful that this form yields anything close to the optimal kernel, in the sense of delivering the lowest quadratic risk. In any case, if we look only at the variance, the sample size will need to be very large for the faster rate to dominate the effect of an inflating constant larger than 10: for instance, in the case of the local polynomial Whittle, the constant doubles from $q = 2$ to $q = 4$, and the rate increases from $n^{8/9}$ to $n^{16/17}$. They break even around $n = 50000$.

4 Adaptive Estimation of Long Memory

We now consider an alternative to plug-in estimation which does not presume prior knowledge of the degree of smoothness of $h(\lambda)$ in the neighbourhood of zero and achieves the optimal rates up to logarithmic factors.

The key to the adaptive estimation procedures is to select a bandwidth m as a function of an adaptively chosen level of smoothness $\hat{s}$, first obtained by Giraitis, Robinson and Samarov (2000), who refer to a method first proposed by Lepskii (1990) in a different nonparametric setting.

Roughly speaking, $\hat{s}$ is chosen as the largest s, such that the estimator based on the bandwidth $m = m(s)$ is not significantly different from all such estimators using $m(\zeta)$, $0 < \zeta < s$; and $m(s)$, instead of balancing variance squared bias, allows squared bias to exceed variance by a factor $\log n$. More precisely, let $\hat{d}(m)$ denote a semiparametric estimator of d based on m harmonics of the periodogram. Let $b(m)$ designate the bias of such an estimator, and $\sigma(m)$ the highest order of magnitude component of the stochastic fluctuation term of $\hat{d}(m)$. A value m of the bandwidth is called "admissible" if, for some positive constant K, chosen large enough,

$$|\hat{d}(m') - \hat{d}(m)| \leq K\sigma(m')(\log n)^{\frac{1}{2}} \text{ for all } m' \leq m. \tag{64}$$

The adaptive value of m, called $\hat{m}$, is chosen as the largest admissible bandwidth (i.e. the largest integer that satisfies (64)). The principle of this adaptive estimation method is then to show that $\hat{m}$ thus chosen is sufficiently close with sufficiently high probability to the value of the bandwidth which balances $b(m)$ and $\sigma(m)(\log n)^{1/2}$. This requires the exponential inequalities (3.11) in Giraitis, Robinson and Samarov (2000) for instance.

For the class of spectral densities $\mathcal{F}(s, a, \delta, K)$ where $a_j = 0$, all $j \geq 1$, Corollary 7.1 of Moulines and Soulier (2003) shows that a tapered version of the LP achieves the rate

$$\hat{m} = n^{s/1+2s}(\log n)^{1/2} \tag{65}$$

when K satisfies $K > 6$ (required by their exponential inequality (T3)). Tapering is needed for better control of the bias, and is achieved by replacing the DFT's in (5) by

$$\phi_h(\lambda) = \left(2\pi \sum_{t=1}^{n} h_t^2\right)^{-\frac{1}{2}} \sum_{t=1}^{n} h_t^2 X_t e^{it\lambda}, \tag{66}$$

where the tapering sequence h_t can be taken equal, in the case of the "cosine bell taper," to

$$h_t = \frac{1}{2}(1 - \cos\lambda_t),\ t = 1, \ldots, n. \tag{67}$$

The earlier adaptive estimator proposed by Giraitis, Robinson and Samarov (2000) was built on a variant of Lepskii's method in the following way: the unknown degree of smoothness s is bounded above and below by s^* and s_* respectively. The interval $[s_*, s^*]$ is endowed with the net

$$B = \{s \geq 0 : s = s^* - \frac{k}{\log n},\ k = 1, \ldots\} \tag{68}$$

and the smoothness estimator is defined as

$$\hat{s} = \sup_{s \in B} \left\{ |\hat{d}(m(s')) - \hat{d}(m(s))| \leq \frac{2(s^* - s')}{(1 + 2s')^2} \left(\frac{\log^2 n}{n}\right)^{\frac{s}{1+2s}},\ s' \in B \right\}, \tag{69}$$

where

$$m(s) = (\log n)^{\frac{2}{1+2s}} n^{\frac{2s}{1+2s}} \tag{70}$$

thereby defining an adaptive estimator that achieves the rate

$$m(\hat{s}) \propto \left(\frac{n}{\log n}\right)^{\frac{s}{1+2s}}. \tag{71}$$

Both rates (65) and (71) reach the optimal rate of convergence identified by Theorem 1 up to logarithmic factors. However, both fail to reach the optimal adaptive rate lower bound identified by the following theorem, proved in Giraitis, Robinson and Samarov (2000), with a restricted to the case $a_j = 0$, all $j \geq 1$:

Theorem 2.

$$\liminf_n \inf_{\hat{d}} \sup_{s_* \leq s \leq s^*} \left(\frac{n}{\log n}\right)^{-\frac{2s}{1+2s}} \sup_{f \in \mathcal{F}(s,a,\delta,K)} E_f(\hat{d} - d)^2 > 0. \tag{72}$$

It is conjectured that this lower bound is the adaptive minimax rate for the class $\mathcal{F}(s, a, \delta, K)$ (including the case where a is unrestricted), but so far, the upper bounds in the literature do not coincide with the lower bound of Theorem 2. In the case of smooth models ($s \geq 2$) with a unrestricted, the method of choosing the largest admissible bandwidth in (64) can still be applied, but needs to be modified to take account of the fact that when s reaches even integers, the estimator changes (to a higher-order kernel version in Robinson and Henry (2003), or a higher-order polynomial version in Andrews and Guggenberger (2003) and Andrews and Sun (2004)) to achieve bias reduction, and so does the first order of magnitude $\sigma(m)$ in the stochastic fluctuation term.

Andrews and Sun (2004) construct an adaptive version of their local polynomial Whittle, i.e. the estimator minimizing (58), in a way that mimics very closely the construction in Giraitis, Robinson and Samarov (2000), and which may also be used with higher-order kernel m-estimators. Call $\sigma_q(m)$ the first order of magnitude term of the stochastic fluctuation term for an estimator $\hat{d}_q(m)$ based on a kernel of order q (Robinson and Henry, 2003) or a local polynomial of order $2q$ (Andrews and Sun, 2004). Now the two user chosen parameters are the bandwidth $m(s)$ and the kernel or polynomial order $q(s)$. Set

$$m(s) = n^{\frac{2s}{1+2s}} \tag{73}$$

$$q(s) = u, s \in (2u, 2(u+1)], \tag{74}$$

and select $\hat{s}$ as the largest admissible value, i.e.

$$\hat{s} = \sup_{s \geq 1} \left\{ |\hat{d}_{q(s')}(m(s')) - \hat{d}_{q(s)}(m(s))| \leq K\zeta_n \sigma_{q(s')}(m(s')) \right\}, \tag{75}$$

with $\zeta_n = \log n (\log \log n)^{\frac{1}{2}}$.

References

1. Andrews, D.W.K. and Guggenberger, P. (2003). A bias reduced log–periodogram estimator for the long memory parameter. *Econometrica*, **71**, 675–712.
2. Andrews, D.W.K. and Sun, Y. (2004). Adaptive local polynomial Whittle estimation of long range dependence. *Econometrica*, **71**, 569–614.
3. Clark, P.K. (1973). A subordinated stochastic process model with fixed variances for speculative prices. *Econometrica*, **41**, 135–156.
4. Deo, R. and Hurvich, C. (2001). On the log–periodogram estimator of the memory parameter in long memory stochastic volatility models. *Econometric Theory*, **17**, 686–710.
5. Geweke, J. and Porter-Hudak, S. (1983). The estimation and application of long memory time series models. *Journal of Time Series Analysis*, **4**, 221–238.
6. Giraitis, L. and Robinson, P.M. (2003a). Edgeworth expansions for semiparametric Whittle estimation of long memory. *The Annals of Statistics*, **31**, 1325–1375.
7. Giraitis, L. and Robinson, P.M. (2003b). Parametric estimation under long range dependence, in Doukhan, P., Oppenheim G. and Taqqu, M.S. (Eds.), *Theory and Applications of Long Range Dependence*. Birkhauser, Boston, pp. 229–250.
8. Giraitis, L., Robinson, P.M. and Samarov, V. (1997). Rate optimal semiparametric estimation of the memory parameter of the Gaussian time series with long range dependence. *Journal of Time Series Analysis*, **18**, 49–60.

9. Giraitis, L., Robinson, P.M. and Samarov, V. (2000). Adaptive semiparametric estimation of the memory parameter. *Journal of Multivariate Analysis*, **72,** 183–207.
10. Henry, M. (2001). Robust automatic bandwidth for long memory. *Journal of Time Series Analysis*, **22,** 293–316.
11. Henry, M. and Robinson, P.M. (1996). Bandwidth choice in Gaussian semiparametric estimation of long range dependence, in Robinson, P.M. and Rosenblatt, M. (Eds), *Athens Conference on Applied Probability and Time Series Analysis, vol II: Time Series Analysis, In Memory of E.J. Hannan.* Springer Verlag, New York, pp. 220-232.
12. Henry, M. and Zaffaroni, P. (2003). The long range dependence paradigm for macroeconomics and finance, in Doukhan, P., Oppenheim G. and Taqqu, M.S. (Eds.), *Theory and Applications of Long Range Dependence.* Birkhauser, Boston, pp. 417–438.
13. Hurvich, C. and Beltrao, K. (1993). Asymptotics for the low frequency ordinates of the periodogram of a long memory time series. *Journal of Time Series Analysis*, **14,** 455–472.
14. Künsch, H. (1987). Statistical properties of self–similar processes, in *Proceedings of the First World Congress of the Bernoulli Society*, VNU Science Press, pp. 67–74.
15. Lepskii, O.V. (1990). On the problem of adaptive estimation in Gaussian white noise. *Theory of Probability and its Applications*, **35,** 454–466.
16. Moulines, E. and Soulier, P. (2003). Semiparametric spectral estimation for fractional processes, in Doukhan, P., Oppenheim G. and Taqqu, M.S. (Eds.), *Theory and Applications of Long Range Dependence.* Birkhauser, Boston, pp. 251–302.
17. Robinson, P.M. (1995a). Gaussian semiparametric estimation of long range dependence. *The Annals of Statistics*, **23,** 1630–1661.
18. Robinson, P.M. (1995b). Log periodogram regression of time series with long range dependence. *The Annals of Statistics*, **23,** 1048–1072.
19. Robinson, P.M. and Henry, M. (1999). Long and short memory conditional heteroskedasticity in estimating the memory parameter of levels. *Econometric Theory*, **15**, 299–336.
20. Robinson, P.M. and Henry, M. (2003). Higher–order kernel semiparametric M–estimation of long memory. *Journal of Econometrics*, **114,** 1–27.

Wavelet Analysis of Nonlinear Long–Range Dependent Processes. Applications to Financial Time Series*

Gilles Teyssière[1] and Patrice Abry[2]

[1] Laboratoire de Statistique Appliquée et MOdélisation Stochastique (SAMOS), Université Paris 1. stats@gillesteyssiere.net

[2] Laboratoire de Physique, École Normale Supérieure de Lyon, CNRS UMR 5672. Patrice.Abry@ens-lyon.fr

Summary. We present and study the performance of the semiparametric wavelet estimator for the long–memory parameter devised by Veitch and Abry (1999). We compare this estimator with two semiparametric estimators in the spectral domain, the local Whittle (LW) estimator developed by Robinson (1995a) and the "log–periodogram" (LP) estimator by Geweke and Porter–Hudak (1983). The wavelet estimator performs well for a wide range of nonlinear long–memory processes in the conditional mean and the conditional variance, and is reliable for discriminating between change–points and long–range dependence in volatility. We also address the issue of selection of the range of octaves used as regressors by the weighted least squares estimator. We will see that using the feasible optimal bandwidths for either the LW and LP estimators, respectively studied by Henry and Robinson (1996) and Henry (2001), is a useful rule of thumb for selecting the lowest octave. We apply the wavelet estimator to volatility series of high frequency (intra–day) Foreign Exchange (FX) rates, and to the volatility and volume of stocks of the Dow Jones Industrial Average Index.

1 Introduction

The occurrence of long–range dependence, or strong dependence or long–memory, in economics is documented by numerous research works. Although it is widely accepted that squared or absolute returns on financial assets display long–range dependence, we are still unsure that what is observed is either genuine long–memory or a statistical artefact, as statistical tools used for the study of long–range dependent processes made the assumption that the process under investigation is homogeneous and stationary. Thus, it has

* We thank Carlos Velasco and Olsen & Associates for providing some of the data used in this chapter, Liudas Giraitis for a very careful reading and Gabriel Lang for useful discussions.

been claimed, see e.g., Mikosch and Stărică (1999, 2003, 2004a, 2004b) that the observed long–range dependence is spurious and the consequence of non–stationarity in the data.

Discriminating long–range dependence from single (or multiple) change–point(s) is a very active research area in statistics, see e.g., Lavielle (1999), Berkes, Horváth, Kokoszka and Shao (2003). Change–points and long–range dependence might coexist, and finding the optimal number of change–points and their location requires a more elaborated approach than Vostrikova's (1981) binary segmentation procedure, see Lavielle and Teyssière (2005), although most econometrics research papers still resort to the binary segmentation methods, or a mix of this algorithm and the tests by Lavielle (1999) and Lavielle and Moulines (2000), a combination giving incorrect results.

We consider here a statistical method for the analysis of long–range dependence, based on wavelets, which does not require these strong stationarity assumptions and is robust to some departures from the previous hypotheses. Empirical research works by Kokoszka and Teyssière (2002) and Teyssière (2003) reached the conclusion that the intensity of strong dependence in volatility processes measured with wavelet based estimators is lower than what is usually found using estimators in either the time or the frequency domain, and then volatility processes mix a moderate level of long–range dependence and change–points.

As empirical evidence suggests that financial time series are highly nonlinear, in particular volatility processes exhibit a combination of nonlinearity, long–range dependence and change–points, we study in this chapter the estimation of the scaling coefficient for some nonlinear long–memory processes used in the statistical and econometrics literature, and compare the performance of the wavelet estimator with the one of two semiparametric estimators in the frequency domain: the local Whittle (LW) estimator by Künsch (1987) and Robinson (1995a), and the "log–periodogram" (LP) estimator by Geweke and Porter–Hudak (1983). These semiparametric estimators are of interest for researchers dealing with real data, as the Data Generating Process (DGP) of the observed data is unknown, the estimation of a misspecified parametric model might lead to serious biases in estimation. Since one has to resort to semiparametric methods, it is then useful to know the performance of these estimators for a wide range of nonlinear dependent processes.

We consider some nonlinear long–memory processes that were not studied in Abry *et al.* (2003): the linear ARCH (LARCH) process, the long–memory stochastic volatility (LMSV) process, the nonlinear moving average (NLMA) process, and some nonlinear transformations of fractionally integrated processes. We also study the performance of the wavelet estimator for the case of change–point processes, i.e., the non–stationary GARCH process, a process with non constant coefficients. The relevance of wavelet analysis for dealing with the issue of change–point and spurious strong dependence in GARCH processes has been conjectured in previous works, see e.g., Kokoszka and Teyssière (2002), Teyssière (2003), but never systematically analyzed. We also consider the case of a volatility process mixing both strong dependence and change in regimes, the non homogeneous LMSV process, and the case of dependent processes with polynomial and broken trends.

We also address a standard issue in semiparametric estimation of strongly dependent processes, the cutoff between short–range and long–range dependence. The bandwidth selection problem has been studied for the LW and LP estimators by Henry and Robinson (1996), Hurvich, Deo and Brodsky (1998) and Henry (2001). For the wavelet estimators, we have to select the range of octaves used as regressors by the weighted least squares estimator. On the basis of some simulations, we suggest some choices for the lowest octave, and we will see that using the optimal bandwidth for the LW estimator is a useful rule of thumb for selecting this lowest octave.

We also report some simulation results for the use of the feasible optimal bandwidth for the LP estimator, and will see that this data–driven bandwidth works well for a large variety of nonlinear long–range dependent processes, and often better than the fixed bandwidths $[T^b]$, with $b \in (0,1)$, usually considered.

We will see that although the wavelet estimator has been devised for the standard Gaussian case, it still works for a broader class of nonlinear processes, provided in the case of highly nonlinear processes, that the sample size is large enough for disentangling the long–memory component from the other nonlinear ones. Since the performance of the wavelet estimator is satisfactory enough, we consider an application of this estimator to high–frequency (intra–day) foreign exchange (FX) rates, and to trading volume.

This chapter is organized as follows: section 2 presents the long–range dependent processes that we will study with the wavelet estimator, while section 3 presents the wavelet and semiparametric spectral estimators. Section 4 compares the performance of the wavelet estimator with other standard semiparametric estimators in the spectral domain for a large variety of nonlinear strongly dependent processes. Section 6 gives several applications to high–frequency Foreign Exchange (FX) rates and to trading volume of stocks of the Dow Jones Industrial Average Index.

2 Long–Memory or Long–Range Dependence

2.1 Definition and Consequence

Definition 1. *Let $\{Y_t, t \in \mathbb{R}\}$ be a second–order stationary process. This process is a long-memory process if its spectrum $f_Y(\lambda)$ is such that in a close positive neighborhood of the zero frequency,*

$$f_Y(\lambda) \sim c_f \lambda^{-\alpha}, \quad \lambda \to 0_+, \quad c_f \in (0, \infty), \tag{1}$$

or equivalently, if its autocorrelation function $\rho_Y(k)$ has the following hyperbolic rate of decay[3]

[3] $x_k \asymp y_k$ means that $\exists$ two constants C_1, C_2 such that $C_1 y_k \leqslant x_k \leqslant C_2 y_k$, $k \to \infty$.

$$\rho_Y(k) \asymp k^{\alpha-1}, \tag{2}$$

with $\alpha \in (0,1)$.

As a consequence of equation (2), the autocorrelations of a long–memory process are not summable, i.e.,

$$\sum_{k=-\infty}^{\infty} \rho_Y(k) = \infty. \tag{3}$$

Long–memory received an interesting non standard exposition based on the covariance under aggregation of the sum of covariances in Gefferth *et al.* (2003).

2.2 Standard Linear Long–Range Dependent Processes

- **Fractionally Integrated Process**

The Fractionally Integrated process, denoted as either FI(d) or I(d), defined as

$$(1-L)^d(Y_t-\mu) = \varepsilon_t, \quad \varepsilon_t \sim N(0,\sigma_\varepsilon^2), \tag{4}$$

where μ is the unknown mean of the process, L stands for the lag operator, i.e., $LY_t = Y_{t-1}$, the fractional difference operator $(1-L)^d$ associated with the degree of fractional integration $d \in (-0.5, 0.5)$, is defined as

$$(1-L)^d = \sum_{j=0}^{\infty} b_j L^j,\ b_0 = 1,\ b_j = \prod_{k=1}^{j}\left(1 - \frac{1+d}{k}\right), \tag{5}$$
$$b_j \sim -\frac{1}{\Gamma(-d)} j^{-(1+d)} \qquad \longrightarrow \infty,$$

where $\Gamma(\cdot)$ denotes the Gamma function. When $d \in (0, 0.5)$ (respectively, $d \in (-0.5, 0)$) the process is said to be persistent (respectively, anti-persistent). For this process, one has:

$$\alpha = 2d. \tag{6}$$

- **Fractional ARIMA (FARIMA) Process**

The FI(d) process is nested into the class of Fractional ARIMA processes, defined as

$$\phi(L)(1-L)^d(Y_t-\mu) = \theta(L)\varepsilon_t, \quad \varepsilon_t \sim N(0,\sigma_\varepsilon^2), \tag{7}$$

where $\phi(L) = 1 - \sum_{j=1}^{p}\phi_j L^j$ and $\theta(L) = 1 + \sum_{j=1}^{q}\theta_j L^j$ are lag polynomials of respective orders p and q with root outside the unit circle. This model, which is denoted as FARIMA(p, d, q), generalizes the class of ARIMA(p, d, q) models with integer degree of differentiation.

The spectrum of a FARIMA(p, d, q) process is equal to

$$f_Y(\lambda) = \frac{\sigma_\varepsilon^2}{2\pi} \frac{|\theta(\exp(-i\lambda))|^2}{|\phi(\exp(-i\lambda))|^2} |1 - \exp(-i\lambda)|^{-2d}. \tag{8}$$

With

$$\alpha - 2d, \tag{9}$$

then the spectrum of a FARIMA(p, d, q) process is of the form of equation (1).

• Estimation Issues

Under the hypothesis of Gaussian error terms, the parameters $\zeta = \{\theta_i, \phi_j, \mu, d\}$ are estimated by maximizing the log–likelihood function, in either the time or the frequency domain; see Beran (1994), Robinson (1994) and Hauser (1999) for a survey. The parameters $\hat{\zeta}$ have the rate of convergence equal to $T^{1/2}$, where T denotes the sample size, except the parameter μ, the rate of convergence of which is equal to $T^{1/2-d}$. This illustrates the difficulty to disentangle long–range dependence from changes in the mean parameter μ of the process.

• Empirical Volatility: The Need for Nonlinear Long–Memory Processes

While returns on asset prices, defined as $r_t = \log(P_t/P_{t-1})$, where P_t denotes the asset price at time t, are uncorrelated, empirical evidence from the series of absolute returns $|r_t|$ and squared returns r_t^2 has shown that the spectrum and the autocorrelation function, henceforth ACF, for both series behave like equations (1) and (2). Both the series $|r_t|$ or r_t^2 are commonly used as empirical measures for the volatility, which hence appear to be strongly dependent. Figures 1 and 2 below display the periodogram and the ACF of absolute returns on dollar-deutschmark FX rate.

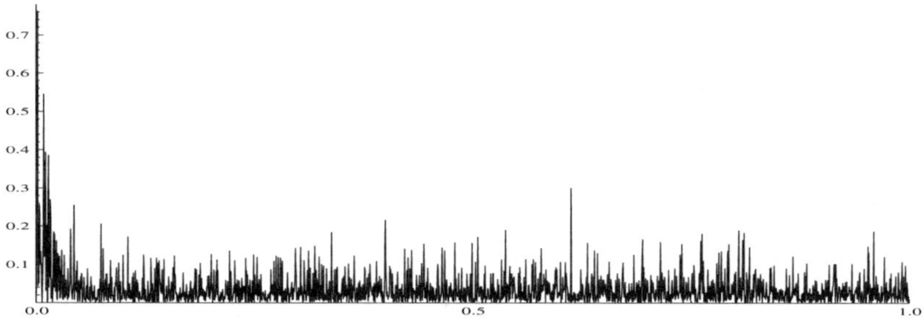

Fig. 1. Periodogram of absolute returns on dollar-deutschmark FX rate

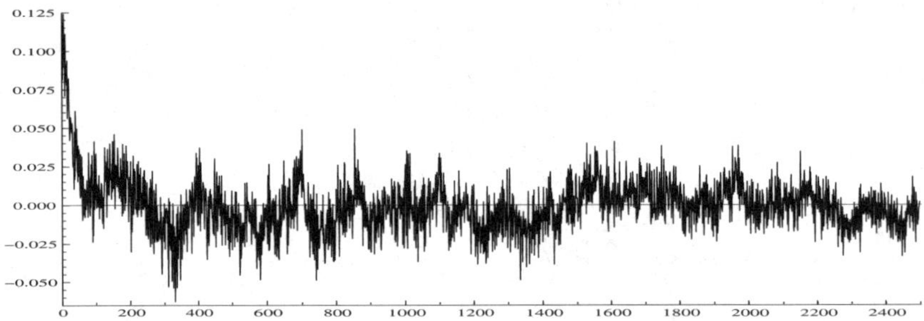

Fig. 2. ACF of absolute returns on dollar-deutschmark FX rate

The rich dynamics of these volatility processes cannot be parsimoniously fitted by standard FARIMA processes. Moreover, as emphasized by Granger (2000, 2002) absolute and squared returns do not display trends, unlike FARIMA processes. Thus, there is a need to resort to long–range dependent nonlinear processes.

2.3 Nonlinear Long–Range Dependent Volatility Processes

• ARCH(∞) Processes

Robinson (1991) introduced the class of ARCH(∞) processes, further developed by Granger and Ding (1995) and other authors. The ARCH(∞) is defined as

$$r_t = \sigma_t \varepsilon_t, \quad \varepsilon_t \sim D(0,1), \quad \sigma_t^2 = \omega + \varphi(L)\varepsilon_t^2, \tag{10}$$

where $D(0,1)$ is a zero-mean unit-variance random variable, and $\varphi(L) = \sum_{i=1}^{\infty} \varphi_i L^i$ is an infinite order lag polynomial the coefficients of which are positive and have asymptotically the following hyperbolic rate of decay

$$\varphi_j = O\left(j^{-(1+d)}\right). \tag{11}$$

However, the existence of a stationary solution to the equation (10) defining an ARCH(∞) process imply $\sum_{i=1}^{\infty} \varphi_i < \infty$, then the model has moderate memory; see Giraitis, Kokoszka and Leipus (2000) for further details.

With these assumptions, the autocorrelation function (ACF) of the sequence of squared returns $\{r_t^2\}$ satisfies

$$\mathrm{Cov}(r_0^2, r_k^2) \asymp k^{2d-1}. \tag{12}$$

Setting $\alpha = 2d$, and $Y_t = r_t^2$, the squared returns process defined by a stationary ARCH(∞) has moderate memory. In the applied econometrics and financial econometrics literature, the conditions for the existence of a stationary solution are disregarded, and it is wrongly claimed that the ARCH(∞) defined by equation (10) has long–memory.

• Linear ARCH

Robinson (1991) introduced the linear ARCH, henceforth LARCH, further studied by Giraitis, Robinson and Surgailis (2000), and defined as

$$r_t = \sigma_t \varepsilon_t, \quad \varepsilon_t \sim D(0,1), \quad \sigma_t = \omega + \beta(L) r_t, \tag{13}$$

where $\beta(L) = \sum_{i=1}^{\infty} \beta_i L^i$ is an infinite order lag polynomial the coefficients of which have asymptotically the following hyperbolic rate of decay

$$\beta_j = O\left(j^{d-1}\right), \tag{14}$$

for some $d \in (0, 1/2)$. For instance, the lag polynomial $\beta(L)$ can be the one of the moving average form of a FARIMA process.

Giraitis, Robinson and Surgailis (2000) proved the following theorem:

Theorem 1. *Suppose* $E\varepsilon_0^4 < \infty$ *and* $L(E\varepsilon_0^4)^{1/2} \sum_{j=1}^{\infty} \beta_j^2 < 1$, *where* $L = 7$ *if the* $\{\varepsilon_k\}$ *is Gaussian and* $L = 11$ *in the other cases. Then, there is a stationary solution to equations* (13) *and* (14) *given by orthogonal Volterra series*

$$r_t = \sigma_t \varepsilon_t, \quad \sigma_t = \omega \sum_{l=0}^{\infty} \sum_{j_1,\ldots,j_l=1}^{\infty} \beta_{j_1} \cdots \beta_{j_l} \varepsilon_{t-j_1} \cdots \varepsilon_{t-j_1-\cdots-j_l}. \tag{15}$$

The sequence $\{r_t^2\}$ *is covariance stationary and*

$$\mathrm{Cov}(r_0^2, r_k^2) \asymp k^{2d-1}. \tag{16}$$

Setting $\alpha = 2d$, and $Y_t = r_t^2$, the ACF of the squared returns process has the hyperbolic decay of a long–memory process given by equation (2). See the chapter by Giraitis, Leipus and Surgailis (2005) in this volume for further details. Giraitis, Kokoszka, Leipus and Teyssière (2000) considered "pox–plot" based estimators and the local Whittle estimator for the estimation of the scaling parameter α. We consider in section 4.1 the estimation of the scaling parameter with wavelet methods.

• The Long–Memory Stochastic Volatility Process

The long–memory stochastic volatility process, proposed by Breidt *et al.* (1998), is defined as

$$r_t = \sigma_t \zeta_t, \quad \zeta_t \sim N(0,1), \tag{17}$$

$$\sigma_t = \sigma \exp(X_t/2), \quad X_t \sim \mathrm{FARIMA}(p,d,q), \tag{18}$$

where σ is a scale parameter, the processes $\{X_t\}$ and $\{\zeta_t\}$ are independent. The process $\{r_t^2\}$ is linearized as follows

$$\log r_t^2 = \log \sigma^2 + E(\log \zeta_t^2) + X_t + \left(\log \zeta_t^2 - E(\log \zeta_t^2)\right) = \mu + X_t + \varepsilon_t, \tag{19}$$

where $\{\varepsilon_t\}$ is i.i.d, $E(\varepsilon_t) = 0$ and $\text{Var}(\varepsilon_t) = \sigma_\varepsilon^2$. Since $\zeta \sim N(0,1)$, then $\log \zeta^2$ is distributed as the logarithm of a χ_1^2 random variable, thus $E(\log \zeta^2) = 1.27$ and $\sigma_\varepsilon^2 = \pi^2/2$.

The spectral density of the process $\{\log r_t^2\}$ is given by:

$$f(\lambda) = \frac{\sigma_e^2}{2\pi} \frac{|\theta(\exp(-i\lambda))|^2}{|\phi(\exp(-i\lambda))|^2} |1 - \exp(-i\lambda)|^{-2d} + \frac{\sigma_\varepsilon^2}{2\pi}, \quad \lambda \in (-\pi, \pi), \tag{20}$$

where σ_e^2 denotes the variance of the innovations of the FARIMA process $\{X_t\}$. Thus, the spectral density of the process $\{\log r_t^2\}$ is the sum of the spectral density of a FARIMA(p,d,q) process and the spectral density of a white noise process. Setting $\alpha = 2d$ and $Y_t = \log r_t^2$, then the spectrum $f_Y(\lambda)$ given by equation (20) satisfies equation (1).

Deo and Hurvich (2003) studied the estimation of the long–memory parameter using the semiparametric log–periodogram (LP) estimator in the spectral domain by Geweke and Porter–Hudak (1983). In section 4.2, we compare the estimations of the long–memory parameter obtained from the wavelet and LP estimators using Henry's (2001) feasible optimal bandwidths.

- **The Nonlinear Moving Average Process**

The nonlinear moving average process, proposed by Robinson and Zaffaroni (1997), is an extension of Robinson's (1977) short range dependent nonlinear moving average process:

$$r_t = \mu + \sigma_{t-1}\varepsilon_t, \quad \sigma_{t-1} = \rho + \sum_{i=1}^{\infty} \beta_i \varepsilon_{t-i}, \quad \sum_{i=1}^{\infty} \beta_i^2 < \infty, \tag{21}$$

where the innovation process $\{\varepsilon_t\}$ is i.i.d and satisfy the following conditions:

$$\begin{aligned} E(\varepsilon_t) &= E(\varepsilon_t^3) = 0, \\ E(\varepsilon_t^2) &= \sigma^2, \quad 0 < \sigma^2 < \infty, \\ E(\varepsilon_t^4) &= \kappa + 3\sigma^4, \end{aligned} \tag{22}$$

where κ is the fourth cumulant of the process $\{\varepsilon_t\}$. If the process $\{\varepsilon_t\}$ is Gaussian, $\kappa = 0$.

The autocorrelation function of the process $\{r_t^2\}$ is given by

$$\begin{aligned} \gamma_{r^2}(k) = {} & 2\sigma^8 \beta_{|k|}^2 \delta_{\beta\beta}(0) + 2\sigma^8 \delta_{\beta\beta}(k) + 4\rho^2\sigma^6 \delta_{\beta\beta}(k) \\ & + \sigma^4 \kappa \left(\beta_{|k|}^2 \delta_{\beta\beta}(0) + \delta_{\beta^2\beta^2}(k) \right) + 4\sigma^4 \mu \beta_{|k|} \left(\rho^2 + \sigma^2 \delta_{\beta\beta}(k) \right) \\ & + 2\rho^2 \sigma^6 \beta_{|k|}^2 + \nu_y \delta(k, 0), \quad k = 0, \pm 1, \ldots, \end{aligned} \tag{23}$$

where ν_r is a strictly positive constant, $\delta_{\beta\varrho} = \sum_{i=1}^{\infty} \beta_i \varrho_{i+u}$, $u = 0, \pm 1, \ldots$ for any square summable sequence $\{\beta_i\}$ $\{\varrho_i\}$, $\delta(\cdot, \cdot)$ is the Kronecker delta.

When the sequence $\{\beta_i\}$ of the MA form of equation (21) verifies the convergence rate given by equation (14) and $\rho \neq 0$, setting $\alpha = 2d$ and $Y_t = r_t^2$, then the autocorrelation function has the rate of decay given by equation (2).

2.4 Nonlinear Transformations of Fractionally Integrated Processes

Another approach for modeling non–Gaussian LRD processes consists in taking nonlinear transformations of Gaussian LRD processes satisfying (2); see Surgailis (1981), the third chapter by Beran (1994), Giraitis and Surgailis (1985, 2005) and references therein for further details.

We consider here the case of a Gaussian process $\{X_t\}$ and its transformation by a function $G(\cdot)$ such that $EG^2(X_t) < \infty$, then this function can be expanded in series of Hermite polynomials:

$$G(x) = \sum_{k=0}^{\infty} \frac{c_k}{k!} H_k(x), \tag{24}$$

where

$$c_k = E[G(X)H_k(X)] = \frac{1}{\sqrt{2\pi}} \int_{-\infty}^{\infty} G(x)H_k(x)e^{-x^2/2}dx, \tag{25}$$

so that $H_0(x) = 1$, $H_1(x) = x$, $H_2(x) = x^2 - 1$, $H_3(x) = x^3 - 3x$, $H_k(x) = (-1)^k e^{(x^2/2)}(e^{-x^2/2})^{(k)}$, $k \geqslant 0$.

If $\{X_t\}$ is a Gaussian LRD process with scaling parameter α, satisfying equation (2), then the expansion

$$Y_t = G(X_t) = \sum_{J=1}^{\infty} \frac{c_J}{J!} H_J(X_t), \tag{26}$$

has the following properties:

- The processes $\{H_J(X_t), t \in \mathbb{Z}\}$ are orthogonal,
- For $1 \leqslant J \leqslant 1/(1-\alpha)$, the process $\{H_J(X_t)\}$ satisfies equation (2) as

  $$\mathrm{Cov}(H_J(X_0), H_J(X_k)) = J!\mathrm{Cov}(X_0, X_k)^J \asymp J! c_\gamma^J k^{J(\alpha-1)}, \tag{27}$$

  thus, the memory parameter of $\{H_J(X_t), t \in \mathbb{Z}\}$ decreases with J.

The index of the lowest nonzero coefficient c_k, $k \geqslant 1$, of the transformation $G(\cdot)$ is called the Hermite rank of the transformation, denoted by R. For a transformation $G(X_t)$ of rank 1,

$$Y_t = G(X_t) = c_1 H_1(X_t) + \sum_{J \geqslant 2} c_J H_J(X_t)/J! = c_1 X_t + S_t, \tag{28}$$

the linear term $c_1 X_t$ dominates the memory of the nonlinear term S_t. As a consequence, Y_t has the same scaling parameter of X_t: the transformation $G(X_t)$ does not increase the memory of X_t.

The intensity of long–range dependence of the nonlinear transformed process $\{Y_t\}$ coincides with the scaling parameter of the lowest nonzero term $k \geqslant 1$, as for a transformation $G(X_t)$ of Hermite rank R,

$$Y_t = G(X_t) = c_R H_R(X_t) + \sum_{J \geqslant R+1} c_J H_J(X_t)/J!, \tag{29}$$

the memory of the leading term $c_R H_R(X_t)$ dominates the memory of the remainder term $\sum_{J \geqslant R+1} c_J H_J(X_t)/J!$.

Dittman and Granger (2002) considered the particular case $X_t \sim \text{FI}(d)$, then $\{Y_t = G(X_t), t \in \mathbb{Z}\}$ is a long–memory process with long–memory parameter $\tilde{d} = \max\{0, (d-0.5)R+0.5\}$. Here, $\alpha = 2\tilde{d}$. We consider in section 4.4 the estimation of the scaling parameter for some nonlinear transformations of FI(d) processes.

3 Estimation of the Long–Memory Parameter

We compare the statistical performance of the wavelet estimator with the two standard semiparametric estimators in the spectral domain used in the statistical and econometric literature for estimating the scaling parameter α of the processes introduced in section 2 and of real data, namely the local Whittle estimator and the LP estimator. The remainder of this section recalls the definitions and compare the key properties of the wavelet, local Whittle and LP estimators (hereafter denoted by $\hat{\alpha}_W$, $\hat{\alpha}_{LW}$ and $\hat{\alpha}_{LP}$).

Remark 1. All three estimators are sharing similar features:

- They are based on the same semiparametric hypothesis: in the neighborhood of the zero frequency, the spectrum of the process under investigation satisfies equation (1).
- They imply the estimation of a second-order statistical quantity from the observed times series (the periodogram $I_Y(\lambda_j)$ or the scalogram $S_Y(j)$).
- They involve the choice of a range of *frequencies* or *scales* over which the estimation is to be performed. This crucial step will be addressed with care as this is one of the main practical issue that strongly controls the actual performance of these estimators.
- For the LP and LW estimators in the spectral domain, we will select the range of frequencies using plug–in methods; see Henry (2001), Moulines and Soulier (2003) for a detailed presentation. For the wavelet estimator, we will derive the lowest octave using the optimal bandwidth provided by these plug–in methods.

3.1 Local Whittle Estimator

• **Definition.** The local Whittle estimator $\hat{\alpha}_{LW}$ has been proposed by Künsch (1987) and further developed by Robinson (1995a). It is defined as

$$\hat{\alpha}_{LW} = \arg\min_{\alpha} G(\alpha, m) = \left\{ \ln\left(\frac{1}{m} \sum_{j=1}^{m} \frac{I_Y(\lambda_j)}{\lambda_j^{-\alpha}} \right) - \frac{\alpha}{m} \sum_{j=1}^{m} \ln(\lambda_j) \right\}, \quad (30)$$

where $I_Y(\lambda_j)$ is the periodogram evaluated on a set of m Fourier frequencies $\lambda_j = 2\pi j/T, j = 1, \ldots, m \ll [T/2]$, where $[\cdot]$ denotes the integer part, the bandwidth parameter m tends to infinity with the sample size T but more slowly since $1/m + m/T \to 0$ as $T \to \infty$. The process does not need to be Gaussian, but its spectrum is differentiable near the zero frequency, and the process has a moving average representation.

• **Performance.** The LW estimator has the following asymptotic distribution

$$\sqrt{m}(\hat{\alpha}_{LW} - \alpha) \xrightarrow{d} N(0, 1), \quad (31)$$

where $\xrightarrow{d}$ means convergence in distribution. The normality result comes from the assumption that the spectrum

$$f_Y(\lambda) = C\lambda^{-\alpha}\left[1 + E_\beta(\alpha)\lambda^\beta + o(\lambda^\beta)\right], \quad 0 < |E_\beta(\alpha)| < \infty, \quad (32)$$

as $\lambda \to 0^+$, with $\beta \in (0, 2]$ controls the smoothness of the spectrum near the zero frequency, C is a strictly positive constant, and the bandwidth m satisfying

$$\frac{1}{m} + \frac{m^{2\beta+1}\log^2 m}{T^{2\beta}} \longrightarrow 0, \quad T \longrightarrow \infty, \quad (33)$$

see Robinson (1995a) and Henry (2001) for further details.

Remark 2. In this framework, Giraitis, Robinson and Samarov (1997) have demonstrated that the best attainable rate for an estimator $\hat{\alpha}$ is $T^{-r(\beta)}$, with $r(\beta) = \beta/(1 + 2\beta) < 1/2$, thus slower than the rate $T^{-1/2}$ of parametric estimators. For the LW estimator, the rate of convergence is $T^{-r(\beta)}M_n$ where $(\log m)^{-1/(1+2\beta)}M_n = o(1)$.

• **Range of Frequencies.** In this framework, i.e., when $f_Y(\lambda)$ satisfies (32) the optimal bandwidth in the sense of minimization of the mean square error (MSE) has been provided by Henry and Robinson (1996) as

$$m_{LW}^{opt} = \left[\frac{(\beta+1)^4}{2\beta^3 E_\beta(\alpha)^2 (2\pi)^{2\beta}}\right]^{1/(1+2\beta)} T^{2\beta/(1+2\beta)}. \quad (34)$$

With the additional assumption that $\beta = 2$, i.e., the smoothest case for the spectrum, we obtain the optimal bandwidth m_{LW}^{opt} with the iterative procedure:

$$\hat{\alpha}^{(k)} = \arg\min_{\alpha} G(\alpha, m^{(k)}), \tag{35}$$

$$m_{LW}^{opt\,(k+1)} = \left(\frac{3T}{4\pi}\right)^{4/5} \left|E_2(\alpha^{(k)})\right|^{-2/5}, \tag{36}$$

this iterative procedure starting at $m^{(0)} = T^{4/5}$, with $\alpha \in (0,1)$. Without additional knowledge on $E_2(\alpha^{(k)})$, this procedure is infeasible. If we further assume the following specification for the spectrum

$$f_Y(\lambda) = |2\sin(\lambda/2)|^{-\alpha} f_\star(\lambda), \tag{37}$$

where $f_\star(\cdot)$ is twice continuously differentiable and positive at $\lambda = 0$, then a feasible optimal bandwidth is obtained with this approximation (see Delgado and Robinson, 1996):

$$E_2(\alpha^{(k)}) = \frac{f_\star''(0)}{2f_\star(0)} + \frac{\alpha}{24} \tag{38}$$

where $f_\star(0)$ and its second derivative $f_\star''(0)$ are the first and last coefficients of the regression of the periodogram $I(\lambda_j)$ against $|1-\exp(i\lambda)|^{-\alpha(0)}(1, \lambda_j, \lambda_j^2/2)$ for the range of Fourier frequencies λ_j for $j = 1, \ldots, m^{(0)}$. This iterative procedure defined by equations (35) and (36) converges very quickly.

Remark 3. The choice for this optimal bandwidth is motivated by two reasons:

- This bandwidth, the theory of which has been developed for linear LRD processes, is also robust to the presence of (long–memory) conditional heteroscedasticity in the process; see Henry (2001).
- This bandwidth works well, even for nonlinear LRD processes, such as the LARCH process; see Giraitis, Kokoszka, Leipus and Teyssière (2000).

3.2 Log–Periodogram Regression

• **Definition.** Geweke and Porter–Hudak (1983) proposed an estimator $\hat{\alpha}_{LP}$ for α that consists in performing a least squares regression on the log–periodogram (LP) over a range of frequencies λ_j:

$$\log I_Y(\lambda_j) = \log c_f - \alpha \log \lambda_j + e_j, \quad j = 1, \ldots, m, \tag{39}$$

where the sequence $\{e_j\}$ is interpreted as error terms; see Henry (2005) in this volume for more details.

• **Performance.** The LP estimator $\hat{\alpha}_{LP}$ has the following asymptotic distribution

$$\sqrt{m}(\hat{\alpha}_{LP} - \alpha) \xrightarrow{d} N\left(0, \frac{\pi^2}{6}\right). \tag{40}$$

Remark 4. This central limit theorem requires that the bandwidth m satisfies equation (33). As mentioned in the remark 2 above, this estimator has the slow rate of convergence $m^{1/2}$. The best rate of convergence $T^{-r(\beta)}$, with $r(\beta) = \beta/(1+2\beta)$ is attained with Robinson's (1995b) version of the LP estimator.

• **Regression Range.** Usually, to select m, one chooses $m = [T^{4/5}]$ for long–memory processes in the conditional mean, e.g., FARIMA processes or nonlinear transformations of FI(d) processes, see e.g., Dittman and Granger (2002), and $m = [T^{1/3}]$ for the LMSV process, see e.g., Deo and Hurvich (2003). For the LMSV process, this usual choice yields estimates with a slightly lower bias than he one obtained with $m = [T^{4/5}]$, while the Root Mean Squared Error of the estimator is increased by a factor equal to 2. For other nonlinear processes, like the ones considered in this chapter, this choice yields strongly biased and then unreliable estimates. Under the restriction that the spectrum has the specification given by equation (37), Hurvich *et al.* (1998) proposed an optimal bandwidth m_{LP}^{opt} minimizing the MSE for the LP estimator

$$m_{LP}^{opt} = \left[\frac{27(2f_\star(0))^2}{512\pi^2(f_\star''(0))^2}\right]^{1/5} T^{4/5}. \tag{41}$$

A feasible bandwidth is obtained by estimating $f_\star(0)$ and $f_\star''(0)$ as above for the case of the optimal bandwidth for the LW estimator; see the useful paper by Henry (2001) for further details.

Remark 5. The choice $m = cT^{r(\beta)}$ gives the best rate, and with a suitable choice of c minimizes the MSE. However, this choice does not imply the central limit theorem of equation (40).

Remark 6. Unlike for the optimal bandwidth of the LW estimator, the properties of robustness of $\hat{\alpha}_{LP}$ to heteroskedasticity have to be established, so that we tend to prefer $\hat{\alpha}_{LW}$.

3.3 Wavelet Based Estimator

A Short Introduction to Wavelet Analysis

For complete and thorough introductions to wavelet analysis and decompositions, the reader is referred to e.g., the books by Daubechies (1992) and Mallat (1998).

• **Mother–Wavelet.**

Definition 2. *A wavelet is a function $\psi(\cdot)$ defined on $\mathbb{R}$ such that*

$$\int_{\mathbb{R}} \psi(t)\, dt = 0, \tag{42}$$

i.e., satisfies the admissibility condition.

We also assume that $\psi(t)$ satisfies some integrability condition, i.e., $\psi(t) \in L^1(\mathbb{R}) \cap L^2(\mathbb{R})$. The wavelet function is then a band–pass function, i.e., a small "wave" the support of which is almost limited in both the time and frequency domains. To perform a wavelet analysis, one choose a reference wavelet, called the *mother–wavelet*, hereafter denoted ψ_0.

• **Wavelet–Basis.** From this ψ_0, a entire family of wavelets is designed using two operators:

1. A times–shift operator: $(\mathcal{T}_\tau \psi_0)(t) = \psi_0(t - \tau)$,
2. A dilation (or change of scale) operator: $(\mathcal{D}_a \psi_0)(t) = \frac{1}{\sqrt{a}} \psi_0 \left(\frac{t}{a} \right)$.

For the particular purpose of this work, we consider here only a particular type of wavelet decomposition, the so-called *Discrete Wavelet Transform* (DWT). From the time-shift and dilation operators above, the specific collection of translated and dilated templates of ψ_0, defined as

$$\left\{ \psi_{j,k} = 2^{-j/2} \psi_0(2^{-j} t - k), \quad j \in \mathbb{Z},\ k \in \mathbb{Z} \right\}, \tag{43}$$

forms (a possibly orthonormal) basis of $L^2(\mathbb{R})$.

• **Wavelets Coefficients of the Discrete Wavelet Transform.** The wavelet coefficients of the DWT for an analyzed process or function Y_t are labeled $d_Y(j,k)$ and defined as

$$d_Y(j,k) = \langle Y, \psi_{j,k} \rangle = \int_{\mathbb{R}} Y(t) \psi_{j,k}(t)\, dt, \tag{44}$$

where $\psi_{j,k}$ is the wavelet basis defined in (43). It is worth noting that the $d_Y(j,k)$s can be computed at a very low cost (of the order of that of a FFT) from a recursive pyramidal algorithm on the condition that the mother-wavelet ψ_0 is chosen to belong to a *Multiresolution analysis*; see e.g., Daubechies (1992) or Mallat (1998).

• **Number of Vanishing Moments and Polynomial Trends.** The mother wavelet ψ_0 is further characterized by an integer N, called the number of vanishing, or zero, moments and defined as:

$$\int_{\mathbb{R}} t^k \psi_0(t) dt \equiv 0, \quad \forall k = 0, \ldots, N-1. \tag{45}$$

Obviously, from the admissibility condition, equation (42) above, one has $N \geq 1$. This integer N constitutes a key degree of freedom in wavelet analysis that can be freely chosen by the user and tuned to a given purpose. We will show that the use of this degree of freedom plays a central role in the analysis of long–range dependent processes as well as in the discrimination between genuine strong dependence, and non–stationarities such as those existing in change–points processes or trended processes. Indeed, by definition of N, the wavelets coefficients of any polynomial of order $P < N$ will be strictly null $d_P(j,k) \equiv 0$. By linearity of the wavelet transform, it implies that wavelet coefficients are identical for the processes $\{Y_t\}$ and $\{Y_t + P_t\}$. This implies, for instance, that when Y_t is a zero-mean long–range dependent process, the potential superposition of a non–stationary polynomial mean will not affect the measure of the long–memory parameter as long as the degree of the polynomial remain strictly lower than the number of vanishing moment, that can be varied by the user. It has been shown elsewhere, see e.g., Abry and Veitch (1999) or Veitch and Abry (2001), that increasing N reduces the impact of the superimposed trend even if it is not strictly a polynomial one. This will be further discussed in Section 5. The counterpart to the increase of N however usually lies in the corresponding increase in the time support of the wavelet, an important practical drawback as discussed below.

Wavelet Analysis of Long–Range Dependent Processes

Let Y_t denote a second order stationary random process. It can be shown that its wavelet coefficients constitute a zero-mean random field that satisfies:

$$Ed_Y(j,k)d_Y(j',k') = \int_{\mathbb{R}} f_Y(\lambda)2^{j/2}\Psi_0(2^j\lambda)2^{j'/2}\Psi_0^*(2^{j'}\lambda)\exp(-\imath 2\pi(2^jk - 2^{j'}k'))d\lambda, \quad (46)$$

where $\Psi_0(\lambda)$ denotes the Fourier transform of the mother wavelet ψ_0.

From this general general result, it can be shown that when Y_t is a second order stationary long–range dependent process, the coefficients $d_Y(j,k)$ possess the two following key properties:

P1: The process $\{d_Y(j,k),\ k \in \mathbb{Z}\}$ is stationary if $N \geq (\alpha - 1)/2$, and its variance reproduces in the limit of large scales the power law behavior of the spectrum of Y_t,

$$2^j \to +\infty, \quad Ed_Y(j,\cdot)^2 \simeq 2^{j\alpha}c_f C(\alpha,\psi_0), \quad C(\alpha,\psi_0) = \int |\lambda|^{-\alpha}|\Psi_0(\lambda)|^2 d\lambda, \quad (47)$$

P2: The process $\{d_Y(j,k),\ k \in \mathbb{Z}\}$ is stationary and short–range dependent, if $N > \alpha/2$. The residual correlations between the elements of the sequence $\{d_Y(j,k),\ k \in\}$ is in reverse relationship with N; see Flandrin (1989):

$$Ed_Y(j,k)d_Y(j,k') \approx C\,|k-k'|^{\alpha-1-2N}, \quad |k-k'| \longrightarrow \infty. \quad (48)$$

Properties **P1** and **P2** above constitute the main rationale for the wavelet analysis of long–range dependent data. Indeed, **P1** yields

$$\log_2\left(Ed_Y(j,\cdot)^2)\right) = j\alpha + \log_2(c_f C(\alpha,\psi_0)). \quad (49)$$

This invites to perform a linear regression in a log-log plot. However, the expectation $Ed_Y(j,\cdot)^2$ needs to be estimated from a (single, finite duration) observation. This is where **P2** plays a key role: the ensemble average can be efficiently replaced with the time average $1/n_j \sum_{k=1}^{n_j} d_Y(j,k)^2$.

Estimation of the Long–Memory Parameter

This estimator was first proposed and studied in Abry *et al.* (1995) and Abry and Veitch (1998) and further developed and analyzed in Veitch and Abry (1999).

• **Definition.** Let $S_Y(j) = 1/n_j \sum_{k=1}^{n_j} d_Y(j,k)^2$, where n_j denotes the number of wavelet coefficients $d_Y(j,k)$ available at octave j. Roughly, n_j varies as $2^{-j}T$, where T denotes the number of observed samples. The wavelet-based estimator consists of a (weighted) linear regression in the so-called logscale diagram: $\log_2 S_Y(j)$ versus $\log_2 2^j = j$. Precisely, it is defined as:

$$\hat{\alpha}_W = \sum_{j_1}^{j_2} w_j(\log_2 S_Y(j) - (\psi(n_j/2)/\log 2 - \log_2(n_j/2))), \tag{50}$$

where $\psi(z) := \Gamma'(z)/\Gamma(z)$ is the Psi function and the weights are chosen to satisfy the two constraints:

$$\sum_{j_1}^{j_2} jw_j \equiv 1, \qquad \sum_{j_1}^{j_2} w_j \equiv 0. \tag{51}$$

A natural form for the w_js reads:

$$w_j = \frac{1}{a_j}\frac{S_0 j - S_1}{S_0 S_2 - S_1^2}, \tag{52}$$

where $S_p = \sum_{j_1}^{j_2} j^p/a_j$, $p = 0, 1, 2$. The ranges of octaves $j \in [j_1, j_2]$ is assumed to be a priori chosen; this will be discussed below. The a_js are arbitrary coefficients used to weight the estimation according to the confidence given to $\log_2 S(j)$. Precisely, for the estimation of the LRD parameter, we chose $a_j \equiv \zeta(2, n_j/2)$, with $\zeta(2,z) := \sum_{n=0}^{\infty} 1/(z+n)^2$ a generalized Riemann Zeta function, that provides us with an approximation of $\mathrm{Var} \log_2 S(j)$. $\zeta(2, n_j/2)$ turns out to be asymptotically proportional to $1/n_j$ (see below). It implies that the larger j, the smaller n_j, the less $\log_2 S(j)$ contributes to the estimate of α.

• **Performance.** To study approximately but analytically the performance of this estimator, the three following assumptions are assumed to hold:

H1: For each j, the sequences $\{d_Y(j,\cdot)\}$ are stationary and uncorrelated,
H2: The processes $\{d_Y(j,\cdot)\}$ and $\{d_Y(j',\cdot)\}$, $j \neq j'$ are uncorrelated,
H3: The processes $\{d_Y(j,\cdot)\}$, $j \in [j_1, j_2]$, are Gaussian.

The two first points constitute idealizations of the decorrelation property **P2**. The third point is obviously satisfied when $\{Y_t\}$ is a Gaussian process.

Using these assumptions, it can be shown that

$$E\log_2 S(j) = \log_2 Ed_Y(j,\cdot)^2 + (\psi(n_j/2)/\log 2 - \log_2(n_j/2)), \tag{53}$$
$$\mathrm{Var}\log_2 S(j) = \zeta(2, n_j/2)/\log^2 2. \tag{54}$$

From this, one shows that

$$E\hat{\alpha}_W = \alpha, \tag{55}$$
$$\mathrm{Var}\hat{\alpha}_W = \frac{(1-2^{-J})/\log^2 2}{1-2^{-(J+1)}(J^2+4)+2^{-2J}}\frac{1}{T}, \tag{56}$$

where $J = j_2 - j_1 + 1$ is the width of the regression range. Veitch and Abry (1999) conjectured the asymptotic approximation of the wavelet estimator

$$\sqrt{T}(\hat{\alpha}_W - \alpha) \approx N\left(0, \frac{1}{\ln^2(2)2^{1-j_1}}\right), \tag{57}$$

where j_1 is the lowest octave, the long–memory behavior being captured by the octaves larger than j_1.

This analytical yet only approximative performance turns out to be extremely close to the actual ones, even when Y_t are non Gaussian processes. In practice, $\hat{\alpha}_W$ has a negligible bias and presents a Gaussian statistic with a variance (approximately) known without the estimation of any quantity. An effective confidence interval can hence be constructed.

Under some assumptions, i.e., $n_{j_1} \to \infty$, but more slowly than the sample size T, Bardet, Lang, Moulines and Soulier (2000) proved the asymptotically Normal distribution of this estimator

$$\sqrt{n_{j_1}}(\hat{\alpha}_W - \alpha) \xrightarrow{d} N(0, K_v(\alpha)), \tag{58}$$

where the expression of $K_v(\alpha)$ depends on the unknown scaling parameter α.

• **Regression Range.** Let us turn now to the choice of the regression range $j \in [j_1, j_2]$. One might first use the graphical representation of the logscale diagram, i.e., the graphical plot of the y_j against the octaves j from which is estimated the scaling parameter $\hat{\alpha}_W$. A graphical analysis might help to select the upper and lower octaves that have a too high leverage effect on the regression line; see Belsley *et al.* (1980).

Obviously, as LRD implies a power law behavior in the limit of large scales, $2^j \to +\infty$, the upper bound j_2 for the octaves has to be chosen as large as allowed by the observation length. We chose

$$j_2 = [\log_2 T - \log_2(2N+1)], \tag{59}$$

where $[\cdot]$ denotes the integer part, so that borders effects do not affect the estimation of the variance. This is equivalent to the fact that the range of frequencies for the LP and LW estimators starts at $\lambda_j = \pi/T$.

The choice for j_1, i.e., the cutoff between short–range dependence and long–range dependence, is similar to the bandwidth selection problem in semiparametric/nonparametric statistics and is exactly equivalent to the choice of the m parameter for the local Whittle and log-periodogram estimators introduced above. This choice consists in optimizing a bias-variance trade-off: a too small j_1 should induce bias as the regression range would include octaves with departures from the power law due to short-range dependencies, conversely, a too large j_1 decreases the bias but implies an increase of the variance, as shown in relation (57). Abry *et al.* (2003) discussed the issue of the selection for the lower octave j_1 in connexion with the bandwidth selection issue for the local Whittle (LW) estimator, which has been addressed by Robinson and Henry (1996) and Henry (2001, 2005); see equation (36). The frequency cutoff, associated with the optimal bandwidth m_{LW}^{opt} for the LW estimator, is then equal to m_{LW}^{opt}/T, and corresponds to the scale 2^{-j_1}. Thus, using m_{LW}^{opt} we could define the optimal lower scale as

$$j_1^{opt} = \left\lceil \frac{\log T - \log m_{LW}^{opt}}{\log 2} \right\rceil . \tag{60}$$

A similar decision rule is possible using the feasible optimal bandwidth m_{LP}^{opt} for the log–periodogram (LP) estimator; see equation (39). Indeed, estimation results given by equation (60) with m_{LW}^{opt} and m_{LP}^{opt} are very close, with a slight advantage for m_{LW}^{opt} as the RMSE is slightly lower with this bandwidth.

In Veitch, Abry and Taqqu (2003), it has been proposed that the choice of the lowest octave j_1 might be guided using a goodness–of–fit function, such as the generalized Pearson statistic defined as

$$Q = \sum_{j=j_1}^{j_2} \frac{(y_j - \hat{\alpha}_W j - \hat{a})^2}{\sigma_j^2}, \tag{61}$$

where $\hat{a}$ is the unbiased estimator for $\log_2(\widehat{c_f C})$, and $\sigma_j^2 = \mathrm{Var}(y_j)$. Under the null hypothesis of Gaussian residuals, the statistic $Q \sim \chi^2(J-2)$, where $J = j_2 - j_1 + 1$. The statistic Q can be viewed as a function of j_1, denoted $Q(j_1)$. Studying the evolution of Q as a function of j_1, Veitch, Abry and Taqqu (2003) proposed an empirical criterion to choose automatically j_1. It turns out that this procedure amounts for most cases to select the j_1 that ensures the lowest RMSE, and hence the bias-variance trade-off. Bardet *et al.* (2000) provided an analytical expression for the choice of an asymptotically optimal j_1 minimizing the MSE, and conjectured that a feasible approximation of this optimal bandwidth using estimates of α, $f_\star(0)$ and $f_\star''(0)$ as for the LW and LP estimators might be obtained. This procedure is beyond the scope of this chapter.

• **Choosing the Number of Vanishing Moments.** On one hand, equation (48) may lead to think that the larger N, the weaker the correlation amongst wavelet coefficients, however we saw that the LRD is turn in SRD as soon as $N > \alpha/2$. On the other hand, as already mentioned, increasing N induces an increase of the wavelet size and hence of the border effects. In turn, this produces a decrease of j_2, as summarized in equation (59). Therefore, an optimal practical choice for the estimation of the parameter α of true LRD processes is $N = 2$ or $N = 3$, as will be seen on numerical simulations conducted in next sections. However, in disentangling LRD from non–stationarities, the possibility of varying N and using larger N is fruitful.

• **Discrete Time.** By definition, the DWT applies to continuous time processes, see equation (44). However, in most practical cases, only a collection of discrete samples $\{y_k, k = 1, \ldots, n\}$ is available. This specific difficulty was carefully addressed in Veitch, Taqqu and Abry (2001). Numerical simulations such as those proposed in the next two sections taking into account this point are under investigation; see Abry and Teyssière (2005).

4 Estimation of the Long–Memory Parameter for Nonlinear Long–Range Dependent Processes

• **Protocol.** In this section, we compare by means of numerical simulations, the performance of the local Whittle, log-periodogram and wavelet estimators for the long–memory parameter. Because wavelet based estimation of the long–memory parameter for linear FARIMA processes has been considered in Veitch and Abry (1999), we consider in this section the estimation of the scaling parameter of the nonlinear processes described in section 2 .

For each process of interest, 5000 replications, with sample sizes T were numerically synthetised. In general, we used $T = 10000$, in some cases, we went up to $T = 20000$, for having a better idea on the asymptotic bias for some highly nonlinear processes. For each simulation, bias and Root Mean Squared Error (RMSE) defined respectively as $B = \langle\hat{\alpha}\rangle - \alpha$ and RMSE $= \sqrt{B^2 + \langle\langle\hat{\alpha}\rangle\rangle}$, where $\langle\cdot\rangle$ and $\langle\langle\cdot\rangle\rangle$ denote respectively the sample mean and sample variance estimators computed from the 5000 replications.

For the wavelet-based estimator, we used the Daubechies wavelets with the number of vanishing moments ranging from $N = 2, \ldots, 10$. We considered the range of lowest octaves $j_1 = 1, \ldots, 7$ and denoted by $\hat{\alpha}_W^{(j_1)}$ the corresponding estimates. We computed the "optimal" j_1 according to equation (60), $\hat{\alpha}_W^m$ denotes the wavelet estimator with the lowest octave j_1 derived from equation (60). We also chose the j_1 and N that minimize the Root Mean Squared Error (RMSE) of $\hat{\alpha}_W$. We denote by j_1^{RMSE} and $\hat{\alpha}_W^{RMSE}$ the corresponding j_1 and estimate. Systematically, we compare this RMSE to that obtained with the choice $j_1 = 6$ and $N = 2$. As we will see below, the RMSE for this choice of (j_1, N) is not too far from the minimum one and the bias is lower. On the basis of simulation results for $T = 20000$, we also conjecture that for some of these nonlinear processes we could get rid off the bias with very large samples, so that we could select a higher value for j_1 which minimizes the RMSE. For each simulation, Tables are reporting the results and logscale diagrams are displayed.

For the local Whittle and the log–periodogram estimators, we respectively denote by $\hat{\alpha}_{LW}^{opt}$ and $\hat{\alpha}_{LP}^{opt}$ those obtained with the optimal bandwidth m^{opt} as in equations (36) and (41).

4.1 The Long–Memory Linear ARCH Process

The long–memory LARCH process is defined as:

$$r_t = \sigma_t \varepsilon_t, \quad \varepsilon_t \sim D(0,1), \quad \sigma_t = \omega + \beta(L) r_t, \tag{62}$$

where the coefficients β_j of the lag polynomial $\beta(L)$ have the rate of decay $\beta_j = O\left(j^{\alpha/2-1}\right)$, for some $0 < \alpha < 1$. We choose here three parameterizations for $\beta(L)$ from the moving average representation of a FARIMA(0,d,0), a FARIMA(1,d,0) and a FARIMA(0,d,1) processes, with $\alpha = 2d$, that we respectively denote as **LARCH A**, **LARCH B** and **LARCH C**. For **LARCH B** and **LARCH C**, we respectively set $\phi = -0.20$ and $\theta = 0.20$.

We consider the following values for the scaling parameter $\alpha = 0.10, 0.15, 0.20, \ldots, 0.60$. For values of α over 0.60, the sequence of coefficients $\{\beta_j\}$ must be largely rescaled for satisfying the stationarity condition of theorem 1, i.e., $L(E\varepsilon_0^4)^{1/2} \sum_{j=1} \beta_j^2 < 1$. We estimate the scaling parameter using the wavelet estimator and the two spectral estimators LW and LP. We report in Table 2 the LP results only as they are very close to the LW estimator, although with a slightly higher RMSE as theoretically expected, and so far there are no results for the estimation results of the LARCH process with the LP estimator, while Giraitis, Kokoszka, Leipus and Teyssière (2000) already report estimation results for LARCH processes with the LW estimator for $T = 3000, 6000$. Tables 1 and 2 below report estimation results for the LARCH processes for $T = 10,000$ while Tables 3 and 4 give the results for $T = 20,000$. For $T = 20,000$, estimation results for the LW estimator are reported on Table 4.

Figures 3 to 5 below display the logscale diagrams for **LARCH A** processes, with $T = 10000$ and $T = 20000$. We can see that the "correct" LRD behavior is captured by the octaves j greater than 5. When T increases, the number of octaves that can be used for the wavelet regression is obviously larger, so that wavelet estimates are more reliable.

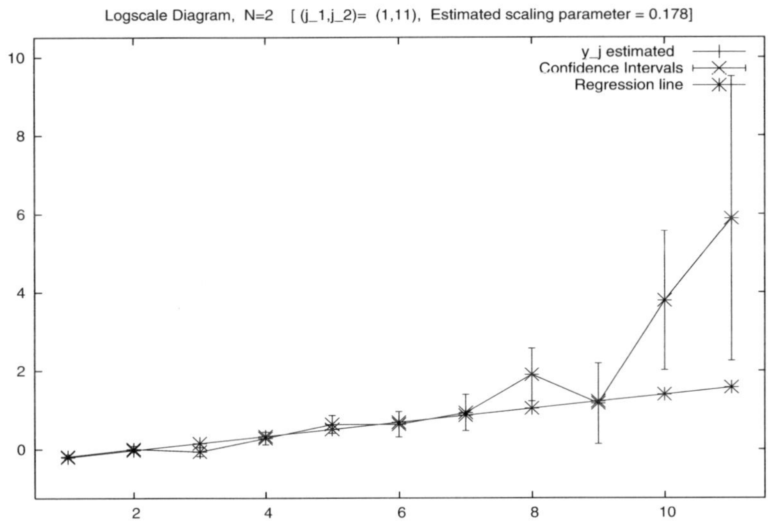

Fig. 3. Logscale diagram for the realization of a **LARCH A** process, with $\alpha = 0.40$, $T = 10000$. We select here $j_1 = 1$, $j_2 = 11$, $N = 2$

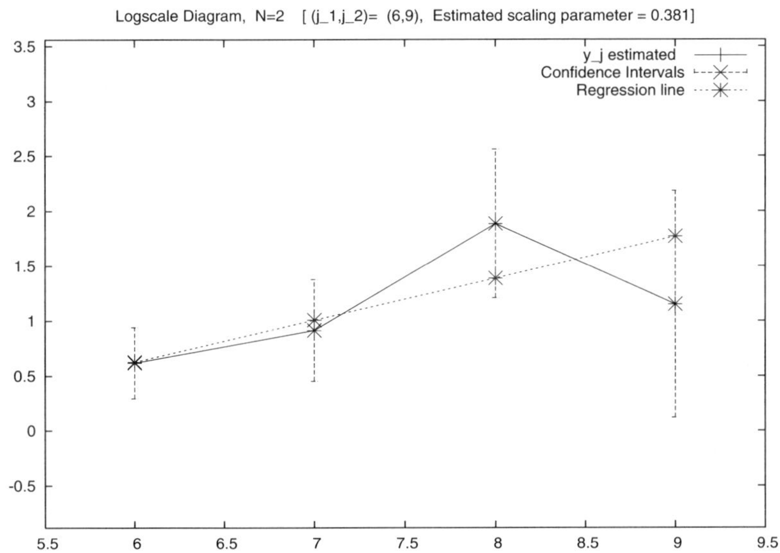

Fig. 4. Logscale diagram for the realization of a **LARCH A** process, with $\alpha = 0.40$, $T = 10000$. We select here $j_1 = 6$, $j_2 = 9$, $N = 2$

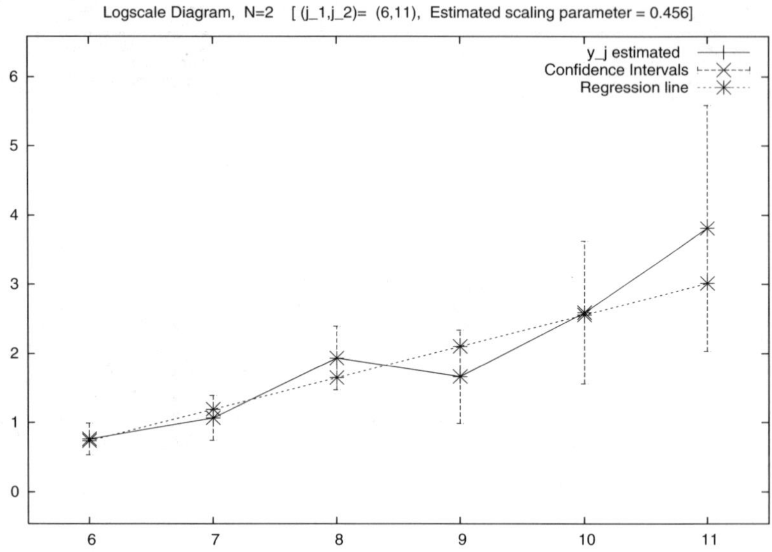

Fig. 5. Logscale diagram for the realization of a **LARCH A** process, with $\alpha = 0.40$, $T = 20000$. We select here $j_1 = 6$, $j_2 = 11$, $N = 2$

From the results in Tables 1 to 4, we can conclude that even for $T = 10,000$, both the LW and LP estimators perform slightly better than the wavelet estimator as both their bias and RMSE are slightly lower. For this sample size, choosing the wavelet estimator with $j_1 = 6$ and $N = 2$ reduces the bias, which becomes smaller than the one of the LW estimator, but increases the variance. However, for $T = 20,000$, the RMSE of the estimates for $j_1 = 6$ is reduced and becomes closer to the one of the LW estimator. For $T = 20,000$, j_1^{RMSE} is often equal to 5, the wavelet estimates are closer to the LW estimates, and we conjecture that for larger samples $j_1 = 6$ would minimize the RMSE, so that the performance of the wavelet, LP and LW estimator would become very close.

We also note that the estimations for the wavelet estimator with j_1 chosen according to equation (60) gives interesting results, not as good as those given by j_1^{RMSE}, but that can be used as a first approximation.

Table 1. Estimation of the scaling parameter for LARCH processes. $T = 10000$, $N = 2$

Model	α	$\hat{\alpha}_W^{RMSE}$	j_1^{RMSE}	$E\hat{\alpha}_W^{RMSE} - \alpha$	RMSE	$\hat{\alpha}_W^{(6)}$	$E\hat{\alpha}_W^{(6)} - \alpha$	RMSE
LARCH A	0.10	0.0255	3	-0.0745	0.0824	0.0445	-0.0555	0.1350
	0.15	0.0690	4	-0.0810	0.0971	0.0966	-0.0534	0.1357
	0.20	0.1199	4	-0.0801	0.0980	0.1622	-0.0378	0.1330
	0.25	0.1785	4	-0.0715	0.0937	0.2325	-0.0175	0.1329
	0.30	0.2394	4	-0.0606	0.0891	0.3046	0.0046	0.1381
	0.35	0.3010	4	-0.0490	0.0891	0.3618	0.0118	0.1487
	0.40	0.3349	4	-0.0650	0.1028	0.3994	-0.0006	0.1549
	0.45	0.3880	5	-0.0620	0.1252	0.4228	-0.0272	0.1585
	0.50	0.4060	5	-0.0940	0.1442	0.4463	-0.0537	0.1663
	0.55	0.4234	5	-0.1266	0.1677	0.4697	-0.0803	0.1778
	0.60	0.4928	6	-0.1072	0.1924	0.4928	-0.1072	0.1923
LARCH B	0.10	0.0125	1	-0.0875	0.0898	0.0039	-0.0961	0.1554
	0.15	0.0128	2	-0.1372	0.1399	0.0314	-0.1185	0.1704
	0.20	0.0633	5	-0.1367	0.1581	0.0824	-0.1176	0.1706
	0.25	0.1222	5	-0.1278	0.1514	0.1531	-0.0969	0.1586
	0.30	0.1938	5	-0.1062	0.1354	0.2351	-0.0649	0.1441
	0.35	0.2712	5	-0.0788	0.1180	0.3190	-0.0310	0.1367
	0.40	0.3483	5	-0.0517	0.1069	0.3977	-0.0023	0.1397
	0.45	0.4202	5	-0.0298	0.1061	0.4669	0.0169	0.1504
	0.50	0.4249	4	-0.0751	0.1117	0.5216	0.0216	0.1640
	0.55	0.4901	5	-0.0599	0.1281	0.5375	-0.0125	0.1641
	0.60	0.4987	5	-0.1013	0.1520	0.5530	-0.0470	0.1709
LARCH C	0.10	0.1014	1	0.0014	0.0306	0.0725	-0.0275	0.1332
	0.15	0.1369	1	-0.0131	0.0375	0.1232	-0.0268	0.1382
	0.20	0.1477	1	-0.0523	0.0635	0.1641	-0.0359	0.1426
	0.25	0.1567	1	-0.0933	0.1004	0.2032	-0.0468	0.1479
	0.30	0.1979	4	-0.1021	0.1243	0.2406	-0.0594	0.1545
	0.35	0.2500	5	-0.1000	0.1413	0.2766	-0.0734	0.1624
	0.40	0.2803	5	-0.1197	0.1570	0.3114	-0.0886	0.1717
	0.45	0.3092	5	-0.1408	0.1745	0.3451	-0.1049	0.1824
	0.50	0.3368	5	-0.1632	0.1939	0.3777	-0.1223	0.1945
	0.55	0.4093	6	-0.1407	0.2081	0.4093	-0.1407	0.2081
	0.60	0.4398	6	-0.1602	0.2230	0.4398	-0.1602	0.2230

Table 2. Estimation of the scaling parameter for LARCH processes. $T = 10000$, $N = 2$. The optimal bandwidth m_{LP}^{opt} is used for the LP estimator

Model	α	$\hat{\alpha}_W^m$	$E\hat{\alpha}_W^m - \alpha$	RMSE	$\hat{\alpha}_{LP}^{opt}$	$E\hat{\alpha}_{LP}^{opt} - \alpha$	RMSE
LARCH A	0.10	0.0289	-0.0711	0.0847	0.0323	-0.0677	0.0821
	0.15	0.0654	-0.0845	0.0981	0.0778	-0.0722	0.0869
	0.20	0.1149	-0.0851	0.1011	0.1379	-0.0621	0.0826
	0.25	0.1756	-0.0744	0.0967	0.2085	-0.0415	0.0754
	0.30	0.2412	-0.0588	0.0934	0.2792	-0.0208	0.0749
	0.35	0.3097	-0.0403	0.0964	0.3482	-0.0018	0.0785
	0.40	0.3485	-0.0515	0.1090	0.3889	-0.0111	0.0902
	0.45	0.3672	-0.0827	0.1310	0.4138	-0.0362	0.1014
	0.50	0.4285	-0.0715	0.1317	0.4421	-0.0579	0.1089
	0.55	0.4081	-0.1419	0.1788	0.4668	-0.0831	0.1313
	0.60	0.4284	-0.1715	0.2041	0.4935	-0.1064	0.1495
LARCH B	0.10	0.0001	-0.0999	0.1102	-0.0019	-0.1019	0.1119
	0.15	0.0130	-0.1370	0.1448	0.0196	-0.1304	0.1379
	0.20	0.0415	-0.1585	0.1658	0.0605	-0.1395	0.1480
	0.25	0.0876	-0.1624	0.1711	0.1216	-0.1284	0.1408
	0.30	0.1507	-0.1493	0.1617	0.2042	-0.0958	0.1178
	0.35	0.2314	-0.1186	0.1417	0.2905	-0.0595	0.1015
	0.40	0.3228	-0.0772	0.1205	0.3803	-0.0197	0.0899
	0.45	0.4041	-0.0459	0.1127	0.4571	0.0071	0.0964
	0.50	0.4688	-0.0312	0.1153	0.5192	0.0192	0.1056
	0.55	0.4813	-0.0687	0.1318	0.5381	-0.0119	0.1075
	0.60	0.4921	-0.1079	0.1565	0.5569	-0.0431	0.1178
LARCH C	0.10	0.0640	-0.0360	0.0661	0.0631	-0.0339	0.0596
	0.15	0.1066	-0.0434	0.0751	0.1094	-0.0406	0.0643
	0.20	0.1380	-0.0620	0.0889	0.1473	-0.0527	0.0758
	0.25	0.1675	-0.0825	0.1063	0.1842	-0.0658	0.0875
	0.30	0.1971	-0.1029	0.1248	0.2206	-0.0794	0.1022
	0.35	0.2256	-0.1244	0.1459	0.2574	-0.0926	0.1160
	0.40	0.2542	-0.1458	0.1674	0.2940	-0.1060	0.1302
	0.45	0.2824	-0.1676	0.1893	0.3308	-0.1192	0.1440
	0.50	0.3113	-0.1887	0.2110	0.3662	-0.1338	0.1598
	0.55	0.3404	-0.2096	0.2327	0.4022	-0.1478	0.1736
	0.60	0.3694	-0.2306	0.2541	0.4383	-0.1617	0.1874

Table 3. Estimation of the scaling parameter for LARCH processes. $T = 20000$, $N = 2$

Model	α	$\hat{\alpha}_W^{RMSE}$	j_1^{RMSE}	$E\hat{\alpha}_W^{RMSE} - \alpha$	RMSE	$\hat{\alpha}_W^{(6)}$	$E\hat{\alpha}_W^{(6)} - \alpha$	RMSE
LARCH A	0.10	0.0315	4	-0.0685	0.0767	0.0475	-0.0525	0.0925
	0.15	0.0858	5	-0.0642	0.0821	0.1022	-0.0478	0.0910
	0.20	0.1464	5	-0.0536	0.0757	0.1701	-0.0299	0.0850
	0.25	0.2132	5	-0.0368	0.0675	0.2419	-0.0081	0.0830
	0.30	0.2802	5	-0.0198	0.0642	0.3107	0.0107	0.0880
	0.35	0.3433	5	-0.0067	0.0680	0.3726	0.0226	0.0972
	0.40	0.3809	5	-0.0191	0.0746	0.4113	0.0113	0.1003
	0.45	0.4007	5	-0.0493	0.0880	0.4361	-0.0139	0.1018
	0.50	0.4202	5	-0.0798	0.1085	0.4612	-0.0388	0.1092
	0.55	0.4862	6	-0.0638	0.1213	0.4862	-0.0638	0.1213
	0.60	0.5110	6	-0.0890	0.1371	0.5110	-0.0890	0.1371
LARCH B	0.10	0.0120	1	-0.0880	0.0891	0.0065	-0.0935	0.1208
	0.15	0.0249	5	-0.1251	0.1347	0.0368	-0.1132	0.1368
	0.20	0.0913	6	-0.1087	0.1335	0.0913	-0.1086	0.1335
	0.25	0.1653	6	-0.0847	0.1157	0.1653	-0.0847	0.1157
	0.30	0.2491	6	-0.0509	0.0955	0.2491	-0.0509	0.0955
	0.35	0.3356	6	-0.0164	0.0855	0.3356	-0.0164	0.0855
	0.40	0.3612	5	-0.0388	0.0726	0.4120	0.0120	0.0895
	0.45	0.4337	5	-0.0163	0.0697	0.4808	0.0308	0.1009
	0.50	0.4943	5	-0.0057	0.0764	0.5357	0.0357	0.1122
	0.55	0.5059	5	-0.0441	0.0885	0.5533	0.0033	0.1074
	0.60	0.5705	6	-0.0295	0.1121	0.5705	-0.0295	0.1121
LARCH C	0.10	0.1019	1	0.0019	0.0218	0.0743	-0.0257	0.0851
	0.15	0.1435	1	-0.0065	0.0245	0.1270	-0.0230	0.0878
	0.20	0.1562	1	-0.0438	0.0504	0.1696	-0.0304	0.0915
	0.25	0.1910	5	-0.0590	0.0862	0.2104	-0.0396	0.0966
	0.30	0.2259	5	-0.0741	0.1243	0.2494	-0.0506	0.1033
	0.35	0.2869	6	-0.0631	0.1113	0.2869	-0.0631	0.1113
	0.40	0.3232	6	-0.0768	0.1210	0.3232	-0.0768	0.1210
	0.45	0.3585	6	-0.0915	0.1321	0.3584	-0.0915	0.1321
	0.50	0.3928	6	-0.1072	0.1446	0.3928	-0.1072	0.1446
	0.55	0.4261	6	-0.1239	0.1585	0.4261	-0.1239	0.1585
	0.60	0.4583	6	-0.1717	0.1737	0.4583	-0.1417	0.1786

Table 4. Estimation of the scaling parameter for LARCH processes. $N = 2$, $T = 20000$. The optimal bandwidth m_{LW}^{opt} is used for the LW estimator

Model	α	$\hat{\alpha}_W^m$	$E\hat{\alpha}_W^m - \alpha$	RMSE	$\hat{\alpha}_{LW}^{opt}$	$E\hat{\alpha}_{LW}^{opt} - \alpha$	RMSE
LARCH A	0.10	0.0299	-0.0701	0.0774	0.0349	-0.0651	0.0736
	0.15	0.0687	-0.0813	0.0888	0.0826	-0.0674	0.0771
	0.20	0.1239	-0.0761	0.0872	0.1470	-0.0530	0.0692
	0.25	0.1926	-0.0574	0.0780	0.2229	-0.0271	0.0584
	0.30	0.2678	-0.0322	0.0684	0.2990	-0.0010	0.0563
	0.35	0.3355	-0.0145	0.0694	0.3663	0.0163	0.0635
	0.40	0.3754	0.0246	0.0765	0.4086	0.0086	0.0659
	0.45	0.3969	-0.0531	0.0907	0.4372	-0.0128	0.0690
	0.50	0.4178	-0.0822	0.1108	0.4671	-0.0329	0.0776
	0.55	0.4381	-0.1119	0.1350	0.4975	-0.0525	0.0900
	0.60	0.4573	-0.1427	0.1620	0.5285	-0.0715	0.1040
LARCH B	0.10	0.0002	-0.0088	0.1050	-0.0003	-0.1003	0.1055
	0.15	0.0155	-0.1345	0.1385	0.0237	-0.1263	0.1308
	0.20	0.0471	-0.1529	0.1570	0.0686	-0.1314	0.1369
	0.25	0.0999	-0.1501	0.1564	0.1375	-0.1125	0.1226
	0.30	0.1800	-0.1200	0.1333	0.2305	-0.0695	0.0904
	0.35	0.2751	-0.0749	0.0972	0.3278	-0.0222	0.0647
	0.40	0.3592	-0.0408	0.0748	0.4141	0.0141	0.0652
	0.45	0.4326	-0.0174	0.0704	0.4862	0.0362	0.0782
	0.50	0.4931	-0.0069	0.0789	0.5429	0.0429	0.0789
	0.55	0.5057	-0.0433	0.0915	0.5652	0.0152	0.0790
	0.60	0.5190	-0.0810	0.1160	0.5883	-0.0117	0.0806
LARCH C	0.10	0.0650	-0.0350	0.0525	0.0650	-0.0350	0.0525
	0.15	0.1066	-0.0434	0.0751	0.1094	-0.0406	0.0640
	0.20	0.1425	-0.0575	0.0736	0.1523	-0.0477	0.0629
	0.25	0.1756	-0.0744	0.0898	0.1921	-0.0579	0.0739
	0.30	0.2094	-0.0906	0.1066	0.2321	-0.0679	0.0848
	0.35	0.2439	-0.1061	0.1229	0.2731	-0.0769	0.0952
	0.40	0.2796	-0.1204	0.1373	0.3140	-0.0860	0.1047
	0.45	0.3137	-0.1363	0.1529	0.3551	-0.0949	0.1143
	0.50	0.3460	-0.1540	0.1696	0.3955	-0.1045	0.1239
	0.55	0.3756	-0.1744	0.1888	0.4351	-0.1149	0.1343
	0.60	0.4032	-0.1968	0.2098	0.4743	-0.1257	0.1451

4.2 The Long–Memory Stochastic Volatility Process

We consider the long–memory stochastic volatility (LMSV) process defined by equation (17) with the scale parameter $\sigma = 0.8$. The process $\{X_t\}$ is a FARIMA(0,d,0) process generated using the Durbin–Levinson algorithm. Tables 2 and 6 report the estimation results for the long–memory parameter $\alpha = 2d$ obtained from the local Whittle, wavelet and log–periodogram estimators. The comparison with the LP estimator is of interest as the use of this estimator for LMSV processes is theoretically justified. Deo and Hurvich (2003) report few simulation results for $m = [T^{0.3}], [T^{0.4}], [T^{0.5}]$: the bias increases with m, while the RMSE decreases. We use both the optimal feasible bandwidth given by (41) and $m = [T^{0.3}]$.

Figures 6 and 7 below display the logscale diagrams for one realization of a LMSV process, with different choices for the range of octaves $[j_1, j_2]$: the long–range dependent behavior of the LMSV process is captured for $j_1 \geq 6$.

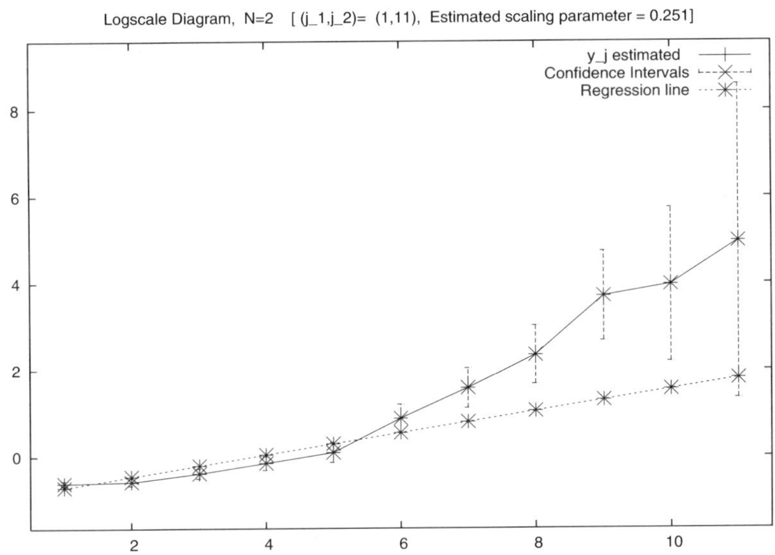

Fig. 6. Logscale diagram for the realization of a LMSV process, with $\alpha = 0.90$. We select here $j_1 = 1$, $j_2 = 11$, $N = 2$

These three estimators give here very similar results, when the optimal bandwidth is used for the LP estimator. This set of simulation results should stimulate further theoretical research on this wavelet estimator for volatility processes. We mention that in that case, the choice for j_1 using equation (60), with either m_{LW}^{opt} or m_{LP}^{opt}, docs not yield the best results.

When using the standard bandwidth $m = [T^{0.3}]$, the mean of the $\hat{\alpha}_{LP}$ estimates are equal to -0.0598, -0.111, 0.0722, 0.1794, 0.2986, 0.4224, 0.5437, 0.6584, 0.7680, for scaling parameters respectively equal to 0.10, 0.20, 0.30, 0.40, 0.50, 0.60, 0.70, 0.80, 0.90, with a RMSE ranging from 0.4897 to 0.5139, i.e., the bias is slightly reduced only for $\alpha > 0.40$, but the RMSE is always very high so that the estimates obtained with $m = [T^{0.3}]$ are not reliable. As we will see for other volatility processes, the choice $m = [T^{0.3}]$ does not appear very sensible.

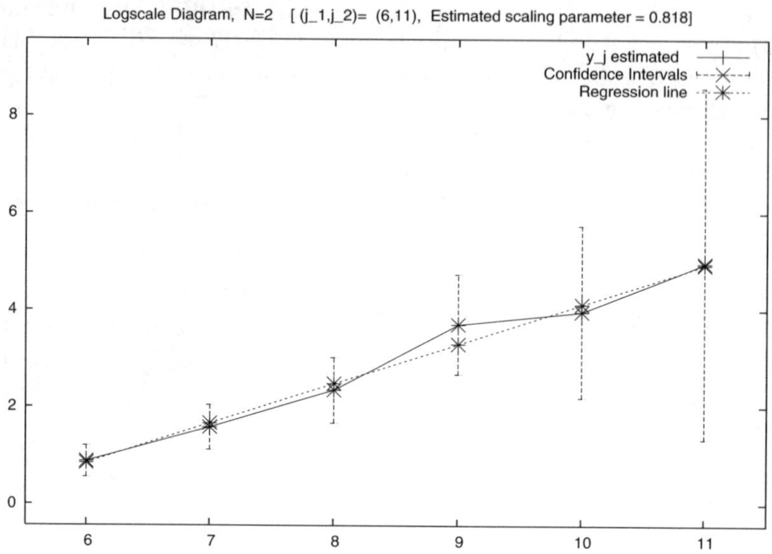

Fig. 7. Logscale diagram for the realization of a LMSV process, with $\alpha = 0.90$. We select here $j_1 = 6$, $j_2 = 11$, $N = 2$

Table 5. Estimation of the scaling parameter for LMSV processes, wavelet estimator. $T = 10000$

α	$\hat{\alpha}_W^{RMSE}$	j_1^{RMSE}	$E\hat{\alpha}_W - \alpha$	RMSE	$\hat{\alpha}_W^{(6)}$	$E\hat{\alpha}_W^{(6)} - \alpha$	RMSE
0.10	0.0183	2	-0.0817	0.0847	0.0213	-0.0787	0.0898
0.20	0.0519	4	-0.1481	0.1559	0.0492	-0.1508	0.1569
0.30	0.1043	5	-0.1957	0.2092	0.0876	-0.2124	0.2170
0.40	0.1872	6	-0.2128	0.2441	0.1872	-0.2128	0.2441
0.50	0.2764	6	-0.2236	0.2536	0.2764	-0.2236	0.2536
0.60	0.3807	6	-0.2193	0.2499	0.3807	-0.2193	0.2499
0.70	0.4959	6	-0.2051	0.2367	0.4959	-0.2051	0.2367
0.80	0.6177	6	-0.1823	0.2184	0.6177	-0.1823	0.2184
0.90	0.7423	6	-0.1577	0.1986	0.7423	-0.1577	0.1986

Remark 7. The results reported here for the wavelet estimator $\hat{\alpha}_W^{RMSE}$ are obtained with $N = 2$. When N increases, both bias and RMSE (slightly) increase, but the octave which minimizes the RMSE does not depend on the choice of N.

Table 6. Estimation of the scaling parameter for LMSV processes, LW and LP estimator. $T = 10000$. The optimal bandwidths m_{LP}^{opt} and m_{LW}^{opt} are respectively used for the LP and LW estimators

α	$\hat{\alpha}_{LP}^{opt}$	$E\hat{\alpha}_{LP}^{opt} - \alpha$	RMSE	$\hat{\alpha}_{LW}^{opt}$	$E\hat{\alpha}_{LW}^{opt} - \alpha$	RMSE
0.10	0.0153	-0.0847	0.1102	0.0186	-0.0814	0.0927
0.20	0.0516	-0.1484	0.1644	0.0508	-0.1492	0.1557
0.30	0.1049	-0.1951	0.2075	0.0971	-0.2029	0.2082
0.40	0.1764	-0.2236	0.2346	0.1622	-0.2378	0.2437
0.50	0.2656	-0.2344	0.2453	0.2525	-0.2475	0.2555
0.60	0.3715	-0.2285	0.2405	0.3679	-0.2321	0.2433
0.70	0.4914	-0.2085	0.2234	0.4978	-0.2022	0.2169
0.80	0.6198	-0.1802	0.1991	0.6313	-0.1687	0.1876
0.90	0.7511	-0.1489	0.1732	0.7634	-0.1366	0.1611

4.3 The Nonlinear Moving Average Process

Full Whittle estimation, i.e., with an exact specification of the spectrum, of the parameters for the Nonlinear Moving Average (NLMA) process has been considered by Robinson and Zaffaroni (1997) and studied by Zaffaroni (2003). Due to the incomplete specification used for the local Whittle estimator, see equation (1), we expect the bias for the LW estimator to be greater.

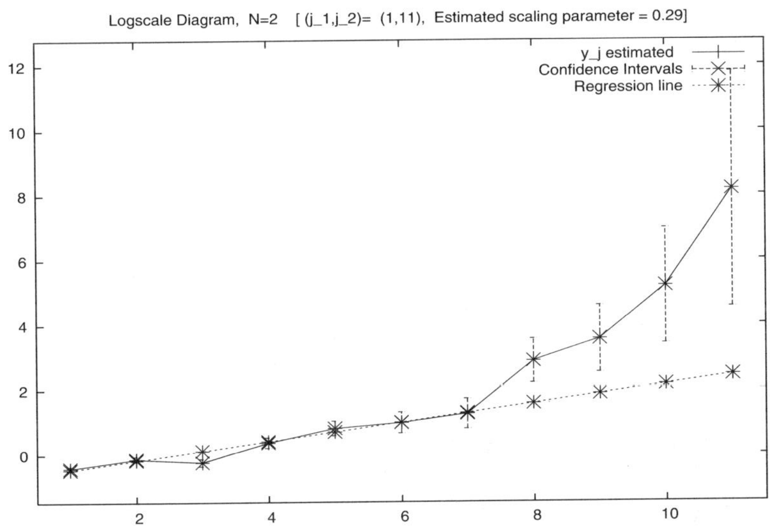

Fig. 8. Logscale diagram for the realization of a NLMA process, with $\alpha = 0.90$. We select here $j_1 = 1$, $j_2 = 11$, $N = 2$

Figures 8 and 9 display the logscale diagrams for the realization of a NLMA process. As for the LMSV process, the long–memory behavior is captured by the octaves $j \geq j_1 = 6$.

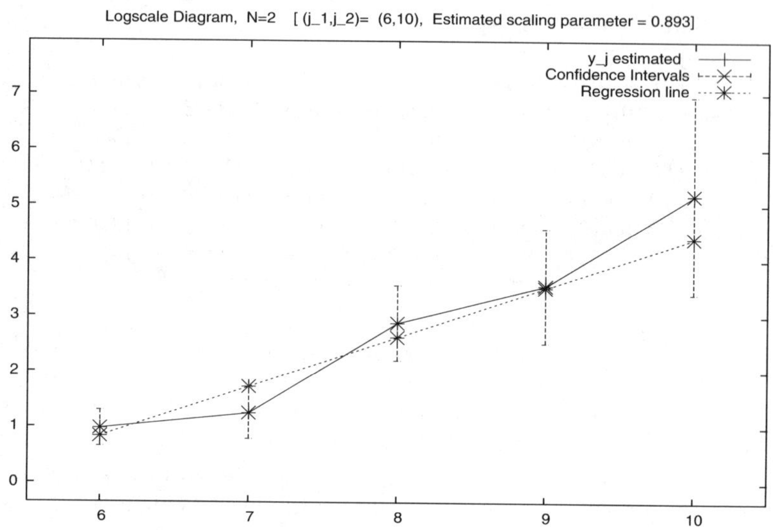

Fig. 9. Logscale diagram for the realization of a NLMA process, with $\alpha = 0.90$. We select here $j_1 = 6$, $j_2 = 10$, $N = 2$

Table 7. Estimation of the scaling parameter for NLMA processes, wavelet estimator. $T = 10000$, $N = 2$

α	$\hat{\alpha}_W^{RMSE}$	j_1^{RMSE}	$E\hat{\alpha}_W^{RMSE} - \alpha$	RMSE	$\hat{\alpha}_W^{(6)}$	$E\hat{\alpha}_W^{(6)} - \alpha$	RMSE
0.10	0.2717	1	0.1717	0.1752	0.3120	0.2120	0.2529
0.20	0.3187	1	0.1187	0.1263	0.2936	0.0936	0.1807
0.30	0.3294	1	0.0294	0.0542	0.2684	0.0316	0.1692
0.40	0.3354	1	-0.0646	0.0793	0.2761	-0.1239	0.2132
0.50	0.3637	2	-0.1363	0.1487	0.3164	-0.1836	0.2565
0.60	0.4028	3	-0.1972	0.2106	0.3816	-0.2184	0.2858
0.70	0.4636	4	-0.2364	0.2554	0.4633	-0.2364	0.3033
0.80	0.5436	5	-0.2564	0.2891	0.5540	-0.2460	0.3130
0.90	0.6205	5	-0.2795	0.3094	0.6469	-0.2531	0.3197

As can be seen from Tables 3 and 8, the wavelet estimator performs slightly better than both the LW and LP estimators, as its bias and RMSE are on average smaller. The wavelet estimator with j_1 selected using equation (60), with either m_{LW}^{opt} or m_{LP}^{opt}, gives results close to the LW estimator and the wavelet estimator with $j_1 = j_1^{RMSE}$. Again in this case, equation (60) might serve as a rule of thumb for selecting j_1.

Table 8. Estimation of the scaling parameter for NLMA processes, local Whittle and log–periodogram estimators. $T = 10000$. The optimal bandwidths m_{LP}^{opt} and m_{LW}^{opt} are respectively used for the LP and LW estimators

α	$\hat{\alpha}_{LP}^{opt}$	$E\hat{\alpha}_{LP}^{opt} - \alpha$	RMSE	$\hat{\alpha}_{LW}^{opt}$	$E\hat{\alpha}_{LW}^{opt} - \alpha$	RMSE
0.10	0.3028	0.2028	0.2124	0.3025	0.2025	0.2109
0.20	0.2899	0.0899	0.1157	0.2892	0.0892	0.1143
0.30	0.2689	-0.0311	0.0942	0.2679	-0.0321	0.0932
0.40	0.2732	-0.1268	0.1620	0.2731	-0.1269	0.1620
0.50	0.3092	-0.1908	0.2204	0.3091	-0.1909	0.2205
0.60	0.3744	-0.2256	0.2556	0.3714	-0.2286	0.2574
0.70	0.4557	-0.2443	0.2756	0.4537	-0.2463	0.2766
0.80	0.5470	-0.2530	0.2845	0.5469	-0.2531	0.2847
0.90	0.6458	-0.2542	0.2895	0.6456	-0.2544	0.2872

When using the standard bandwidth $m = [T^{0.3}]$, the mean of the $\hat{\alpha}_{LP}$ estimates is equal to 0.3024, 0.3161, 0.3170, 0.3350, 0.3672, 0.4071, 0.4483, 0.4847, 0.5105, with a RMSE ranging from 0.0563 to 0.3927. The advantage provided by the optimal bandwidth m_{LP}^{opt} is not as obvious as for the LMSV model, but in the absence of the knowledge on the true Data Generating Process (DGP), using this optimal bandwidth does not yield any significant loss for the estimation of the parameter.

4.4 Nonlinear Transformations of Fractionally Integrated Processes

We also consider some nonlinear transformations of FI(d) processes, i.e., processes $\{Y_t\}$ defined as $Y_t = G(X_t)$, where $X_t \sim$ FI(d), and $G(\cdot)$ can be written as a sum of Hermite polynomials. As in Dittman and Granger (2002), we choose the transformations $G(x) = x^2$, $G(x) = x^3$, $G(x) = x^4$, $G(x) = x^3 - 3x$, and $G(x) = x^4 - 6x^2$, the Hermite rank of which are respectively equal to $R = 2, 1, 2, 3, 4$. We also consider some trigonometric and exponential transformations: $G(x) = \sin(x)$, $G(x) = \cos(x)$, $G(x) = \exp(x)$, and $G(x) = (1 + \exp(-x))^{-1}$. The Hermite rank of these transformations are all equal to 1, except for the cosine transformation, the Hermite rank of which is $R = 2$.

Granger and Dittman (2002) report simulation results for smaller sample sizes and the LP estimator with bandwidth $m = [T]^{4/5}$. The use of the optimal bandwidth m_{LP}^{opt} results for some transformations in a marginal increase of the bias and RMSE, but not very significant, so that we can reliably stick to that optimal bandwidth. We observe that for all nonlinear transformations, the estimators $\hat{\alpha}_{LW}$ and $\hat{\alpha}_{LP}$ give very close results.

The FI(d) processes have been generated using the Durbin–Levinson algorithm. Consider the first series of transformations, see Tables 4 and 10: the wavelet estimator performs generally better than the LP and LW estimators, i.e., its bias and RMSE are most of the times lower than those of the LP and LW estimators, the advantage of the wavelet estimator being obvious for the largest values of $\tilde{\alpha}$. For the highly nonlinear transformations, e.g., $G(x) = x^3 - 3x$ and $G(x) = x^4 - 6x^2$, the bias is huge for all estimators for the lowest values of $\tilde{\alpha}$, and choosing a higher value for j_1, even for samples of size $T = 20,000$, does not yield any significant improvement. Estimation results with j_1 computed from equation (60) are not as good as the ones with either j_1^{RMSE} or the LP and LW estimators, but might be considered as informative for a first analysis.

Remark 8. When N increases, the wavelet estimates do not differ too much in terms of bias and RMSE, except for the case $j_1 = 7$ and $N = 6$ for which the RMSE is very large.

Figures 10 to 18 below display the logscale diagrams for the 9 transformations $G(x)$ considered here.

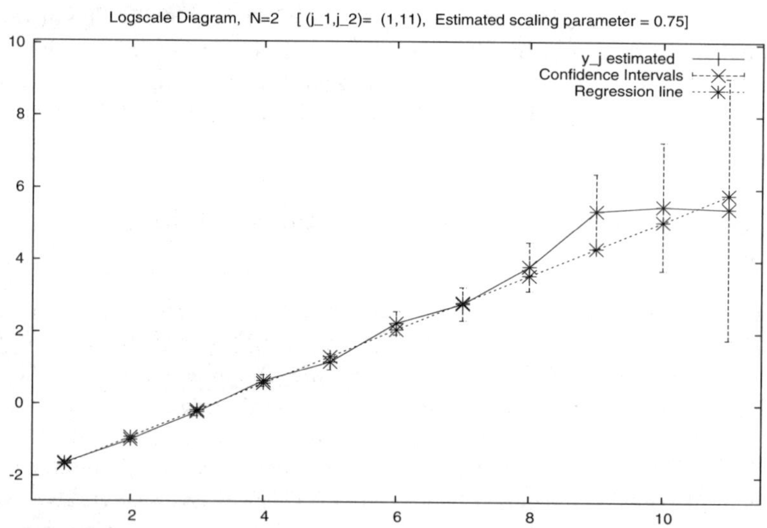

Fig. 10. Logscale diagram for the realization of a nonlinear transformation of a FI(d) process, with $G(x) = x^2$, $d = 0.45$ then $\alpha = 0.80$, $T = 10000$. We select here $j_1 = 1$, $j_2 = 11$, $N = 2$

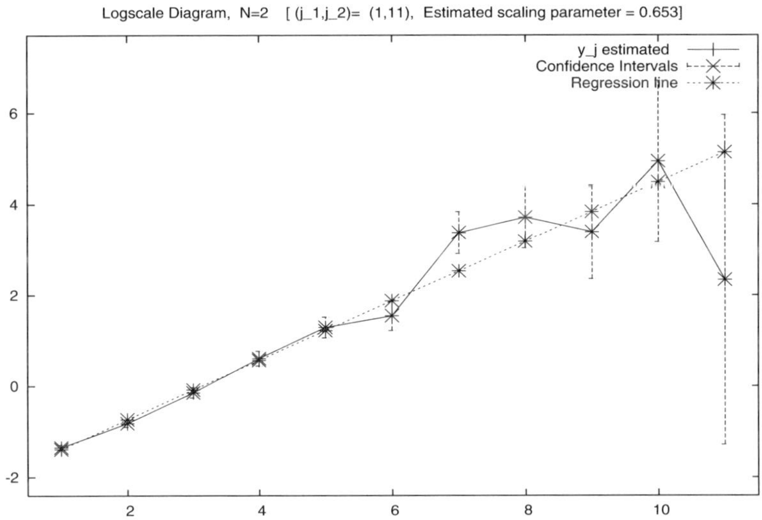

Fig. 11. Logscale diagram for the realization of a nonlinear transformation of a FI(d) process, with $G(x) = x^3$, $d = 0.45$ then $\alpha = 0.90$, $T = 10000$. We select here $j_1 = 1$, $j_2 = 11$, $N = 2$

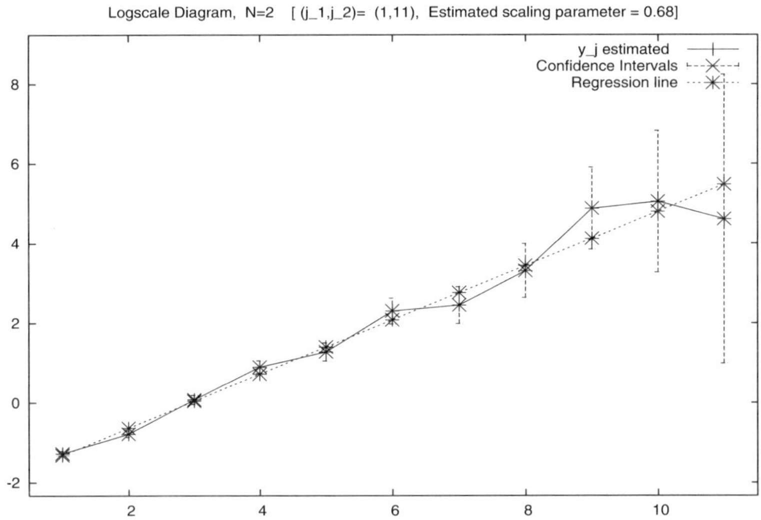

Fig. 12. Logscale diagram for the realization of a nonlinear transformation of a FI(d) process, with $G(x) = x^4$, $d = 0.45$ then $\alpha = 0.80$, $T = 10000$. We select here $j_1 = 1$, $j_2 = 11$, $N = 2$

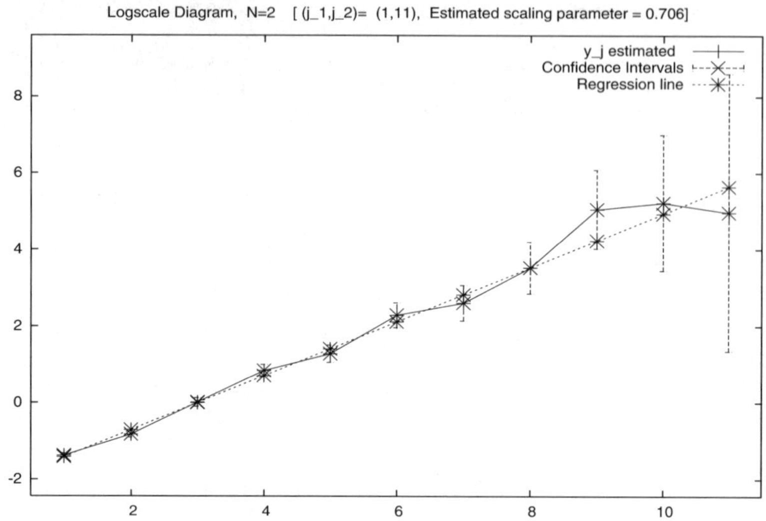

Fig. 13. Logscale diagram for the realization of a nonlinear transformation of a FI(d) process, with $G(x) = x^3 - 3x$, $d = 0.45$ then $\alpha = 0.80$, $T = 10000$. We select here $j_1 = 1$, $j_2 = 11$, $N = 2$

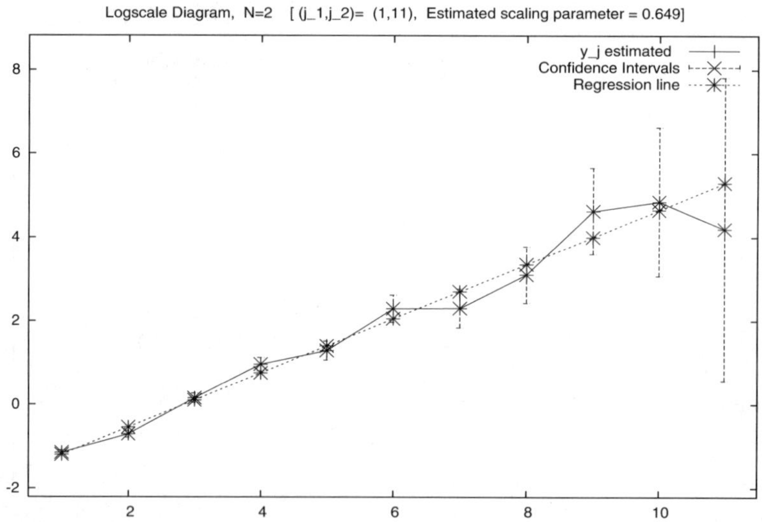

Fig. 14. Logscale diagram for the realization of a nonlinear transformation of a FI(d) process, with $G(x) = x^4 - 6x$, $d = 0.45$ then $\alpha = 0.60$, $T = 10000$. We select here $j_1 = 1$, $j_2 = 11$, $N = 2$

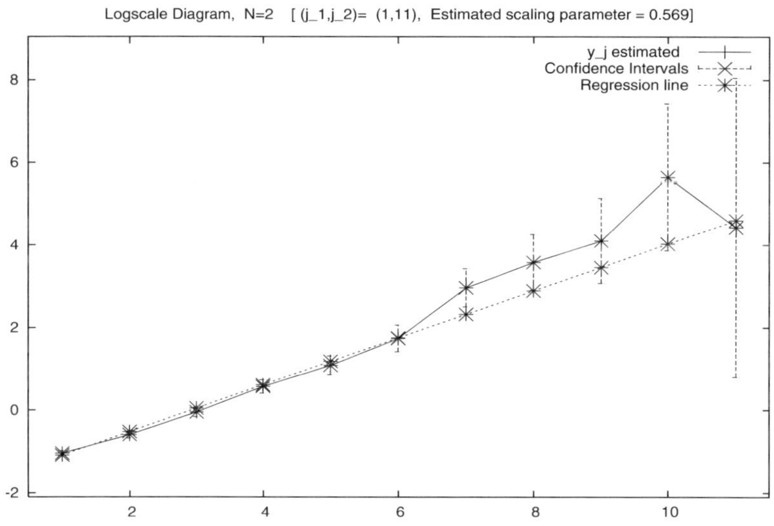

Fig. 15. Logscale diagram for the realization of a nonlinear transformation of a FI(d) process, with $G(x) = \sin(x)$, $d = 0.45$ then $\alpha = 0.90$, $T = 10000$. We select here $j_1 = 1$, $j_2 = 11$, $N = 2$

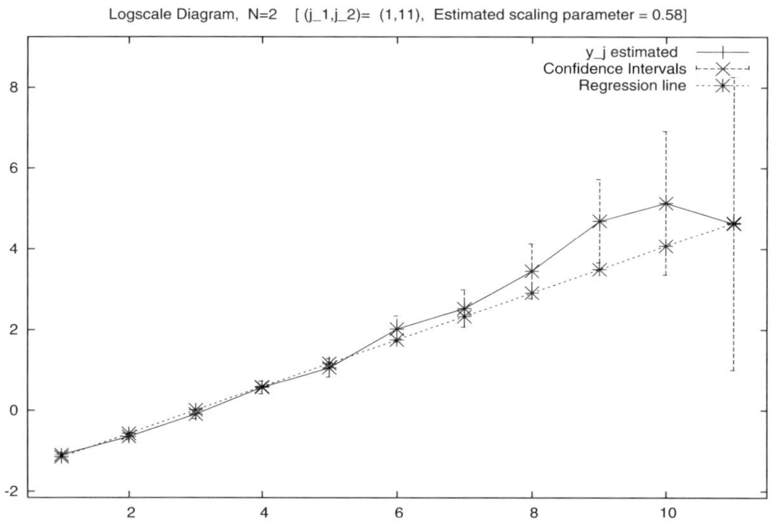

Fig. 16. Logscale diagram for the realization of a nonlinear transformation of a FI(d) process, with $G(x) = \cos(x)$, $d = 0.45$ then $\alpha = 0.80$, $T - 10000$. We select here $j_1 = 1$, $j_2 = 11$, $N = 2$

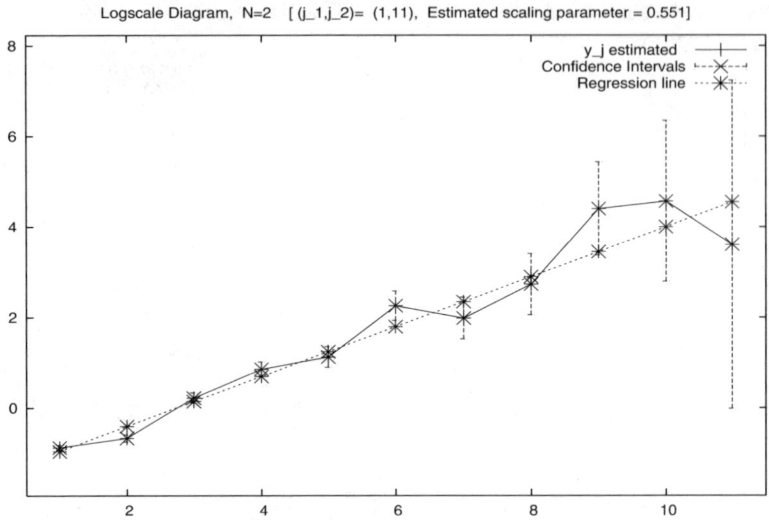

Fig. 17. Logscale diagram for the realization of a nonlinear transformation of a FI(d) process, with $G(x) = \exp(x)$, $d = 0.45$ then $\alpha = 0.90$, $T = 10000$. We select here $j_1 = 1$, $j_2 = 11$, $N = 2$

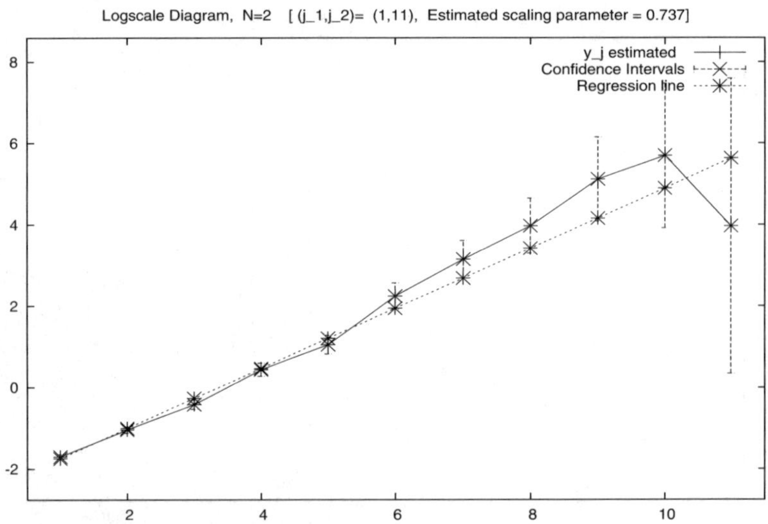

Fig. 18. Logscale diagram for the realization of a nonlinear transformation of a FI(d) process, with $G(x) = (1 + \exp(-x))^{-1}$, $d = 0.45$ then $\alpha = 0.90$, $T = 10000$. We select here $j_1 = 1$, $j_2 = 11$, $N = 2$

Table 9. Estimation of the scaling parameter for nonlinear transformations of FI(d) processes, with $\alpha = 2\tilde{d}$ and $\tilde{d} = \max\{0, (d-0.5)R + 0.5\}$. $N = 2$, $T = 10000$

$G(x)$	α	$\hat{\alpha}_W^{RMSE}$	j_1^{RMSE}	$E\hat{\alpha}_W^{RMSE} - \alpha$	RMSE	$\hat{\alpha}_W^m$	$E\hat{\alpha}_W^m - \alpha$	RMSE
x^2	0.00	0.1176	5	0.1176	0.1488	0.1403	0.1403	0.1527
$(R=2)$	0.20	0.2627	1	0.0627	0.0705	0.2449	0.0449	0.0919
	0.40	0.3964	1	-0.0036	0.0578	0.3786	-0.0214	0.1136
	0.60	0.5769	2	-0.0231	0.0861	0.5396	-0.0604	0.1530
	0.80	0.7447	2	-0.0553	0.0888	0.7145	-0.0855	0.1935
x^3	0.10	0.0608	2	-0.0392	0.0464	0.0653	-0.0347	0.0599
$(R=1)$	0.20	0.1341	3	-0.0659	0.0810	0.1385	-0.0615	0.0790
	0.30	0.2224	4	-0.0775	0.0948	0.2122	-0.0808	0.0962
	0.40	0.3087	4	-0.0913	0.1066	0.3062	-0.0938	0.1097
	0.50	0.3966	4	-0.1034	0.1178	0.3974	-0.1026	0.1213
	0.60	0.4841	4	-0.1159	0.1304	0.4882	-0.1118	0.1335
	0.70	0.5705	4	-0.1295	0.1454	0.5759	-0.1241	0.1493
	0.80	0.6596	4	-0.1404	0.1596	0.6640	-0.1360	0.1712
	0.90	0.7564	3	-0.1436	0.1587	0.7589	-0.1411	0.1971
x^4	0.00	0.1272	1	0.1272	0.1348	0.1129	0.1129	0.1374
$(R=2)$	0.20	0.2094	1	0.0094	0.0522	0.2047	0.0047	0.0961
	0.40	0.3392	2	-0.0608	0.0962	0.3222	-0.0778	0.1466
	0.60	0.4796	2	-0.1204	0.1469	0.4591	-0.1409	0.2143
	0.80	0.6105	2	-0.1600	0.1839	0.6097	-0.1903	0.2803
$x^3 - 3x$	0.00	0.1483	1	0.1483	0.1557	0.1633	0.1633	0.1863
$(R=3)$	0.10	0.2781	1	0.1781	0.1878	0.3188	0.2188	0.2474
	0.40	0.4368	1	0.0368	0.0782	0.4884	0.0884	0.1646
	0.70	0.6703	2	-0.0297	0.0929	0.6651	-0.0349	0.1736
$x^4 - 6x^2$	0.00	0.1738	1	0.1738	0.1949	0.1676	0.1676	0.2268
$(R=4)$	0.20	0.3333	1	0.1333	0.1664	0.3327	0.1327	0.2341
	0.60	0.5350	1	-0.0650	0.1197	0.5270	-0.0730	0.2459

Table 10. Estimation of the scaling parameter for nonlinear transformations of FI(d) processes, with $\alpha = 2\tilde{d}$ and $\tilde{d} = \max\{0, (d - 0.5)R + 0.5\}$. $T = 10000$. The optimal bandwidths m_{LP}^{opt} and m_{LW}^{opt} are respectively used for the LP and LW estimators

$G(x)$	α	$\hat{\alpha}_{LP}^{opt}$	$E\hat{\alpha}_{LP}^{opt} - \alpha$	RMSE	$\hat{\alpha}_{LW}^{opt}$	$E\hat{\alpha}_{LW}^{opt} - \alpha$	RMSE
x^2	0.00	0.1193	0.1193	0.1371	0.1192	0.1192	0.1369
$(R = 2)$	0.20	0.2248	0.0248	0.0975	0.2225	0.0225	0.0960
	0.40	0.3644	-0.0356	0.1315	0.3599	-0.0401	0.1302
	0.60	0.5223	-0.0777	0.1735	0.5190	-0.0810	0.1736
	0.80	0.6953	-0.1047	0.2133	0.6867	-0.1132	0.2149
x^3	0.10	0.0638	-0.0362	0.0580	0.0636	-0.0364	0.0576
$(R = 1)$	0.20	0.1410	-0.0590	0.0754	0.1409	-0.0591	0.0750
	0.30	0.2277	-0.0723	0.0882	0.2276	-0.0724	0.0881
	0.40	0.3217	-0.0783	0.0964	0.3216	-0.0784	0.0964
	0.50	0.4184	-0.0816	0.1029	0.4185	-0.0815	0.1028
	0.60	0.5134	-0.0866	0.1099	0.5134	-0.0866	0.1099
	0.70	0.6019	-0.0981	0.1261	0.6020	-0.0980	0.1245
	0.80	0.6877	-0.1123	0.1495	0.6866	-0.1134	0.1446
	0.90	0.7755	-0.1245	0.1755	0.7728	-0.1272	0.1699
x^4	0.00	0.0981	0.0981	0.1180	0.0978	0.0978	0.1176
$(R = 2)$	0.20	0.1934	-0.0066	0.0936	0.1906	-0.0094	0.0910
	0.40	0.3158	-0.0842	0.1474	0.3157	-0.0843	0.1475
	0.60	0.4609	-0.1391	0.2043	0.4593	-0.1407	0.2048
	0.80	0.6185	-0.1815	0.2566	0.6105	-0.1895	0.2568
$x^3 - 3x$	0.00	0.1856	0.1856	0.2029	0.1850	0.1850	0.2022
$(R = 3)$	0.10	0.3542	0.2542	0.2759	0.3533	0.2533	0.2743
	0.40	0.5286	0.1286	0.1789	0.5259	0.1259	0.1736
	0.70	0.6893	-0.0107	0.1423	0.6866	-0.0134	0.1377
$x^4 - 6x^2$	0.00	0.1736	0.1736	0.2114	0.1732	0.1732	0.2107
$(R = 4)$	0.20	0.3506	0.1506	0.2204	0.3502	0.1502	0.2195
	0.60	0.5487	-0.0513	0.1992	0.5435	-0.0565	0.1934

For the second series of transformations, see Tables 11 and 12, the wavelet estimator performs better than the two other estimators, except for the exponential transformation.

Table 11. Estimation of the scaling parameter for nonlinear transformations of FI(d) processes, with $\alpha = 2\tilde{d}$ and $\tilde{d} = \max\{0, (d-0.5)R+0.5\}$. $N = 2$, $T = 10000$

$G(x)$	α	$\hat{\alpha}_W^{RMSE}$	j_1^{RMSE}	$E\hat{\alpha}_W^{RMSE} - \alpha$	RMSE	$\hat{\alpha}_W^m$	$E\hat{\alpha}_W^m - \alpha$	RMSE
$\sin(x)$	0.10	0.0822	2	-0.0178	0.0352	0.0871	-0.0129	0.0450
$(R=1)$	0.20	0.1694	2	-0.0306	0.0382	0.1777	-0.0223	0.0492
	0.30	0.2652	3	-0.0348	0.0474	0.2704	-0.0296	0.0544
	0.40	0.3556	3	-0.0444	0.0553	0.3642	-0.0358	0.0620
	0.50	0.4577	4	-0.0423	0.0639	0.4585	-0.0415	0.0711
	0.60	0.5413	4	-0.0587	0.0761	0.5496	-0.0504	0.0824
	0.70	0.6353	5	-0.0647	0.1013	0.6235	-0.0765	0.1051
	0.80	0.6648	5	-0.1352	0.1671	0.6561	-0.1439	0.1716
	0.90	0.6606	5	-0.2394	0.2606	0.6556	-0.2444	0.2654
$\cos(x)$	0.00	0.1505	1	0.1505	0.1520	0.1310	0.1310	0.1421
$(R=2)$	0.20	0.2394	1	0.0394	0.0473	0.2333	0.0333	0.0779
	0.40	0.3749	2	-0.0251	0.0519	0.3644	-0.0356	0.1015
	0.60	0.5192	3	-0.0808	0.1043	0.5177	-0.0823	0.1389
	0.80	0.6407	4	-0.1593	0.1751	0.6435	-0.1565	0.1890
$\exp(x)$	0.10	0.0628	3	-0.0372	0.0598	0.0643	-0.0357	0.0648
$(R=1)$	0.20	0.1326	3	-0.0674	0.0826	0.1361	-0.0639	0.0851
	0.30	0.2169	4	-0.0831	0.1035	0.2152	-0.0848	0.1034
	0.40	0.2986	4	-0.1014	0.1202	0.2967	-0.1033	0.1221
	0.50	0.3798	4	-0.1202	0.1389	0.3807	-0.1193	0.1414
	0.60	0.4573	4	-0.1427	0.1626	0.4599	-0.1401	0.1664
	0.70	0.5276	4	-0.1724	0.1953	0.5320	-0.1680	0.2005
	0.80	0.5868	4	-0.2132	0.2420	0.5885	-0.2114	0.2557
	0.90	0.6260	3	-0.2740	0.3022	0.6265	-0.2735	0.3384
$(1+e^{-x})^{-1}$	0.10	0.0871	1	-0.0129	0.0202	0.0980	-0.0020	0.0429
$(R=1)$	0.20	0.1905	2	-0.0095	0.0245	0.1964	-0.0036	0.0439
	0.30	0.2879	2	-0.0121	0.0260	0.2945	-0.0055	0.0464
	0.40	0.3842	2	-0.0158	0.0277	0.3935	-0.0065	0.0531
	0.50	0.4799	2	-0.0201	0.0307	0.4946	-0.0054	0.0614
	0.60	0.5878	3	-0.0121	0.0349	0.5961	-0.0039	0.0697
	0.70	0.6848	3	-0.0152	0.0363	0.6951	-0.0049	0.0816
	0.80	0.7796	3	-0.0204	0.0395	0.7925	-0.0075	0.0958
	0.90	0.8669	3	-0.0331	0.0508	0.8796	-0.0204	0.1196

Table 12. Estimation of the scaling parameter for nonlinear transformations of FI(d) processes, with $\alpha = 2\tilde{d}$ and $\tilde{d} = \max\{0, (d-0.5)R+0.5\}$. $T = 10000$. The optimal bandwidths m_{LP}^{opt} and m_{LW}^{opt} are respectively used for the LP and LW estimators

$G(x)$	α	$\hat{\alpha}_{LP}^{opt}$	$E\hat{\alpha}_{LP}^{opt} - \alpha$	RMSE	$\hat{\alpha}_{LW}^{opt}$	$E\hat{\alpha}_{LW}^{opt} - \alpha$	RMSE
$\sin(x)$	0.10	0.0855	-0.0145	0.0465	0.0856	-0.0144	0.0461
$(R=1)$	0.20	0.1779	-0.0221	0.0518	0.1779	-0.0221	0.0517
	0.30	0.2736	-0.0264	0.0581	0.2736	-0.0264	0.0579
	0.40	0.3709	-0.0291	0.0649	0.3710	-0.0290	0.0647
	0.50	0.4689	-0.0311	0.0728	0.4697	-0.0303	0.0712
	0.60	0.5618	-0.0382	0.0795	0.5620	-0.0380	0.0791
	0.70	0.6417	-0.0583	0.0972	0.6425	-0.0575	0.0950
	0.80	0.6788	-0.1212	0.1582	0.6793	-0.1207	0.1569
	0.90	0.6587	-0.2413	0.2735	0.6603	-0.2397	0.2704
$\cos(x)$	0.00	0.1135	0.1135	0.1303	0.1129	0.1129	0.1296
$(R=2)$	0.20	0.2147	0.0147	0.0883	0.2141	0.0141	0.0881
	0.40	0.3512	-0.0488	0.1243	0.3510	-0.0490	0.1240
	0.60	0.5121	-0.0879	0.1633	0.5108	-0.0892	0.1622
	0.80	0.6441	-0.1559	0.2050	0.6450	-0.1550	0.2026
$\exp(x)$	0.10	0.0617	-0.0383	0.0592	0.0615	-0.0385	0.0588
$(R=1)$	0.20	0.1370	-0.0630	0.0787	0.1369	-0.0031	0.1194
	0.30	0.2203	-0.0797	0.0952	0.2204	-0.0796	0.0951
	0.40	0.3085	-0.0915	0.1089	0.3086	-0.0914	0.1087
	0.50	0.3978	-0.1022	0.1226	0.3979	-0.1021	0.1223
	0.60	0.4818	-0.1182	0.1442	0.4821	-0.1179	0.1406
	0.70	0.5587	-0.1413	0.1671	0.5590	-0.1410	0.1666
	0.80	0.6206	-0.1794	0.2136	0.6201	-0.1799	0.2092
	0.90	0.6627	-0.2273	0.2743	0.6618	-0.2382	0.2727
$(1+\exp(-x))^{-1}$	0.10	0.0962	-0.0038	0.0441	0.0961	-0.0039	0.0439
$(R=1)$	0.20	0.1943	-0.0057	0.0482	0.1942	-0.0058	0.0475
	0.30	0.2922	-0.0078	0.0529	0.2923	-0.0077	0.0529
	0.40	0.3913	-0.0087	0.0636	0.3915	-0.0085	0.0601
	0.50	0.4904	-0.0096	0.0686	0.4905	-0.0095	0.0684
	0.60	0.5885	-0.0115	0.0822	0.5880	-0.0120	0.0766
	0.70	0.6834	-0.0166	0.0903	0.6833	-0.0167	0.0902
	0.80	0.7766	-0.0234	0.1053	0.7764	-0.0236	0.1041
	0.90	0.8614	-0.0386	0.1350	0.8610	-0.0390	0.1245

5 Robustness: Long–Memory Versus Non–Stationarity

This section aims at contributing to the issue of disentangling long–memory and non–stationarity. This can be addressed through two major questions. Can one estimate correctly the long–memory parameter of an actual long–range dependent process when non–stationarities are superimposed to it? Can one figure out that a given time series is non–stationary and has no long–memory? The first question is illustrated in Sections 5.1, 5.2 and 5.4 where trends, change points or both are respectively superimposed to long–memory processes. The second one is considered in Section 5.3 where change points in a short–range memory process may be confused with long–memory.

5.1 Long–Memory Plus Trends: Polynomial Trends Superimposed to Fractionally Integrated Processes

As mentioned earlier, the wavelets coefficients $d_Y(j,\cdot)$ are the same for the processes $\{Y_t\}$ and $\{Y_t + P_t\}$ where P_t is a polynomial of order at most $N-1$, where N is the number of vanishing moments of the mother wavelet ψ_0. Thus, with an appropriate choice for N, we can discriminate between trended and long–range dependent processes. This separation is also possible with Fourier based methods, when replacing the periodogram by a tapered version of it, see e.g., Lobato and Velasco (2000).

We consider here the accuracy of the wavelet based estimator, as we will use this estimator in section 6.2 for estimating the dependence of financial time series with a trend, i.e., transactions volume. We consider the following process

$$Y_t = X_t + T_t, \quad X_t \sim \mathrm{FI}(d), \quad T_t = \sum_{l=0}^{q} \xi_l t^l, \tag{63}$$

which additively mixes a long–range dependent process and a polynomial trend. We consider $q = 1, 2, 3$. Table 13 below reports simulation results for $N = q + 1$ only.

For the relevant choices of N, i.e., those for which $N > q$, the wavelet estimates are the same as those for the FARIMA process without polynomial trend, since we used the same sequence of pseudo–error terms for all the FI(d) process.

Remark 9. In this section, we do not compare the performance of the wavelet estimator with the one of the LP and LW estimators. For the LW estimator, we would have to consider the tapered periodogram, as in Lobato and Velasco (2000), and very likely perform a similar transformation for the LP estimator.

Table 13. Estimation of the scaling parameter for an additive combination of a FI(d) processes and a polynomial trend of order q. $T = 10000$

Order	α	$\hat{\alpha}_W^{RMSE}$	j_1^{RMSE}	$E\hat{\alpha}_W^{RMSE} - \alpha$	RMSE	$\hat{\alpha}_W^{(5)}$	$E\hat{\alpha}_W^{(5)} - \alpha$	RMSE
$q = 1$	0.10	0.0860	2	-0.0140	0.0205	0.1016	0.0016	0.0744
(N=2)	0.20	0.1907	2	-0.0093	0.0235	0.2017	0.0017	0.0745
	0.30	0.2868	2	-0.0132	0.0254	0.3018	0.0018	0.0747
	0.40	0.3833	2	-0.0167	0.0275	0.4020	0.0020	0.0748
	0.50	0.4800	2	-0.0200	0.0297	0.5022	0.0022	0.0751
	0.60	0.5771	2	-0.0229	0.0318	0.6023	0.0023	0.0753
	0.70	0.6934	3	-0.0066	0.0338	0.7025	0.0025	0.0757
	0.80	0.7929	3	-0.0713	0.0341	0.8026	0.0026	0.0761
	0.90	0.8924	3	-0.0076	0.0345	0.9027	0.0027	0.0766
$q = 2$	0.10	0.0875	2	-0.0125	0.0196	0.1022	0.0022	0.0771
(N=3)	0.20	0.1930	2	-0.0070	0.0234	0.2023	0.0023	0.0772
	0.30	0.2899	2	-0.0101	0.0246	0.3023	0.0023	0.0773
	0.40	0.3870	2	-0.0130	0.0259	0.4024	0.0024	0.0776
	0.50	0.4842	2	-0.0158	0.0275	0.5024	0.0024	0.0779
	0.60	0.5817	2	-0.0183	0.0291	0.6024	0.0024	0.0783
	0.70	0.6792	2	-0.0208	0.0308	0.7042	0.0042	0.0788
	0.80	0.7769	2	-0.0231	0.0325	0.8025	0.0025	0.0794
	0.90	0.8747	2	-0.0253	0.0342	0.9026	0.0026	0.0800
$q = 3$	0.10	0.0880	1	-0.0120	0.0194	0.1030	0.0030	0.0817
(N=4)	0.20	0.1937	2	-0.0063	0.0237	0.2030	0.0030	0.0817
	0.30	0.2909	2	-0.0091	0.0247	0.3031	0.0031	0.0817
	0.40	0.3881	2	-0.0119	0.0259	0.4031	0.0031	0.0818
	0.50	0.4855	2	-0.0145	0.0272	0.5030	0.0030	0.0819
	0.60	0.5830	2	-0.0170	0.0287	0.6030	0.0030	0.0820
	0.70	0.6806	2	-0.0194	0.0303	0.7029	0.0029	0.0823
	0.80	0.7783	2	-0.0217	0.0319	0.8028	0.0028	0.0826
	0.90	0.8761	2	-0.0239	0.0335	0.9027	0.0027	0.0830

5.2 Long–Memory Plus Trends Plus Change–Points: Change–Points in Fractionally Integrated Processes

Since economic processes are subject to changes in regime, we consider the extreme case of the following change–point process

$$\begin{aligned} Y_t &= X_t, \quad X_t \sim \mathrm{FI}(d), \quad t \leqslant k, \\ Y_t &= X_t + T_t, \quad X_t \sim \mathrm{FI}(d), \quad T_t = \sum_{l=0}^{q} \xi_l (t-k+1)^l, \quad t > k, \end{aligned} \tag{64}$$

i.e., a process with a broken polynomial trend. In some sense, here two types of difficulties are mixed up together. The non–stationarity superimposed to long–memory results both from trends and change points. However, in that particular extreme case, a visual inspection of the series will suggest to split the series at the break points.

Remark 10. These results reported here obviously hold provided that the polynomial trend series are of the same order of magnitude as the series X_t.

Table 14. Estimation of the scaling parameter for an additive combination of a FI(d) processes and with a broken polynomial trend of order $q=1$, $q=2$ and $q=3$, defined by equation (64). Columns 3 to 6 report estimation results for $k=[T/10]$, while columns 7 to 10 report estimation results for $k=[T/2]$, $T=10000$

N	α	$\hat{\alpha}_W^{RMSE}$	j_1^{RMSE}	$E\hat{\alpha}_W^{RMSE}-\alpha$	RMSE	$\hat{\alpha}_W^{RMSE}$	j_1^{RMSE}	$E\hat{\alpha}_W^{RMSE}-\alpha$	RMSE
$q=1$	0.10	0.0970	2	-0.0030	0.0232	0.0971	2	-0.0029	0.0232
$N=6$	0.20	0.1942	2	-0.0057	0.0238	0.1943	2	-0.0057	0.0238
	0.30	0.2915	2	-0.0085	0.0246	0.2916	2	-0.0084	0.0246
	0.40	0.3889	2	-0.0111	0.0258	0.3889	2	-0.0111	0.0258
	0.50	0.4863	2	-0.0137	0.0270	0.4863	2	-0.0137	0.0271
	0.60	0.5838	2	-0.0162	0.0285	0.5838	2	-0.0162	0.0285
	0.70	0.6814	2	-0.0186	0.0301	0.6814	2	-0.0186	0.0301
	0.80	0.7790	2	-0.0210	0.0317	0.7790	2	-0.0210	0.0317
	0.90	0.8767	2	-0.0233	0.0334	0.8767	2	-0.0233	0.0334
$q=2$	0.10	0.0915	1	-0.0085	0.0178	0.1018	1	0.0018	0.0159
$N=6$	0.20	0.1981	2	-0.0019	0.0232	0.2046	2	0.0046	0.0243
	0.30	0.2941	2	-0.0059	0.0240	0.2991	2	-0.0009	0.0239
	0.40	0.3906	2	-0.0094	0.0251	0.4050	2	0.0050	0.0239
	0.50	0.4874	2	-0.0126	0.0265	0.4996	2	-0.0004	0.0234
	0.60	0.5846	2	-0.0154	0.0281	0.6023	2	0.0023	0.0233
	0.70	0.6819	2	-0.0181	0.0299	0.6960	2	-0.0040	0.0237
	0.80	0.7793	2	-0.0207	0.0315	0.7903	2	-0.0097	0.0255
	0.90	0.8769	2	-0.0231	0.0332	0.8851	2	-0.0149	0.0281
$q=3$	0.10	0.0996	2	-0.0004	0.0231	0.0981	2	-0.0019	0.0162
$N=6$	0.20	0.1960	2	-0.0040	0.0235	0.2018	2	0.0018	0.0244
	0.30	0.2927	2	-0.0073	0.0243	0.3039	2	0.0039	0.0242
	0.40	0.3897	2	-0.0103	0.0255	0.3990	2	-0.0010	0.0238
	0.50	0.4869	2	-0.0131	0.0268	0.5041	2	0.0041	0.0237
	0.60	0.5842	2	-0.0158	0.0283	0.6037	2	0.0037	0.0235
	0.70	0.6816	2	-0.0184	0.0299	0.6972	2	-0.0028	0.0235
	0.80	0.7792	2	-0.0208	0.0315	0.7913	2	-0.0087	0.0251
	0.90	0.8768	2	-0.0232	0.0333	0.8858	2	-0.0142	0.0276

We note that for this particular process, the value of the statistic Q for $j_1 = j_1^{RMSE}$ does not differ too much for the one when j_1 is given following the arguments in Veitch, Abry, Taqqu (2003) when the number of moments N is adequately chosen, i.e., large enough for fitting all possible variations of the process. Figures 19 and 20 display the logscale diagrams for trended processes with change–points.

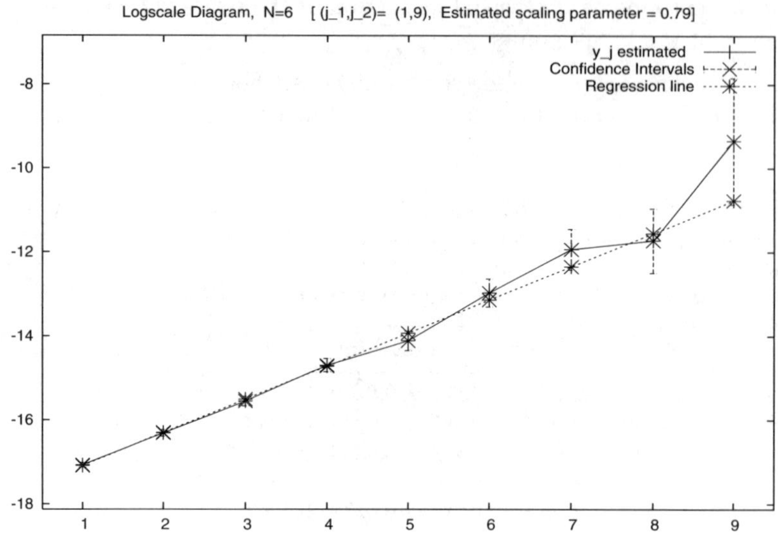

Fig. 19. Logscale diagram for the realization of an additive combination of a FI(d) process ($\alpha = 0.8$) and a broken polynomial trend of order $q = 1$, with $k = [T/10]$. We select here $j_1 = 1$, $j_2 = 9$, $N = 6$

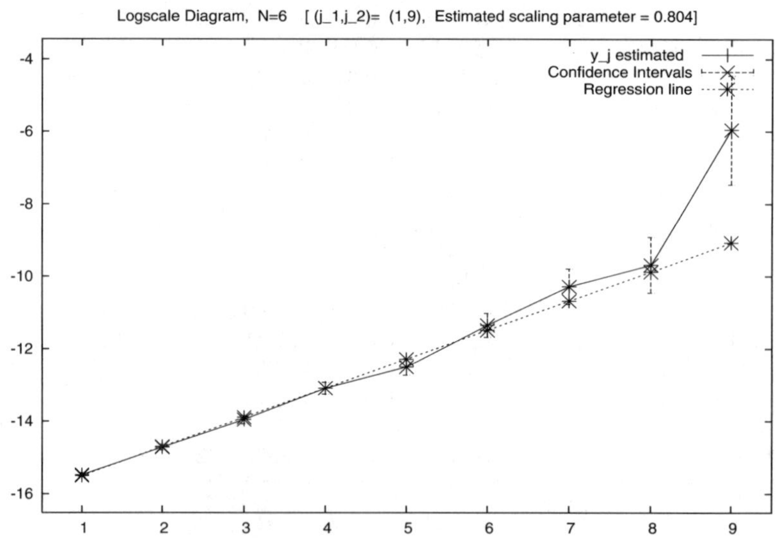

Fig. 20. Logscale diagram for the realization of an additive combination of a FI(d) process ($\alpha = 0.8$) and a broken polynomial trend of order $q = 1$, with $k = [T/2]$. We select here $j_1 = 1$, $j_2 = 11$, $N = 6$

For both cases, provided that N is large enough, $N = 6$ there, the long–memory behavior quite always captured from the lowest octave $j_1 = 2$.

5.3 Short–Memory and Change–Points: Change–Point GARCH Process

Volatility, GARCH Process and Change–Points?

As mentioned in the introduction, a potentially useful application of wavelet methods is the analysis of volatility series, the volatility of asset prices P_t being commonly defined by either the absolute returns $|r_t|$ or the squared returns r_t^2, where the returns r_t are defined as $r_t = \ln(P_t/P_{t-1})$.

Volatility processes are modeled using either GARCH type models, or stochastic volatility models, although GARCH type models are more often used as they are easy to estimate, parsimonious as a simple GARCH(1,1) model with three parameters can fit most of the series of asset returns, and have nice properties for asset pricing, see e.g., Duan (1995) and Kallsen and Taqqu (1998).

A standard GARCH(1,1) process is defined as

$$r_t = \sigma_t \varepsilon_t, \quad \varepsilon_t \sim N(0,1), \quad \sigma_t^2 = \omega + \beta\sigma_{t-1}^2 + \gamma r_{t-1}^2. \tag{65}$$

The literature on the properties of GARCH–type processes is pretty impressive; interested readers are referred to the chapter by Giraitis, Leipus and Surgailis (2005) in this volume.

However, GARCH–type homogeneous processes have several drawbacks and inconsistencies:

1. They cannot fit the heavy tails of the returns r_t,
2. When the sample size increases, the sum of estimated parameters $\hat{\beta} + \hat{\gamma}$ tends to 1, a property called the "IGARCH effect". As a consequence, the variance of the process $\{r_t\}$ is infinite since $\omega > 0$,
3. For large samples, the ACF of the series $|r_t|$ and r_t^2 behave like equation (2), this is the so called long–range dependence in the volatility.

Mikosch and Stărică (1999, 2003, 2004a, 2004b) emphasized in a series of research works that points 2 and 3 are inconsistent, as the ACF of the power transformation of a process with infinite variance is not properly defined. Thus, the GARCH(1,1) model might be the "right" model, provided that it is estimated on the sample for which the parameters of the process are constant, and that IGARCH, long–range dependence and tail effects might be the consequence of the non–stationarity of the GARCH process, and, for instance, of changes in the parameters so that the unconditional variance of the process is not constant. A major issue in practical time series analysis hence consists in being able to decide whether long–memory is truly present in the analyzed data or if a change–point type non–stationary property actually exists and is likely to be misinterpreted as long–memory.

Discriminating Between True Long–Memory and Change–Points?

There is a large number of research works dealing with the issue of change–point detection, see e.g., Berkes *et al.* (2004), Berkes, Horváth and Kokoszka (2004), Kokoszka and Teyssière (2002), Mikosch and Stărică (1999, 2004b), Chu (1995). So far, none of the change–point tests proposed in research papers is based on wavelets. More precisely, the relevance of the use of the wavelet analysis in the context of non–homogeneous GARCH processes has not been studied, since as it is discussed below, is does not fit in the standard framework of change–point in the conditional mean processes. Such an analysis is the purpose of this section.

We consider a GARCH(1,1) process with change–point at time k:

$$\begin{aligned} r_t &= \sigma_t \varepsilon_t, \quad \varepsilon_t \sim N(0,1), \\ \sigma_t^2 &= \omega + \beta \sigma_{t-1}^2 + \gamma r_{t-1}^2 \quad t \leqslant k, \\ \sigma_t^2 &= \omega^\star + \beta^\star \sigma_{t-1}^2 + \gamma^\star r_{t-1}^2, \quad t > k. \end{aligned} \tag{66}$$

with $\omega \neq \omega^\star$ or $\beta \neq \beta^\star$ or $\gamma \neq \gamma^\star$. The parameter of interest is the unconditional variance of the process $\sigma^2 = \omega/(1-\beta-\gamma)$, and we consider two cases: small change and large change in the unconditional variance. We choose $k = [T/2]$ and set for the first part of the process

$$\omega = 0.1, \quad \beta = 0.3, \quad \gamma = 0.3, \quad \sigma^2 = 0.25, \tag{67}$$

while the parameters for the second part of the two processes are:

- **GARCH A** (small change): $\omega^\star = 0.125$, $\beta^\star = 0.6$, $\gamma^\star = 0.1$, $\sigma^2 = 0.4667$,
- **GARCH B** (large change): $\omega^\star = 0.15$, $\beta^\star = 0.65$, $\gamma^\star = 0.25$, $\sigma^2 = 1.5$.

Since financial time series of size $T = 10000$ are likely affected by several changes in regimes, we consider the following GARCH process, denoted by **GARCH C**, with two un-periodic changes in regimes at times k and k':

$$\begin{aligned} r_t &= \sigma_t \varepsilon_t, \quad \varepsilon_t \sim N(0,1), \\ \sigma_t^2 &= 0.15 + 0.65\sigma_{t-1}^2 + 0.25 r_{t-1}^2, \quad t \leqslant k \quad \sigma^2 = 1.5, \\ \sigma_t^2 &= 0.1 + 0.3\sigma_{t-1}^2 + 0.3 r_{t-1}^2, \quad k < t \leqslant k', \quad \sigma^2 = 0.25, \\ \sigma_t^2 &= 0.25 + 0.6\sigma_{t-1}^2 + 0.2 r_{t-1}^2, \quad t > k', \quad \sigma^2 = 1.25. \end{aligned} \tag{68}$$

We set $k = [T/6]$ and $k' = [5T/6]$, i.e., near the end–points of the sample, which is always the most difficult configuration to detect. Let us note that the absolute value and squares of change–point GARCH processes are not genuine change–point processes in the conditional mean, which makes the standard theory for change–point in the conditional mean not applicable.

We consider also the performance of the estimators for the case without change points, i.e., the Data Generating Process (65) with parameters given by equation (67) that we denote as **DGP 0**.

Let us first analyze the logscale diagrams (presented in figures 21–26) for various realizations of the **GARCH A**, **GARCH B** and **GARCH C** processes. One can see that fitting the entire logscale diagram (i.e., starting from $j_1 = 1$) would lead to the conclusion that long-memory exists in the data. However, we also see that the spurious long–memory property caused by the non–stationarity of the process is gotten rid-off by selecting the lowest octave $j_1 \geq 5$.

Remark 11. The optimal choice j_1^{RMSE} is the same for all values of the number of vanishing moments N. Thus, we cannot use lower j_1 by increasing N.

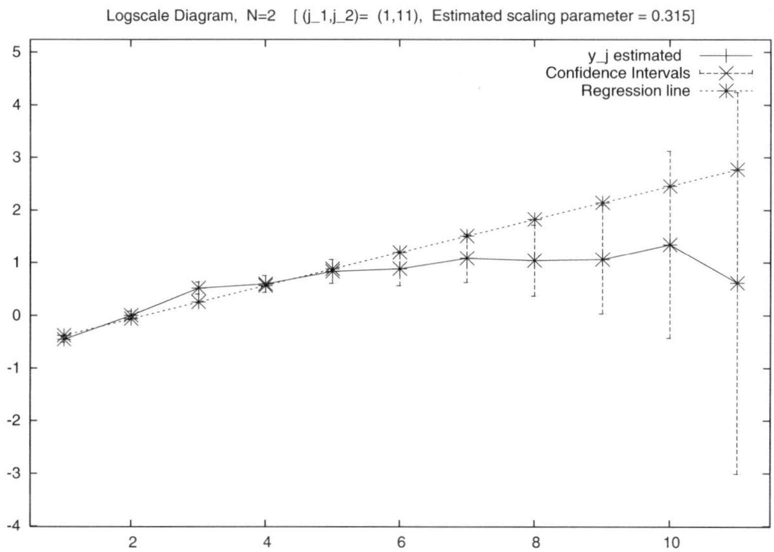

Fig. 21. Logscale diagram for the realization of the absolute value **GARCH A** process. We select here $j_1 = 1$, $j_2 = 11$, $N = 2$

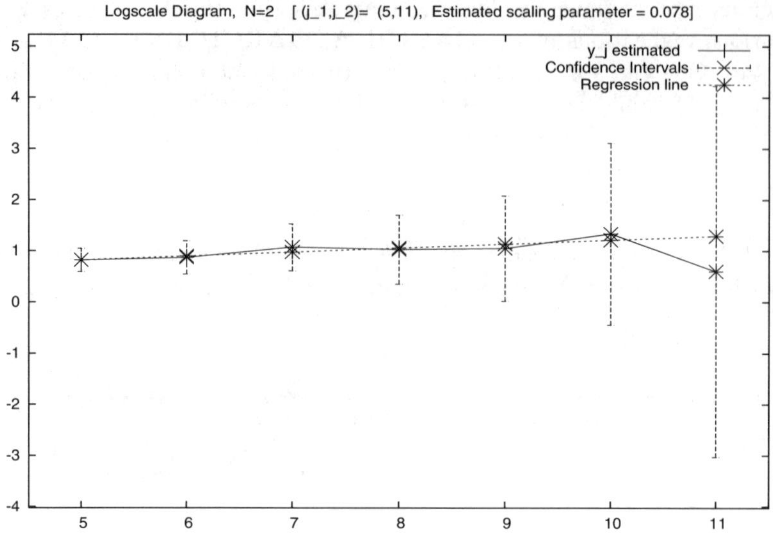

Fig. 22. Logscale diagram for the realization of the squares of a **GARCH A** process. We select here $j_1 = 5$, $j_2 = 11$, $N = 2$

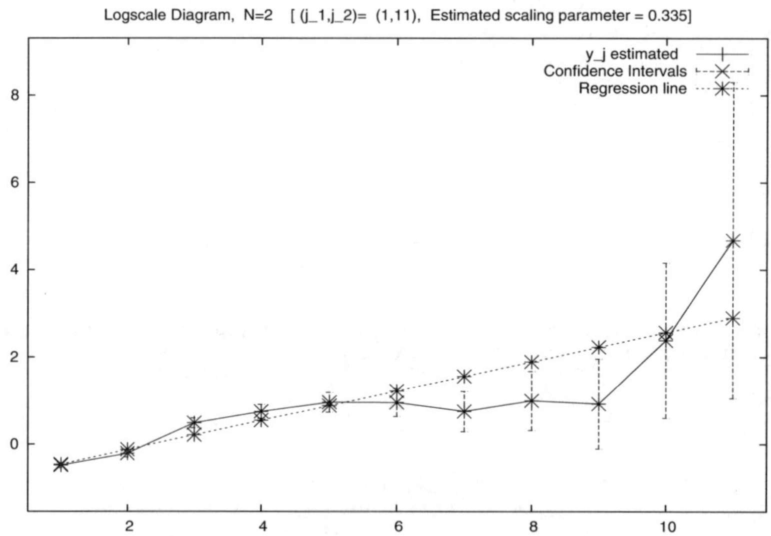

Fig. 23. Logscale diagram for the realization of the squares of a **GARCH B** process. We select here $j_1 = 1$, $j_2 = 11$, $N = 2$

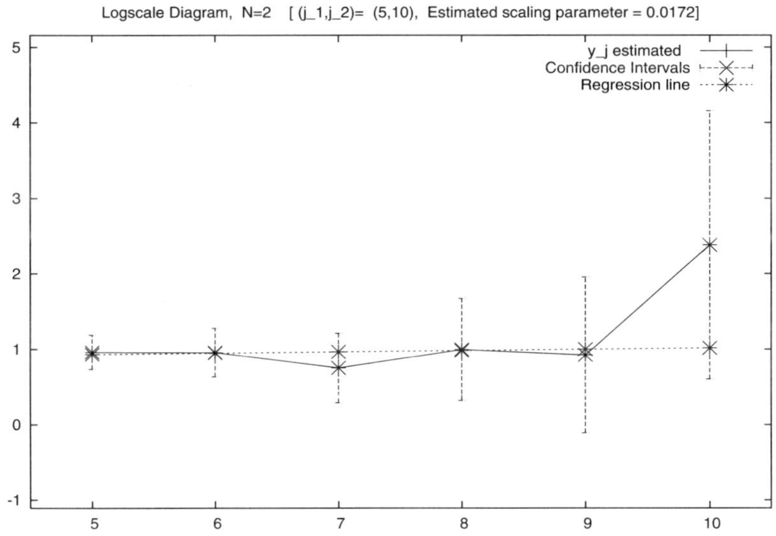

Fig. 24. Logscale diagram for the realization of the squares of a **GARCH B** process. We select here $j_1 = 5$, $j_2 = 10$, $N = 2$

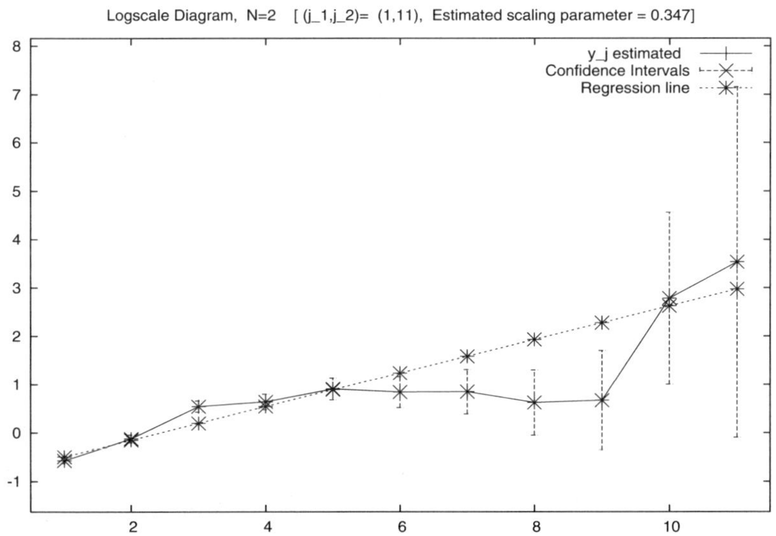

Fig. 25. Logscale diagram for the realization of the squares of a **GARCH C** process. We select here $j_1 = 1$, $j_2 = 11$, $N = 2$

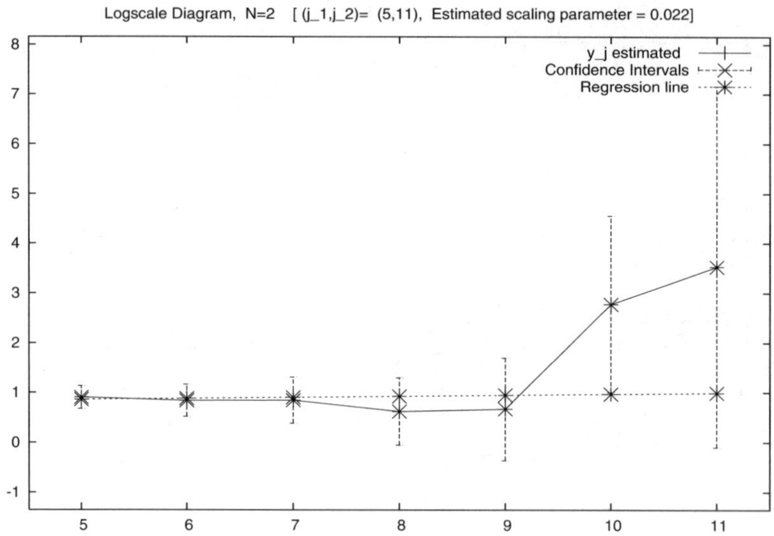

Fig. 26. Logscale diagram for the realization of the squares of a **GARCH C** process. We select here $j_1 = 5$, $j_2 = 11$, $N = 2$

For the **GARCH A** process, i.e., with a constant unconditional variance, both the LW and the LP estimators do not detect any long–range dependence, while the wavelet estimator detects a very moderate intensity of long–memory, although the Gaussian confidence intervals contains the value zero (let us remind that $\alpha \equiv 0$ corresponds to short memory processes). The wavelet estimate with the octave j_1 selected using the optimal bandwidth, see equation (60), gives estimates from $\hat{\alpha}_W = 0.0392$ to $\hat{\alpha}_W = 0.0333$, depending on the number of vanishing moments N.

Table 15. Estimation of the scaling parameter for change–points GARCH processes, wavelet estimators. $T = 10000$, $N = 2$

Model	$\hat{\alpha}_W^{RMSE}$	j_1^{RMSE}	$E\hat{\alpha}_W^{RMSE} - \alpha$	RMSE	$\hat{\alpha}_W^{(5)}$	$E\hat{\alpha}_W^{(5)} - \alpha$	RMSE
GARCH 0	0.0369	4	0.0369	0.0719	0.0151	0.0151	0.0866
GARCH A	0.0495	4	0.0495	0.0791	0.0241	0.0241	0.0900
GARCH B	0.0520	5	0.0520	0.1086	0.0865	0.0865	0.1238
GARCH C	0.0219	4	0.0219	0.0708	0.0682	0.0682	0.1126

There is a significant discrepancy between the estimators for both **GARCH B** and **GARCH C** processes. While the wavelet estimator does detect a very moderate intensity of long–range dependence, both the LW and LP estimators give a high value for $\hat{\alpha}$, with a RMSE far greater than the one of the

Table 16. Estimation of the scaling parameter for change–points GARCH processes, LP and LW estimators. $T = 10000$. The optimal bandwidths m_{LP}^{opt} and m_{LW}^{opt} are respectively used for the LP and LW estimators

Model	$\hat{\alpha}_{LP}^{opt}$	$E\hat{\alpha}_{LP}^{opt} - \alpha$	RMSE	$\hat{\alpha}_{LW}^{opt}$	$E\hat{\alpha}_{LW}^{opt} - \alpha$	RMSE
GARCH 0	0.0038	0.0038	0.0669	0.0015	0.0015	0.0612
GARCH A	0.0043	0.0043	0.4735	0.0065	0.0065	0.0651
GARCH B	0.1921	0.1921	1.4738	0.4015	0.4015	0.4043
GARCH C	0.2440	0.2440	1.4087	0.4188	0.4188	0.4227

wavelet estimator. For the LP, the results obtained with $m = [T^{0.3}]$ are very high, as $\hat{\alpha}_{LP} = 0.0595$ for **GARCH A**, $\hat{\alpha}_{LP} = 1.4454$ for **GARCH B**, and $\hat{\alpha}_{LP} = 1.3801$ for **GARCH C**. This advocates again the use of the optimal bandwidth (41) for volatility processes.

When selecting the lowest octave j_1 that maximizes the goodness–of–fit function (61), the estimated long–memory parameter for the **GARCH B** process is $\hat{\alpha}_W = 0.0539$ for $N = 4$ with a maximum for $\hat{\alpha}_W = 0.0877$ for $N = 2$. For the **GARCH C** process, $\hat{\alpha}_W = 0.0069$ for $N = 4$, with a maximum $\hat{\alpha}_W = 0.0723$ for $N = 2$. This shows that increasing N helps to get rid of spurious long–memory. The wavelet estimates with the lowest octave j_1 given by equation (60) are equal to 0.0053 for the **GARCH A**, 0.0575 for the **GARCH B**, and −0.0281 for the **GARCH C**, in all three cases this "optimal" octave is $j_1 = 7$, which yields a very large variance as well emphasizes the need for further research in this direction.

This example clearly shows that in spite of the octaves selection issue due to the nonlinearity and non Gaussianity of the sequence $\{r_t^2\}$, the wavelet estimator detects a very moderate level of long–range dependence, i.e., $\alpha < 0.0520$ in the worst case, while both the LW and the LP estimators are "fooled" by the occurrence of a large change in the unconditional variance for the short memory GARCH process. This advocates the use of the wavelet estimator, in conjunction with other estimators, for adjudicating between strong dependence and change–point for volatility processes.

5.4 Long–Memory and Change–Points: The Non–Homogeneous Long–Memory Stochastic Volatility Process

We consider here a nonlinear process that displays long–range dependence in conditional variance with a change–point, defined as:

$$\begin{aligned} &r_t = \sigma_t \zeta_t, \quad \zeta_t \sim N(0,1), \quad \sigma_t = \sigma \exp(X_t/2), \quad X_t \sim \text{FARIMA}(p,d,q), \\ &E X_t = 0, \quad t \leqslant k, \quad E X_t = \mu, \quad t > k, \end{aligned} \tag{69}$$

i.e., the process $\{X_t\}$ mixes long–range dependence and a change–point at time k. We consider here samples of size $T = 10000$, $k = [T/2]$, and $\mu = 2.0$.

Remark 12. The magnitude of the jump $\mu = 2.0$ has to be compared with the square root of the variance of the X_t process,

$$\mathrm{Var}(X_t) = \sigma_\epsilon^2 \frac{\Gamma(1-2d)}{\Gamma^2(1-d)}.$$

Here, we set $\sigma_\epsilon^2 = 1$, so that $\sqrt{\mathrm{Var}(X_t)}$ ranges from 1.0022 ($d = 0.05$) to 1.9085 ($d = 0.45$), i.e., the change is of significant magnitude.

Figures 27 and 28 display the logscale diagram for one realization of a non–homogeneous LMSV process for different regression ranges $[j_1, j_2]$. Provided that $j_1 \geq 5$, the long–range dependent behavior is satisfactorily captured.

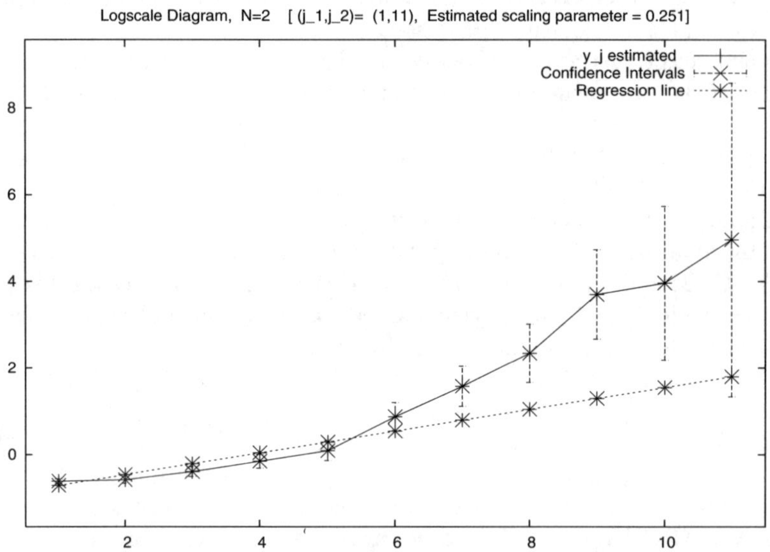

Fig. 27. Logscale diagram for the realization of a change–point LMSV process, with $\alpha = 0.90$. We select here $j_1 = 1$, $j_2 = 11$, $N = 2$

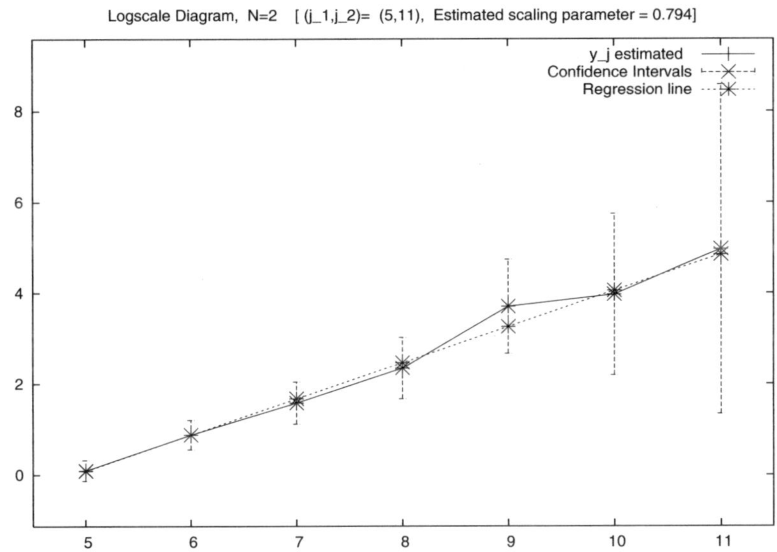

Fig. 28. Logscale diagram for the realization of a change–point LMSV process, with $\alpha = 0.90$. We select here $j_1 = 5$, $j_2 = 11$, $N = 2$

Table 17. Estimation of the scaling parameter for change–point LMSV processes, wavelet estimator. $T = 10000$, $N = 2$

α	$\hat{\alpha}_W^{RMSE}$	j_1^{RMSE}	$E\hat{\alpha}_W^{RMSE} - \alpha$	RMSE	$\hat{\alpha}_W^{(6)}$	$E\hat{\alpha}_W^{(6)} - \alpha$	RMSE
0.10	0.0959	3	-0.0041	0.0317	0.4740	0.3740	0.3878
0.20	0.1777	4	-0.0223	0.0509	0.4748	0.2748	0.2939
0.30	0.3079	5	0.0079	0.0690	0.4820	0.1820	0.2109
0.40	0.3385	5	-0.0605	0.0925	0.4984	0.0984	0.1473
0.50	0.5274	6	0.0274	0.1164	0.5274	0.0274	0.1164
0.60	0.5720	6	-0.0280	0.1120	0.5720	-0.0280	0.1120
0.70	0.6337	6	-0.0663	0.1367	0.6337	-0.0663	0.1367
0.80	0.7122	6	-0.0878	0.1496	0.7122	-0.0878	0.1496
0.90	0.8047	6	-0.0953	0.1546	0.8047	-0.0953	0.1546

Table 18. Estimation of the scaling parameter for change–point LMSV processes, LW and LP estimators. $T = 10000$. The optimal bandwidths m_{LP}^{opt} and m_{LW}^{opt} are respectively used for the LP and LW estimators

α	$\hat{\alpha}_{LP}^{opt}$	$E\hat{\alpha}_{LP}^{opt} - \alpha$	RMSE	$\hat{\alpha}_{LW}^{opt}$	$E\hat{\alpha}_{LW}^{opt} - \alpha$	RMSE
0.10	0.5308	0.4308	0.4365	0.9026	0.8026	0.8034
0.20	0.5377	0.3377	0.3451	0.8931	0.6931	0.6940
0.30	0.5514	0.2514	0.2617	0.8829	0.5829	0.5842
0.40	0.5751	0.1751	0.1899	0.8734	0.4734	0.4751
0.50	0.6103	0.1103	0.1334	0.8657	0.3657	0.3684
0.60	0.6577	0.0577	0.0968	0.8625	0.2625	0.2678
0.70	0.7173	0.0173	0.0826	0.8676	0.1676	0.1795
0.80	0.7877	-0.0123	0.0851	0.8864	0.0064	0.1160
0.90	0.8675	-0.0325	0.0947	0.9251	0.0251	0.0927

Since the sequence of pseudo–error terms is the same for LMSV processes with and without change–point, the results of Tables 17 and 18 can be directly compared to those of Tables 2 and 6. The wavelet estimator is affected by the presence of this change point, but in a far lower extent than both the LW and LP estimators. What is really informative here is the strong discrepancy between the wavelet estimator and the spectral based estimators, which suggests the presence of a break. For that case, the use of the optimal bandwidth for the LP estimator is justified as the mean estimates obtained for $m = [T^{0.3}]$ are between 1.4114 and 1.8578, with a RMSE ranging from 0.6349 to 1.7718.

Remark 13. The best results, i.e., with the lowest RMSE, are obtained with $N = 2$.

6 Financial Time Series

6.1 Intra–day Foreign Exchange (FX) Rates

• **Financial data.** We consider four high–frequency time series on intra–day FX rates provided by Olsen and Associates, i.e., the US dollar–Swiss franc (USD–CHF), the US dollar–Japanese yen (USD–JPY), the US dollar–German deutsche mark (USD–DEM), and the British pound–US dollar (GBP–USD). The data are observed for the whole year 1996 every 30 minutes in a time scale denoted as ϑ–time, where all intra–day periodic components have been removed: time scale with high volatility (activity) are expanded while time scale with low activity are shortened. The ϑ–time scale can be interpreted as a business scale, and then removes the seasonality in the volatility process; see Dacorogna *et al.* (1993) for further details. Since the activity (volatility) is not the same for the series considered, they do not have the same time scale.

We consider here the logarithmic middle price $x(t)$ defined by

$$x_t \equiv \frac{x_t^{bid} + x_t^{ask}}{2} \quad \text{with} \quad x_t^{bid} = \ln(p_t^{bid}), \quad x_t^{ask} = \ln(p_t^{ask}), \tag{70}$$

where p_t^{bid} and p_t^{ask} respectively denote the "bid" and "ask" price at time t. This variable behaves symmetrically when the price is inverted. Next, we define the returns r_t as

$$r_t = x_t - x_{t-\Delta t}, \tag{71}$$

where Δt represents the time interval between two consecutive observations; here, $\Delta t = 30$ minutes. The sample sizes of the series are quite large, as $T =$ 17524, 17529, 17508 and 17520 for the log of returns on USD–DEM, USD–CHF, USD–JPY and GBP–USD respectively.

Figures 29 and 30 below display the ACF of the series of absolute returns $|r_t|$ and squared returns r_t^2 on US dollar–Japanese yen, up to the order 2500. The hyperbolic decay of the ACF is typical or strongly dependent processes, and is slower for the series of absolute returns than for the series of squared returns: this is the so–called "Taylor effect", i.e., the persistence of the series $|r_t|^\delta$ is the strongest for $\delta = 1$. Unlike daily returns, intra–returns are correlated, this negative correlation, which can be modeled by antipersistent processes, is the consequence of market microstructure effects.

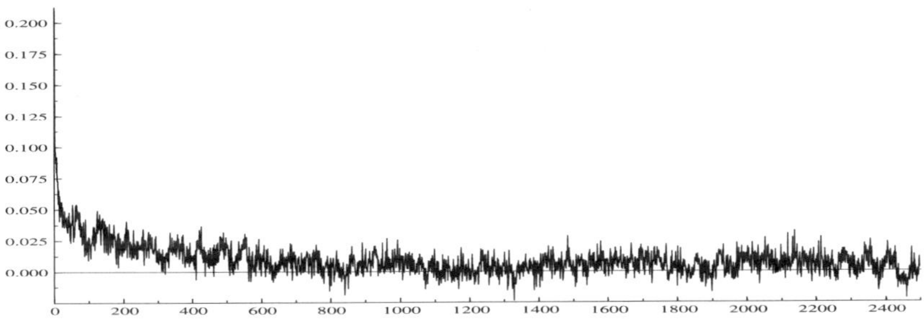

Fig. 29. ACF of absolute returns on US dollar-Japanese yen FX rate

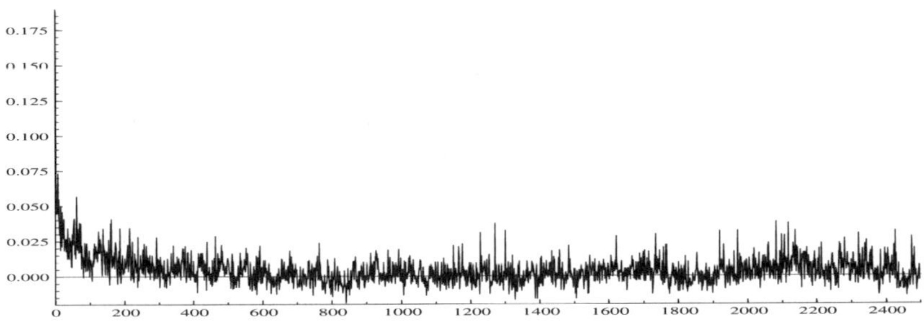

Fig. 30. ACF of squared returns on US dollar-Japanese yen FX rate

• **Long–Memory Analysis.** The estimation of the long–memory parameter for the series of high frequency financial time series has been considered by Müller *et al.* (1990). A wavelet analysis of financial volatilities has been conducted in Kokoszka and Teyssière (2002) and Teyssière (2003). They estimated the degree of long–range dependence of absolute and squared returns on FX rates and equities using the wavelet estimator by Veitch and Abry (1999) and the local Whittle estimator. Their empirical results show that the scaling parameter estimated with the wavelet estimator is far lower than the one obtained with the local Whittle estimator.

In the present work, we estimate the degree of persistence for the series of absolute returns and squared returns for the four intra–day FX rates, denoted by $|r_{usd-chf}|$, $|r_{usd-dem}|$, $|r_{usd-jpy}|$, $|r_{gbp-usd}|$, $r^2_{usd-chf}$, $r^2_{usd-dem}$, $r^2_{usd-jpy}$ and $r^2_{gbp-usd}$. The choices for j_1 and N are guided by the inspection of the logscale diagrams.

We observe that the wavelet-based estimates of the long–memory parameter are significantly lower than the spectral-based estimates, confirming the earlier finding mentioned above. Because the wavelet estimator benefits from robustness against non–stationarities (change points and trends), such discrepancies are in favor of the conjecture that the empirical high intensity of strong dependence in asset price volatilities is a statistical artefact caused by the occurrence of change–point(s). However, the results obtained here also indicate that the estimates significantly depart from 0, i.e., the volatility series do display long–range dependence. This has implications for forecasting purposes, as mentioned by Granger and Hyung (2004) who mix both long–memory and change–point processes for forecasting absolute returns.

Table 19. Estimation of the long–memory parameter for volatilities of intra–day FX rates, $N = 2$. Standard errors are between parentheses, every other line. The optimal bandwidths m_{LP}^{opt} and m_{LW}^{opt} are respectively used for the LP and LW estimators

Series	$\hat{\alpha}_W$	j_1	j_2	$\hat{\alpha}_W^m$	$\hat{\alpha}_{LP}^{opt}$	m_{LP}^{opt}	$\hat{\alpha}_{LW}^{opt}$	m_{LW}^{opt}
$\lvert r_{usd-chf}\rvert$	0.4309	5	9	0.4837	0.5956	549	0.5255	510
	(0.0594)			(0.0541)	(0.0547)		(0.0443)	
$\lvert r_{usd-dem}\rvert$	0.4511	5	11	0.4586	0.5863	557	0.5637	470
	(0.0520)			(0.0804)	(0.0543)		(0.0461)	
$\lvert r_{usd-jpy}\rvert$	0.3581	4	9	0.4017	0.4981	543	0.5597	467
	(0.0381)			(0.0572)	(0.0550)		(0.0463)	
$\lvert r_{gbp-usd}\rvert$	0.2923	5	11	0.3029	0.4126	555	0.5335	447
	(0.0526)			(0.0541)	(0.0544)		(0.0472)	
$r_{usd-chf}^2$	0.3088	3	10	0.3056	0.5808	594	0.5548	604
	(0.0242)			(0.0346)	(0.0526)		(0.0406)	
$r_{usd-dem}^2$	0.2543	4	9	0.2663	0.2583	694	0.3250	749
	(0.0376)			(0.0346)	(0.0486)		(0.0365)	
$r_{usd-jpy}^2$	0.2566	5	12	0.2794	0.4163	577	0.4351	527
	(0.0513)			(0.0572)	(0.0533)		(0.0436)	
$r_{gbp-usd}^2$	0.0982	5	9	0.1204	0.2796	753	0.2859	784
	(0.0603)			(0.0346)	(0.0467)		(0.0357)	

Table 20. Wavelet estimates of the long–memory parameter for volatilities of intra–day FX, for different values of the number of moments N. Standard errors are between parentheses, every other line.

	$N = 3$			$N = 4$			$N = 5$			$N = 6$		
Series	$\hat{\alpha}_W$	j_1	j_2	$\hat{\alpha}_W$	j_1	j_2	$\hat{\alpha}_W$	j_1	j_2	$\hat{\alpha}_W$	j_1	j_2
$\lvert r_{usd-chf}\rvert$	0.4365	4	9	0.4137	4	9	0.4605	5	9	0.4630	5	9
	(0.0380)			(0.0385)			(0.0621)			(0.0631)		
$\lvert r_{usd-dem}\rvert$	0.4924	5	11	0.4277	5	11	0.4371	6	10	0.4328	6	10
	(0.0541)			(0.0573)			(0.0963)			(0.1005)		
$\lvert r_{usd-jpy}\rvert$	0.3591	4	8	0.3449	4	8	0.3776	4	10	0.3879	4	10
	(0.0418)			(0.0426)			(0.0383)			(0.0391)		
$\lvert r_{gbp-usd}\rvert$	0.2569	4	11	0.2125	5	10	0.2078	5	10	0.2126	4	9
	(0.0356)			(0.0581)			(0.0600)			(0.0401)		
$r_{usd-chf}^2$	0.3831	4	10	0.3856	4	10	0.4765	5	10	0.4400	5	10
	(0.0362)			(0.0374)			(0.0599)			(0.0617)		
$r_{usd-dem}^2$	0.2340	3	11	0.2977	5	11	0.2415	5	10	0.2871	5	10
	(0.0240)			(0.0573)			(0.0600)			(0.0617)		
$r_{usd-jpy}^2$	0.2851	4	11	0.2618	4	9	0.2871	4	10	0.2616	4	10
	(0.0380)			(0.0390)			(0.0383)			(0.0391)		
$r_{gbp-usd}^2$	0.1208	5	11	-0.0324	5	11	0.1541	4	10	-0.0217	4	10
	(0.0541)			(0.0573)			(0.0383)			(0.0391)		

Remark 14. We observe that the estimated scaling parameter does not significantly vary for different values of the number of moments N, which confirms the presence of long–range dependence in the volatility series. The series of squared returns $r^2_{gbp-usd}$ on dollar/pound FX rate, is a noticeable exception; in fact, this highly nonlinear series was always very difficult to fit, see e.g., Gallant, Hsieh and Tauchen (1991).

6.2 Application to Trading Volume

• **Bivariate Time Series.** Lobato and Velasco (2000) considered the bivariate process (volatility, log trading volume) of asset prices of stocks constituting the Dow Jones Average Industrial Index, observed from July 1962 to December 1994, i.e., 8176 daily observations, and investigated the issue of common degree of long–range dependence for these two processes.

We consider here few of the series used by Lobato and Velasco (2000), and check whether the wavelet based estimator gives the same results, i.e., the commonality of strong dependence between volatility and log of trading volume.

Figures 31, 32, and 33 below display the log trading volume on three stocks, Eastman Kodak, Chevron and ATT. These three series display strong and clear upward trends.

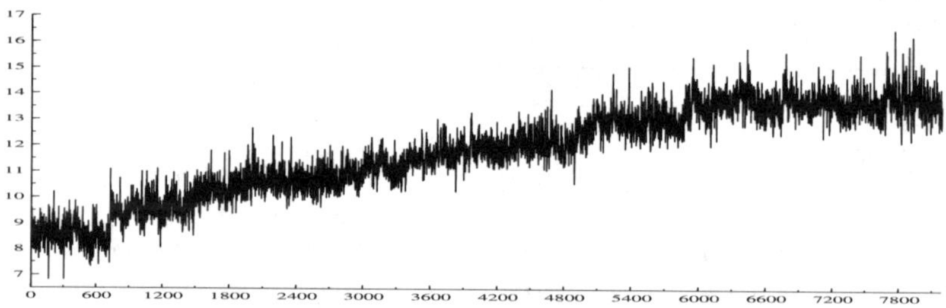

Fig. 31. Logarithm of trading volume on Eastman Kodak stock

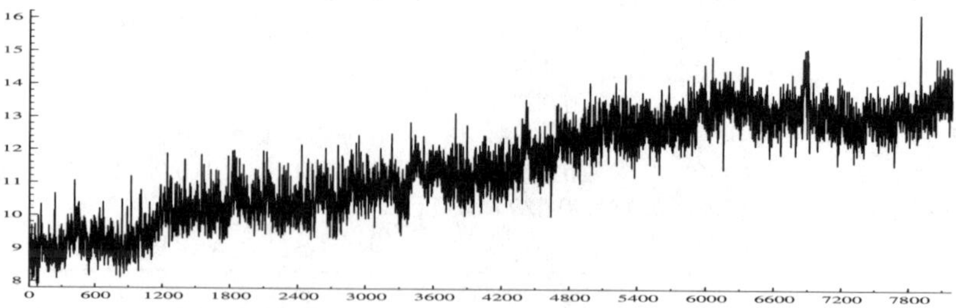

Fig. 32. Logarithm of trading volume on Chevron stock

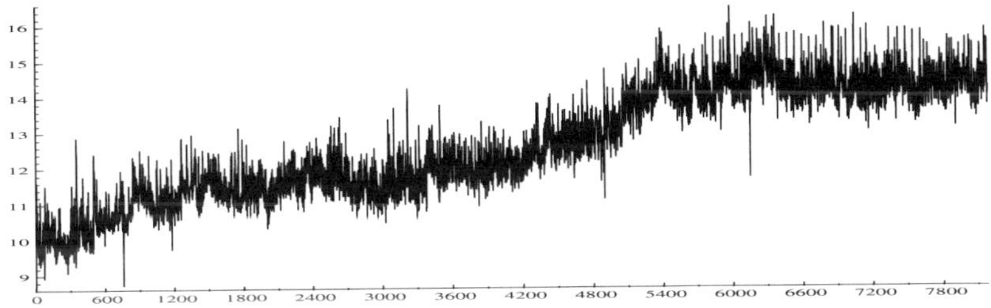

Fig. 33. Logarithm of trading volume on ATT stock

• **Long–Memory Analysis.** We estimate the long–memory parameters for the log of trading volume and the absolute value of returns of six stocks: AT & T Corp., Allied Signal Inc., Chevron Corp., Eastman Kodak Corp., Exxon Corp. and IBM.

Table 21. Estimation of the scaling parameter α for log of trading volume, and absolute value of returns on stocks. Standard errors are between parentheses, every other line. The optimal bandwidth m_{LW}^{opt} is used for the LW estimator

Stocks	$\hat{\alpha}_W$ (Volume)	N	j_1	j_2	$\hat{\alpha}_W$ $(\lvert r_t \rvert)$	j_1	j_2	$\hat{\alpha}_{LW}$ $(\lvert r_t \rvert)$
AT & T	0.7632	6	1	9	0.6422	5	10	0.3465
	(0.0179)				(0.0871)			(0.0533)
Allied Signal	0.6564	6	1	9	0.3126	5	8	0.4546
	(0.0181)				(0.1051)			(0.0546)
Chevron	0.6548	6	1	9	0.3715	7	10	0.5108
	(0.0181)				(0.2530)			(0.0572)
Eastman Kodak	0.7051	6	1	9	0.2484	5	8	0.4146
	(0.0181)				(0.1051)			(0.0516)
Exxon	0.7151	6	1	9	0.5331	6	10	0.3088
	(0.0182)				(0.1431)			(0.0515)
IBM	0.8532	6	1	9	0.3626	4	10	0.2553
	(0.0181)				(0.0556)			(0.0462)

For the series of volume, estimation results no longer differ as soon as $N > 6$, see also Table 22. As we can see from figures 34 and 35, the logscale diagrams consist of straight lines for all octaves, except for Chevron, see Figure 36. Together with the estimated values of the long–memory parameter, this tells that there does exist long–memory in the trading volume times series. The wavelet based estimator is not fooled by the obvious trends that also exist on top of long–memory. We also observe that the estimated long–memory parameters for the series of absolute returns are lower than the estimated scaling parameters for the series of log of volume. In fact, logscale diagrams for the series of absolute returns are more difficult to interpret than the ones of the series of log–volume.

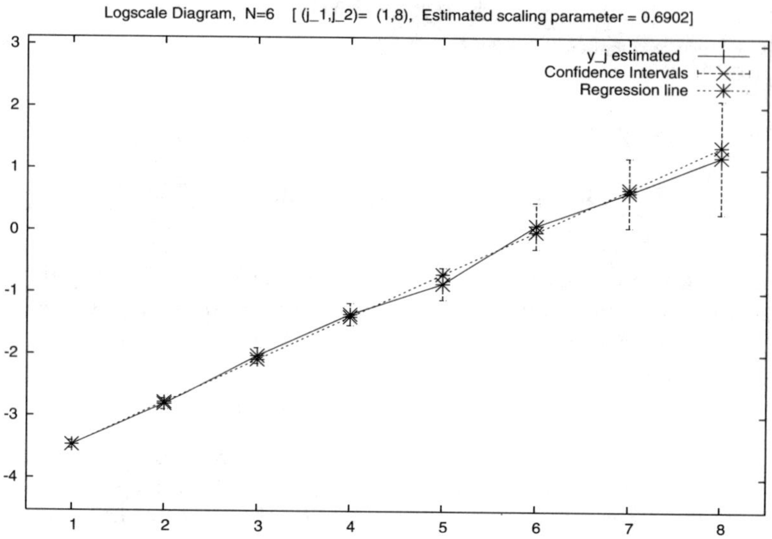

Fig. 34. Logscale diagram for the logarithm of volume on AT& T , $N = 6$

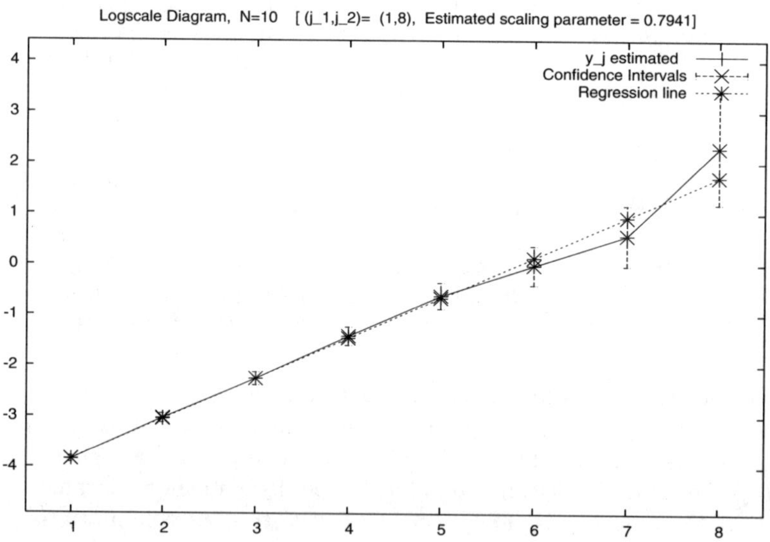

Fig. 35. Logscale diagram for the logarithm of volume on IBM , $N = 10$

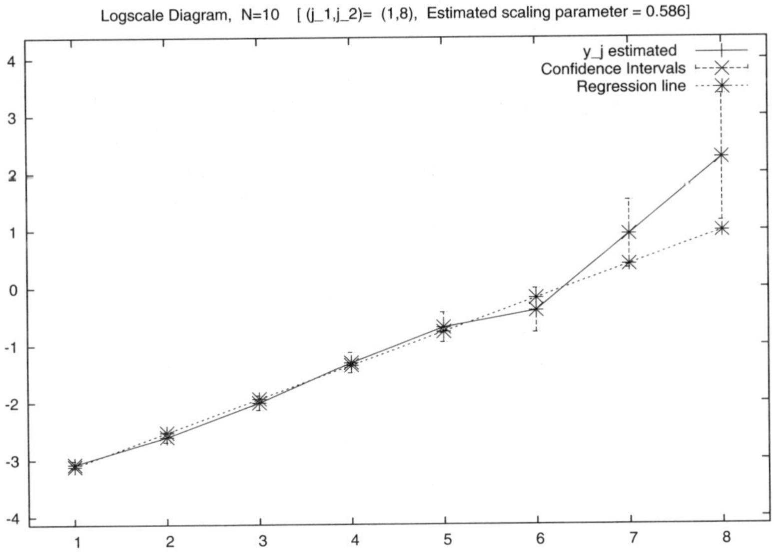

Fig. 36. Logscale diagram for the logarithm of volume on Chevron , $N = 10$

Table 22. Estimation of the scaling parameter α for log of trading volume, and absolute value of returns on stocks, for different values of the number of moments N. Standard errors are between parentheses, every other line

	$N = 7$			$N = 8$			$N = 9$			$N = 10$		
Series	$\hat{\alpha}_W$	j_1	j_2	$\hat{\alpha}_W$	j_1	j_2	$\hat{\alpha}_W$	j_1	j_2	$\hat{\alpha}_W$	j_1	j_2
AT & T	0.7548	1	9	0.7475	1	8	0.7518	1	8	0.7337	1	8
	(0.0180)			(0.0183)			(0.0187)			(0.0189)		
Allied Signal	0.6405	1	9	0.6385	1	8	0.6367	1	8	0.6316	1	8
	(0.0180)			(0.0183)			(0.0187)			(0.0189)		
Chevron	0.6459	1	9	0.6435	1	8	0.6432	1	8	0.6384	1	8
	(0.0180)			(0.0183)			(0.0187)			(0.0189)		
Eastman Kodak	0.7118	1	9	0.7041	1	8	0.6896	1	8	0.6896	1	8
	(0.0180)			(0.0183)			(0.0187)			(0.0189)		
Exxon	0.7070	1	9	0.7099	1	8	0.7046	1	8	0.7089	1	8
	(0.0180)			(0.0183)			(0.0187)			(0.0189)		
IBM	0.8351	1	9	0.8451	1	8	0.8301	1	8	0.8235	1	8
	(0.0180)			(0.0183)			(0.0187)			(0.0189)		

7 Conclusion

In nominal situations, i.e., when the analyzed time series corresponds to an actual non corrupted long–range dependent process, the wavelet estimator for the long–memory parameter works satisfactorily well and this for a large variety of linear and nonlinear strongly dependent processes in conditional mean an in conditional variance. In the same situations, the local Whittle estimator slightly outperforms the log periodogram and wavelet estimators whose performance are mostly comparable. From our point of view, the key advantage of the wavelet based estimator lies in the possibility of varying N which indeed constitutes a key degree of freedom in the wavelet based analysis of long–memory: as long as the logscale diagrams and the corresponding estimates of the long–memory parameter vary with N, it shows that non–stationarities exist in the time series and may impair a correct analysis of the long–memory or confused with long–memory. When, for large enough N, no variation with N are observed, we know we have untangled long–memory from non–stationarity and can accurately estimate the corresponding parameter. Section 5 consistently showed that the wavelet estimator significantly outperformed the two other estimators in situations where non–stationarities were added to long–memory or confused with.

Also, for financial time series, the possibility of varying N enabled to show that the intensity of the long–range dependence of volatility processes is lower than the one measured with spectral based semiparametric estimators. Thus, it reveals that volatility processes mix non–stationarities with long–range dependence. It also showed however that long–memory truly exists both in the volatility and volume time series. The wavelet analysis of the series of volume and volatility finally showed that the dynamics of volatility processes is more complex than the one of volume processes.

References

1. Abry, P., Gonçalvès P. and Flandrin, P. (1995). Wavelets, spectrum analysis and $1/f$ processes. In: Antoniadis, A. and Oppenheim, G. (Eds.), *Wavelets and Statistics*, Lecture Notes in Statistics, Vol. 103. Springer Verlag, New York, pp. 15–30.
2. Abry P. and Veitch, D. (1998). Wavelet analysis of long–range dependent traffic. *IEEE Transactions on Information Theory*, **44,** 2–15.
3. Abry, P., Flandrin, P., Taqqu, M. and Veitch, D. (2000). Wavelets for the analysis, estimation, and synthesis of scaling data. In: Park, K. and Willinger, W. (Eds.), *Self Similar Network Traffic and Performance Evaluation.* Wiley, pp. 39–88.
4. Abry, P., Flandrin, P., Taqqu, M. and Veitch, D. (2003). Self–similarity and long–range dependence through the wavelet lens. In: Doukhan, P., Oppenheim, G. and Taqqu, M.S. (Eds.), *Long–Range Dependence: Theory and Applications.* Birkhauser, Boston, pp. 527–556.

5. Abry, P. and Teyssière, G. (2005). Wavelet Estimation of the Scaling Parameter of Prefiltered Discrete Time Nonlinear LRD Processes. *Preprint.*
6. Bardet, J.M., Lang, G., Moulines, E. and Soulier, P. (2000). Wavelet estimator of long–range dependent processes. *Statistical Inference for Stochastic Processes*, **3,** 85–99.
7. Belsley, D., Kuh, E. and Welsch, R.E., (1980). *Regression Diagnostics*, Wiley, New York.
8. Beran, J. (1994). *Statistics for Long–Memory Processes.* Chapman & Hall, New York.
9. Berkes, I., Horváth, L. and Kokoszka, P.S. (2004). Testing for parameter constancy in GARCH(p,q) models. *Statistics and Probability Letters*, **70,** 263–273.
10. Berkes, I., Gombay, E., Horváth, L. and Kokoszka, P.S. (2004). Sequential change–point detection in GARCH(p,q) models. *Econometric Theory*, **20,** 1140–1167.
11. Berkes, I., Horváth, L., Kokoszka, P.S. and Shao, Q-M. (2003). On discriminating between long–range dependence and changes in mean. *The Annals of Statistics*, forthcoming.
12. Breidt, J., Crato, N. and de Lima, P. (1998). On the detection and estimation of long memory in stochastic volatility. *Journal of Econometrics*, **83,** 325–348.
13. Chu, C.-S.J. (1995). Detecting parameter shift in GARCH models. *Econometric Reviews*, **14,** 241–266.
14. Dacorogna, M.M., Müller, U.A., Nagler, R.J., Olsen, R.B. and Pictet, O.V. (1993). A geographical model for the daily and weekly seasonal volatility in the FX Market. *Journal of International Money and Finance*, **12,** 413–438.
15. Daubechies, I. (1992). *Ten Lectures on Wavelets.* SIAM Philadephia, CBMS–NSF series, vol. **61**.
16. Delgado, M. and Robinson, P. (1996). Optimal spectral bandwidth for long memory. *Statistica Sinica*, **67,** 97–112.
17. Deo, R.S. and Hurvich, C.M. (2003). Estimation of long–memory in volatility. In: Doukhan, P., Oppenheim, G. and Taqqu, M.S. (Eds.), *Long–Range Dependence: Theory and Applications.* Birkhauser, Boston, pp. 313–324.
18. Dittman, I. and Granger, C.W.J. (2002). Properties of nonlinear transformations of fractionally integrated processes. *Journal of Econometrics*, **110,** 113–133.
19. Duan, J.-C. (1995). The GARCH option pricing model. *Mathematical Finance*, **5,** 13–32.
20. Flandrin, P. (1992). Wavelet analysis and synthesis of fractional Brownian motion. *IEEE Transactions on Information Theory*, **38,** 910–917.
21. Flandrin, P. (1989). On the spectrum of fractional Brownian motion. *IEEE Transactions on Information Theory*, **35,** 197–199.
22. Gallant, A.R., Hsieh, D.A. and Tauchen, G. (1991). On fitting a recalcitrant series: the pound/dollar exchange rate, 1974-1983. In: Barnett, W.A., Powell, J. and Tauchen, G. (Eds.), *Nonparametric and Semiparametric Methods in Economics and Statistics.* Cambridge University Press.
23. Gefferth, A., Veitch, D., Maricza, I., Molnár, S. and Ruzsa, I. (2003). The nature of discrete second-order self similarity. *Advances in Applied Probability*, **35,** 395–416.
24. Geweke, J. and Porter–Hudak, S. (1983). The estimation and application of long memory time series models. *Journal of Time Series Analysis*, **4,** 221–238.
25. Giraitis, L. and Surgailis, D. (2005). Lectures on long–memory processes. *In preparation*

26. Giraitis, L. and Surgailis, D. (1985). Central limit theorems and other limit theorems for functionals of Gaussian processes. *Z. Wahrscheinlichkeitstheorie verw. Gebiete* **70,** 191–212.
27. Giraitis, L., Leipus, R. and Surgailis, D. (2005). Recent advances in ARCH modelling. In: Teyssière, G. and Kirman, A. (Eds.), *Long–Memory in Economics.* Springer Verlag, Berlin. Appears in this volume.
28. Giraitis, L., Robinson, P. and Surgailis, D. (2000). A model for long memory conditional heteroskedasticity. *Annals of Applied Probability,* **10,** 1002–1024.
29. Giraitis, L., Kokoszka, P.S. and Leipus, R. (2000). Stationary ARCH Models: Dependence Structure and Central Limit Theorem. *Econometric Theory,* **16** , 3–22.
30. Giraitis, L., Kokoszka, P., Leipus, R. and Teyssière, G. (2000). Semiparametric estimation of the intensity of long–memory in conditional heteroskedasticity. *Statistical Inference for Stochastic Processes,* **3,** 113–128.
31. Giraitis, L., Robinson, P. and Samarov, A. (1997). Rate optimal semiparametric estimation of the memory parameter of the Gaussian time series with long–range dependence. *Journal of Time Series Analysis,* **18,** 49–60.
32. Granger, C.W.J. and Hyung, N. (2004). Occasional structural breaks and long memory with an application to the S&P 500 absolute stock returns. *Journal of Empirical Finance,* **11,** 399–421.
33. Granger, C.W.J. (2002). Long memory, volatility, risk and distribution. *Preprint.*
34. Granger, C.W.J. (2000). Current perspectives on long memory processes. *Academia Economic Papers,* **28,** 1–16.
35. Granger, C.W.J. and Ding, Z. (1995). Some properties of absolute returns, an alternative measure of risk. *Annales d'Économie et de Statistique,* **40,** 67–91.
36. Hauser, M.A. (1999). Maximum likelihood estimators for ARMA and ARFIMA models: a Monte Carlo study. *Journal of Statistical Planning and Inference,* **80,** 229-255.
37. Henry, M. (2005). Bandwidth choice, optimal rates and adaptivity in semiparametric estimation of long memory. In: Teyssière, G. and Kirman, A. (Eds.), *Long–Memory in Economics.* Springer Verlag, Berlin. Appears in this volume.
38. Henry, M. (2001). Robust automatic bandwidth for long–memory. *Journal of Time Series Analysis,* **22,** 293–316.
39. Henry, M. and Robinson, P.M. (1996). Bandwidth choice in Gaussian semiparametric estimation of long–range dependence. In: Robinson, P.M. and Rosenblatt, M. (Eds.), *Athens Conference on Applied Probability and Time Series Analysis, Time Series Analysis, In Memory of E.J. Hannan.* Lecture Notes in Statistics, Vol. 115. Springer Verlag, New York, pp. 220–232.
40. Hurvich, C., Deo, R. and Brodsky, J. (1998). The mean squared error of Geweke and Porter–Hudak's estimator of the memory parameter of a long memory time series. *Journal of Time Series Analysis,* **19,** 19–46.
41. Kallsen, J. and Taqqu, M.S. (1998). Option pricing in ARCH–type models. *Mathematical Finance,* **8,** 13–26.
42. Kokoszka, P.S. and Teyssière, G. (2002). Change–point detection in GARCH models: asymptotic and bootstrap tests. Under revision for the *Journal of Business and Economic Statistics.*
43. Kokoszka, P.S. and Leipus, R. (2003). Detection and estimation of changes in regime. In: Doukhan, P., Oppenheim, G. and Taqqu, M.S. (Eds.), *Long–Range Dependence: Theory and Applications.* Birkhauser, Boston, pp. 325–337.

44. Künsch, H.R. (1987). Statistical Aspects of Self-Similar Processes, in Prohorov, Yu. and Sazanov, V.V. (Eds), *Proceedings of the First World Congress of the Bernoulli Society*, **1,** 67–74. VNU Science Press, Utrecht.
45. Lang, G. and Roueff, F. (2001). Semiparametric estimation of the Hölder exponent of a stationary Gaussian process with minimax rates. *Statistical Inference for Stochastic Processes*, **4,** 283–306.
46. Lavielle, M. (1999). Detection of multiple changes in a sequence of dependent variables. *Stochastic Processes and their Applications*, **83,** 79–102.
47. Lavielle, M. and Moulines, E. (2000). Least–squares estimation of a unknown number of shifts in a time series. *Journal of Time Series Analysis*, **21,** 33–59.
48. Lavielle, M. and Teyssière, G. (2005). Adaptive detection of multiple change–points in asset price volatility. In: Teyssière, G. and Kirman, A. (Eds.), *Long–Memory in Economics*. Springer Verlag, Berlin. Appears in this volume.
49. Lobato, I.N., and Velasco, C. (2000). Long memory in stock market trading volume. *Journal of Business and Economic Statistics*, **18,** 410–427.
50. Mallat, S. (1998). *A Wavelet Tour of Signal Processing*. Academic Press, Boston.
51. Mikosch, T. and Stărică, C. (2004a). Non–stationarities in financial time series: the long–range dependence and the IGARCH effects. *Review of Economics and Statistics*, **86,** 278–290.
52. Mikosch, T. and Stărică, C. (2004b). Change of structure in financial time series and the GARCH model. *REVSTAT Statistical Journal*, **2,** 41–73.
53. Mikosch, T. and Stărică, C. (2003). Long–range dependence effects and ARCH modeling. In: Doukhan, P., Oppenheim, G. and Taqqu, M.S. (Eds.), *Long–Range Dependence: Theory and Applications*. Birkhauser, Boston, pp. 439–459.
54. Mikosch, T. and Stărică, C. (1999). Change of structure in financial time series, long range dependence and the GARCH model. *Preprint*.
55. Moulines, E. and Soulier, P. (2003). Semiparametric spectral estimation for fractional processes. In: Doukhan, P., G. and Taqqu, M.S. (Eds.), *Long–Range Dependence: Theory and Applications*. Birkhauser, Boston, pp. 251–301.
56. Müller, U.A., Dacorogna, M.M., Olsen, R.B., Pictet, O.V., Schwarz, M. and Morgenegg, C. (1990). Statistical study of foreign exchange rates, empirical evidence of a price change scaling law, and intraday analysis. *Journal of Banking and Finance*, **14,** 1189–1208.
57. Robinson, P.M. (1995a). Gaussian semiparametric estimation of long–range dependence. *The Annals of Statistics*, **23,** 1630–1661.
58. Robinson, P.M. (1995b). Log periodogram regression of time series with long range dependence. *The Annals of Statistics*, **23,** 1048–1072.
59. Robinson, P.M. (1994). Time Series with Strong Dependence, in C.A. Sims (Ed), *Advances in Econometrics, Sixth World Congress*. Cambridge University Press, pp. 47–95.
60. Robinson, P.M. (1991). Testing for strong serial correlation and dynamic conditional heteroskedasticity in multiple regression. *Journal of Econometrics*, **47,** 67–84.
61. Robinson, P.M. (1977). The estimation of a nonlinear moving average model. *Stochastic Processes and their Applications*, **5,** 81–90.
62. Robinson, P.M. and Zaffaroni, P. (1997). Modelling nonlinearity and long–memory in time series. *Fields Institute Communications*, **11,** 161–170.
63. Surgailis, D. (1981). Convergence of sums of nonlinear functions of moving averages to self–similar processes. *Soviet Mathematics*, **23,** 247–250.

64. Teyssière, G. (2003). Interaction models for common long–range dependence in asset price volatility. In: Rangarajan, G. and Ding, M. (Eds.), *Processes with Long Range Correlations: Theory and Applications,* Lecture Notes in Physics, Vol. 621. Springer Verlag, Berlin, pp. 251–269.
65. Veitch, D. and Abry, P. (1999). A wavelet based joint estimator of the parameters of long–range dependence. *IEEE Transactions on Information Theory*, **45,** 878–897.
66. Veitch, D., Taqqu, M.S. and Abry, P. (2000). Meaningfull MRA initialisation for discrete time series. *Signal Processing*, **80,** 1971–1983. EURASIP Best Paper Award.
67. Veitch, D. and Abry, P. (2001). A statistical test for the time constancy of scaling exponents. *IEEE Transactions on Signal Processing*, **49,** 2325–2334.
68. Veitch, D., Abry, P. and Taqqu, M.S. (2003). On the automatic selection of the onset of scaling. *Fractals*, **11,** (4), 377-390.
69. Vostrikova, L.Ju. (1981). Detection of "disorder" in multidimensional random processes. *Soviet Mathematics Doklady*, **24,** 55–59.
70. Zaffaroni, P. (2003). Gaussian inference on certain long–range dependent volatility models. *Journal of Econometrics*, **115,** 199–258.

Prediction, Orthogonal Polynomials and Toeplitz Matrices. A Fast and Reliable Approximation to the Durbin–Levinson Algorithm

Djalil Kateb[1], Abdellatif Seghier[2] and Gilles Teyssière[3]

[1] Université de Technologie de Compiègne, L.M.A.C, Centre de Recherche de Royallieu, B.P. 529, 60205 Compiègne. Djalil.Kateb@utc.fr
[2] Université Paris–Sud, Laboratoire de Mathématiques, Bâtiment 425; F-91405 Orsay Cedex. Abdelatif.Seghier@math.u-psud.fr
[3] Statistique Appliquée et de MOdélisation Stochastique, Université Paris 1. stats@gillesteyssiere.net

Summary. Let f be a given function on the unit circle such that $f(e^{i\theta}) =\mid 1-e^{i\theta} \mid^{2\alpha} f_1(e^{i\theta})$ with $\mid \alpha \mid< \frac{1}{2}$ and f_1 a strictly positive function that will be supposed to be sufficiently smooth. We give the asymptotic behavior of the first column of the inverse of $T_N(f)$, the $(N+1)\times(N+1)$ Toeplitz matrix with elements $(f_{i-j})_{0\leq i,j\leq N}$ where $f_k = \frac{1}{2\pi}\int_0^{2\pi} f(e^{i\theta})e^{-ik\theta}\, d\theta$. We shall compare our numerical results with those given by the Durbin-Levinson algorithm, with particular emphasis on problems of predicting either stationary stochastic long–range dependent processes, or processes with a long–range dependent component.

1 Introduction

Let $f \in L^1(\mathbb{T})$ be a function defined on the unit circle with Fourier coefficients

$$f_k = \hat{f}(k) = \frac{1}{2\pi}\int_0^{2\pi} f(e^{i\theta})e^{-ik\theta}\, d\theta,\ k \in \mathbb{Z}, \tag{1}$$

we consider the $(N+1)\times(N+1)$ Toeplitz matrix

$$T_N(f) = \left(\hat{f}(i-j)\right)_{0\leq i,j\leq N}, \tag{2}$$

with symbol f. Such matrices, their corresponding determinants and inverses arise in many applications in mathematics, statistics, mathematical physics and chemistry; see Fisher and Hartwig (1968) and Wu (1966) for more details. The transformation that defines a Toeplitz matrix is the discrete analogue of a convolution and serve naturally to model the action of linear digital filters.

If we suppose f to be the spectral density of a stationary stochastic process $\{X_k\}, -\infty < k < \infty$, of random variables with

$$E(X_i \bar{X}_j) = f_{j-i}, \tag{3}$$

with $f_{-i} = \bar{f}_i$, we can try to estimate the value of X_j from the observed values of neighboring X_i. This is the problem of prediction: we have the forward prediction if the (X_k) precede X_j and the backward prediction if they follow. In the classical backward prediction of X_0 from $X_1, \ldots, X_N$, we have to find, in the subspace spanned by $X_1, \ldots, X_N$, the element that is closest to X_0 in the norm defined by the inner product

$$[v, w] = \langle T_N(f)v, w \rangle. \tag{4}$$

To solve the problem, we have to take the residual error prediction $X_0 + \sum_{k=1}^{N} e_k X_k$ such that

$$[X_0 + \sum_{k=1}^{N} e_k X_k, X_j] = 0, \ 1 \le j \le N, \tag{5}$$

or

$$T_N(f)\,(1, e_1, \ldots, e_N)^T = \mu_N^2\,(1, 0, \ldots, 0, 0)^T, \tag{6}$$

where the prediction error μ_N^2 is given by

$$\mu_N^2 = [E_N, E_N], \tag{7}$$

where $E_N = (, e_1, e_2, \ldots, e_N)^T$. A remarkable aspect is that we need only the first (resp. the last) column inverse for the backward (resp. the forward) prediction. We point out that it is classical to invert $T_N(f)$ through Yule–Walker type equations, with complexity in $O(N^3)$ operations, or through its improved version the Durbin–Levinson algorithm with complexity in $O(N^2)$ elementary operations.
The purpose of our chapter is to show how one can compute best predictors without having to perform any matrix inversions or a recursive scheme. To be more precise, we give an asymptotic behavior of $(T_N(f))^{-1}_{k,1}$, $0 \le k \le N$, for large values of N; the computations of the inverse elements are performed in $O(N)$ operations with a fixed tolerance ε. The idea is very simple : thanks to the classical formula of Wu and Bleher for inverting Toeplitz matrix and the determinant formula for Fisher–Hartwig symbols, straightforward calculations enable us to study the asymptotic behavior of the first elements. The elements of the inverse which are in the "heart" are computed thanks to Rambour and Seghier (2003) and are inspired from the paper by Bleher (1981). Concerning the last elements, we use a general machinery developed by Szegö for solving prediction problems. It uses the theory of orthogonal polynomials. To give an idea, we know how the first and last column are connected to the coefficients of some orthogonal polynomials Φ_N of degree N on the unit circle with respect to

the nonsingular measure $d\mu(\theta) = f(e^{i\theta})\frac{d\theta}{2\pi}$. Thanks to the Szegö recurrences satisfied by the orthogonal polynomials, we know how to deduce the asymptotic behavior of the remaining last terms $(T_N(f))^{-1}_{N-k+1,1}$ from the elements $(T_{N-m}(f))^{-1}_{N-m+1,1}$, $0 \leq m \leq k$ and $(T_{N-m-1}(f))^{-1}_{k-m+1,1}$, $0 \leq m \leq k$.

To illustrate our theory, we consider the modeling of the volatility of asset prices. The volatility of asset prices P_t is defined by either the absolute value of returns $|r_t|$, or the squared returns r_t^2, where the returns are defined as $r_t = \log(P_t) - \log(P_{t-1})$. The series of absolute returns and squared returns display a very rich dynamics, which mixes long–range dependence and nonlinearity. Recent works by Granger and Hyung (2004) propose to forecast the volatility by the sum of a fractionally integrated process and a change point process. The issue of volatility forecasting is of interest in finance as option pricing formulas rely on a "plug–in" estimate of the volatility.

Before we state our main theorem, let us introduce some preliminaries notations and known results.

2 Notations and Definitions. Preliminary Results

2.1 Wiener–Hopf Factorization

The Wiener–Levy theorem, see Zygmund (1959), on analytic functions of Fourier series ensures the existence of the Wiener–Hopf factorization of f_1:

$$f_1 = (f_1)_- GM(f_1)(f_1)_+, \tag{8}$$

where

$$GM(f_1) = \exp\left(\int_0^{2\pi} \log f_1(\chi(\theta))\,\frac{d\theta}{2\pi}\right), \tag{9}$$

is the geometric mean of f_1, and where

$$(f_1)_+(\chi) = \exp\left(\sum_{n=1}^{\infty} \chi^n \widehat{\log f_1}(n)\right), \tag{10}$$

and

$$(f_1)_-(\chi) = \exp\left(\sum_{n=-\infty}^{-1} \chi^n \widehat{\log f_1}(n)\right). \tag{11}$$

The notation χ stands for $\chi(\theta) = e^{i\theta}$.

2.2 The Szegö Function

When the function is of form $f(\chi) = e^{L(\chi)}$ with $L \in L^1$, Szegö defines a function $g(z)$, defined on the open unit disk $\mathbb{D}$:

$$g(z) = \exp\left(\frac{1}{2}\int_0^{2\pi} \frac{\chi(\theta)+z}{\chi(\theta)-z} \log f(e^{i\theta})\, d\theta\right), \quad |\, z \,|< 1, \tag{12}$$

which belongs to the Hardy space $H^2(\mathbb{D})$. Furthermore, we know that

$$\lim_{r\uparrow 1} g(re^{i\theta}) \equiv g(e^{i\theta}),$$

exists for a.e θ and that

$$|\, g(e^{i\theta}) \,|^2 = f(e^{i\theta}), \tag{13}$$

the function g is called the outer function of f. The Cauchy's integral formula learns us that

$$\hat{g}(0) = \sqrt{GM(f)}. \tag{14}$$

Before we state the main result, we need some notations. We denote by $(\beta_k^{(\alpha)})_{k\in\mathrm{N}}$ the Fourier coefficients of $\frac{1}{g}$

$$\frac{1}{g}(\chi) = \sum_{k=0}^{\infty} \beta_k^{(\alpha)} \chi^k,$$

and by $(g_k^{(\alpha)})$ the Fourier coefficients of $(1-\chi)^{-\alpha}$. A straightforward calculation shows that

$$\beta_k^{(\alpha)} = \sum_{p=0}^{k} g_k^{(\alpha)} \widehat{\frac{1}{g_1}}(k-p),$$

where g_1 is the outer function of f_1. From the elementary fact that

$$g_k^{(\alpha+1)} = \sum_{p=0}^{k} g_p^{(\alpha)},$$

it comes that $\beta_k^{(\alpha+1)} = \sum_{p=0}^{k} \beta_p^{(\alpha)}$.
Here is the main theorem

Theorem 2.1 *Let α be a real such that $|\, \alpha \,|< \frac{1}{2}$ and k be a nonnegative integer such that $\frac{k}{N} \to 0$ when $N \to \infty$. If the generating function f is such that $f(\chi) =|\, 1-\chi \,|^{2\alpha} f_1(\chi)$ where f_1 is a sufficiently smooth and strictly positive on the unit circle then*

i)

$$((T_N(f))^{-1})_{k+1,1} = \sqrt{GM(\frac{1}{f_1})}\left(\bar{\beta}_k^{(\alpha)} - \frac{\alpha^2}{N}\bar{\beta}_k^{(\alpha+1)}\right) + o(\frac{1}{N}), \tag{15}$$

when $N \to \infty$.

ii) Given $0 < x < 1$, *we have*

$$((T_N(f))^{-1})_{[Nx]+1,1} = \sqrt{GM(\frac{1}{f_1})} N^{\alpha-1} K_\alpha(x) + o\left(N^{\alpha-1}\right), \tag{16}$$

where

$$K_\alpha(x) = \frac{1}{\Gamma(\alpha)} x^{\alpha-1}(1-x)^\alpha. \tag{17}$$

iii) We have

$$((T_N(f))^{-1})_{N+1-k,1} = \left(\sqrt{GM(\frac{1}{f_1})} \frac{(f_1)_-(1)}{(f_1)_+(1)} \bar{\beta}_k^{(\alpha)}\right) \alpha N^{-1} + o(N^{-1}). \tag{18}$$

3 A Brief Overview on Orthogonal Polynomials on the Unit Circle

In his investigation into the inversion of Toeplitz matrices, Szegö introduced orthogonal polynomials on the circle. If $d\mu(t) = f(e^{it})\, dt$ is a measure without singular part, we denote the scalar product

$$\langle h_1, h_2 \rangle_f := \int_{\mathbb{T}} h_1(\chi)\overline{h_2}(\chi) d\mu(t),$$

and the associated norm as $\|\ \|$. It is natural to consider the monic orthogonal polynomials $\Phi_n(\chi) = \Phi_n(\chi, \mu)$ defined by

$$\Phi_n(z) = z^n + \sum_{k=0}^{n-1} \Phi_{k,n} z^k,$$

$$\langle \Phi_n(\chi), \chi^j \rangle_f = 0, \ j = 0, \ldots, n-1, \tag{19}$$

if Φ_n^* denotes the reversed polynomials,

$$\Phi_n^*(z) = z^n \overline{\Phi_n(\frac{1}{\bar{z}})}, \tag{20}$$

we know, thanks to Szegö and Atkinson, the following theorem

Theorem 3.1 *There exists a sequence of numbers* $(\rho_{n,f})_{n=0}^\infty$ *so that*

$$\Phi_n(z) = z\Phi_{n-1}(z) + \rho_{n,f}\Phi_{n-1}^*(z). \tag{21}$$

Moreover, the sequence of $(\rho_{n,f})$ *belongs to the open unit disk and*

$$\frac{\| \Phi_n \|^2}{\| \Phi_{n-1} \|^2} = 1 - | \rho_{n,f} |^2, \tag{22}$$

and

$$\| \Phi_n \| = \hat{f}(0) \prod_{i=1}^n \left(1 - | \rho_i |^2\right). \tag{23}$$

We know that the last column of $(T_N(f))^{-1}$ is made of the coefficients of $\frac{\|\Phi_N\|}{\|\Phi_N\|^2}$, indeed we have

$$T_N(f)\,(\Phi_{N,0},\Phi_{N,1},\ldots,\Phi_{N,N-1},1)^T = \left(0,0,\ldots,0,\ldots,0,\|\,\Phi_N\,\|^2\right)^T, \quad (24)$$

where $\Phi_N(z) = z^N + \sum_{k=0}^{N-1}\Phi_{N,k}z^k$. We know also that the first column of the inverse is made of the coefficient of the polynomial $\frac{\Phi_N^*}{\|\Phi_N\|^2}$:

$$T_N(f)\left(1,\overline{\Phi_{N,N-1}},\ldots,\overline{\Phi_{N,1}},\Phi_{N,N}\right)^T = \left(\|\,\Phi_N\,\|_N^2,0,0,\ldots,0\right)^T, \quad (25)$$

with $\Phi_{N,N} = \Phi_N(0)$. The sequence $\left(1,\overline{\Phi_{N,N-1}},\ldots,\overline{\Phi_{N,1}},\Phi_{N,N}\right)$ will be denoted as $(1,p_{N,1},\ldots,p_{N,N-1},p_{N,N})$.

3.1 The Determinant of Toeplitz Matrices

An interesting aspect is that we know how to connect the determinant $D_n(f)$ to the sequence of reflection coefficients $(\rho_{n,f}) = (\Phi_n(0))$. Indeed , we have

Proposition 3.1

$$D_n(f) = \hat{f}(0)\prod_{i=1}^{n}\mu_{i,f}^2 = (\hat{f}(0))^{n+1}\prod_{i=1}^{n}\prod_{j=1}^{i}(1-\mid\rho_{j,f}\mid^2). \quad (26)$$

Let us point out that the sequence of coefficients $(1,p_{N,1},\ldots,p_{N,N-1},p_{N,N} = \Phi_N(0))$ of Φ_N^* satisfy an interesting relation that will be crucial for the proof of the main theorem. We have:

Lemma 3.1 *Given an integer $k\neq 0$, we have*

$$p_{N,N-k} = \sum_{m=0}^{k} p_{N-m,N-m}p_{k-m,N-m-1}. \quad (27)$$

Proof From the relation

$$\Phi_s^*(z) = \Phi_{s-1}^*(z) + p_{s,s}z\Phi_{s-1}(z),$$

we get after identification

$$p_{N,N-m} = p_{N-1,N-m} + p_{N,N}p_{N-1,m}, \quad (28)$$

for a fixed k and all $m\leq k$ we obtain

$$p_{N-m,N-k} = p_{N-m-1,N-k} + p_{N-m,N-m}p_{N-m-1,k-m}, \quad (29)$$

and then

$$\sum_{m=0}^{k}\left(p_{N-m,N-k} - p_{N-m-1,N-k}\right) = \sum_{m=0}^{k} p_{N-m,N-m}p_{N-m-1,k-m}.$$

To end the proof, it suffices to remark that since $p_{N-k-1,0} = 1$, the telescopic sum $\sum_{m=0}^{k}\left(p_{N-m,N-k} - p_{N-m-1,N-k}\right)$ is reduced to $p_{N,N-k}$.

4 The Inversion of a Toeplitz Matrix

4.1 Some Preliminary Results and Estimates

We denote by $(\beta_k^{(\alpha)})_{k\in\mathbb{N}}$ the Fourier coefficients of the outer function $\frac{1}{g}$ of $\frac{1}{f}$, by $(g_k^{(\alpha)})_{k\in\mathbb{N}}$ the Fourier coefficients of $(1-e^{i\theta})^{-\alpha}$ and by $(\gamma_k^{(\alpha)})_{k\in\mathbb{Z}}$ the Fourier coefficients of $\frac{g}{\bar{g}}$. We recall some useful results, the proofs of which can be found in Zygmund (1959).

Lemma 4.1 *When $N \to +\infty$, we have*

i)

$$g_N^{(\alpha)} = \frac{N^{\alpha-1}}{\Gamma(\alpha)}\left(1+o(1)\right),$$

ii)

$$\beta_N^{(\alpha)} = \frac{N^{\alpha-1}}{g_1(1)\Gamma(\alpha)} + o(N^{\alpha-1}),$$

where g_1 is the outer function of f_1.

iii)

$$\gamma_N^{(\alpha)} = \frac{g_1(1)}{\bar{g}_1(1)}\frac{\sin(\pi\alpha)}{\pi(\alpha-N)} + o(N^{-1}).$$

4.2 The Bleher–Wu Inversion Formula

The next results need some notations: for $p \geq 1$ $H^p(\mathbb{T})$ denotes the Hardy space of those $f \in L^p(\mathbb{T})$ with $\hat{f}(k) = 0$ for all $k < 0$. We shall denote $H^{2-} = L^2(\lambda) \ominus H^2$. We introduce the orthogonal projectors

$$\begin{aligned}
&\Pi_+ \ : \ L^2(\mathbb{T}) \mapsto H^2(\mathbb{T}),\\
&\Pi_N \ : \ L^2(\mathbb{T}) \mapsto \mathcal{P}_N = span\{\chi_0, \chi_1, \ldots, \chi_N\},\\
&\Pi_N^+ \ : \ L^2(\mathbb{T}) \mapsto \overline{span\{\chi_{N+1}, \chi_{N+2}, \ldots, \}},\\
&\Pi_N^- = I - \Pi_N^+.
\end{aligned}$$

Given $N \geq 1$, let

$$\phi_N = \frac{g}{\bar{g}}\chi^{N+1}, \tag{30}$$

and the Hankel operator $H_{\phi_N} : H^2 \mapsto H^{2-}$ be as

$$H_{\phi_N}(\psi) = \Pi_-(\phi_N\psi), \ \psi \in H^2, \tag{31}$$

its adjoint $H_{\phi_N}^*$ satisfies

$$H_{\phi_N}^*(\psi) = \Pi_+(\bar{\phi}_N\psi), \ \psi \in H^{2-}. \tag{32}$$

We are ready to give a result enabling to compute the elements of the inverse. We have the following theorem:

Theorem 4.1 *Given some integers $0 \leq k, l \leq N$, we have*

$$(T_N^{-1}(f))_{k+1,l+1} = A^{(N)}_{k+1,l+1} - B^{(N)}_{k+1,l+1}, \tag{33}$$

where

$$A^{(N)}_{k+1,l+1} = \langle \Pi_+(\frac{\chi^k}{\bar{g}}), \Pi_+(\frac{\chi^l}{\bar{g}}) \rangle, \tag{34}$$

and where

$$B^{(N)}_{k+1,l+1} = \langle \sum_{s=0}^{\infty} (H^*_{\phi_N} H_{\phi_N})^s \Pi_+(\overline{\phi_N} \Pi_+(\frac{\chi^k}{\bar{g}})), \overline{\phi_N} \Pi_+(\frac{\chi^l}{\bar{g}}) \rangle. \tag{35}$$

Since we are interested by the first column of the inverse, we have to compute explicitly the two terms $A^{(N)}_{k+1,1}$ and $B^{(N)}_{k+1,1}$. We have

Lemma 4.2 *Given $0 \leq k \leq N$ we have*

$$A^{(N)}_{k+1,1} = \beta_0^{(\alpha)} \bar{\beta}_k^{(\alpha)}.$$

Concerning $B^{(N)}_{k+1,1}$ where $0 \leq k \leq N$, we have

Lemma 4.3 *i) Let $N_\alpha = N + 1 + \alpha$. When $N \to \infty$, we have*

$$B^{(N)}_{k+1,1} = \sum_{m=0}^{\infty} \left(\sum_{u=0}^{k} \bar{\beta}^{(\alpha)}_{k-u} F^{(\alpha)}_{2m,N}(u) \right) \left(\frac{\sin(\pi\alpha)}{\pi} \right)^{2m+2} + o\left(\frac{1}{N}\right), \tag{36}$$

where

$$F^{(\alpha)}_{2m,N}(u) = \sum_{j_0}^{\infty} \frac{1}{j_0 + N_\alpha} \sum_{j_1=0}^{\infty} \frac{1}{j_0 + j_1 + N_\alpha} \cdots \sum_{j_{2m-2}=0}^{\infty} \frac{1}{j_{2m-2} + j_{2m-1} + N_\alpha},$$

$$\sum_{j_{2m}=0}^{\infty} \frac{1}{j_{2m} + j_{2m-1} + N_\alpha} \frac{1}{j_{2m} + N_\alpha - u}.$$

We get the following explicit asymptotic formula for inverting.

Corollary 4.1 *Let k be an integer $0 \leq k \leq N$. When $N \to \infty$ we have*

$$(T_N(f))^{-1}_{k+1,1} = \beta_0^{(\alpha)} \left(\bar{\beta}_k^{(\alpha)} - \sum_{m=0}^{\infty} \left(\sum_{u=0}^{k} \bar{\beta}_u^{(\alpha)} F^{(\alpha)}_{2m,N}(k-u) \right) \left(\frac{\sin(\pi\alpha)}{\pi} \right)^{2m+2} \right) + o\left(\frac{1}{N}\right).$$

5 The Determinant of Toeplitz Matrices with a Fisher–Hartwig Symbol

The large behavior of the determinant of a Toeplitz matrix have seen many physical applications from the phase transition in Ising models, from combinatorics and string theory. The Fisher–Hartwig asymptotics refer to the asymptotic behavior of the determinant $D_N(f)$ of a class of $(N+1) \times (N+1)$ Toeplitz matrix $T_N(f)$. When the symbol has no zeros nor jumps in the unit disk, is strictly positive on the circle and satisfies some additional regularity assumptions , the asymptotic behavior is known and described by the celebrated classical Szegö limit theorem. Explicitly, let

$$\ln f(e^{i\theta}) = \sum_{p=-\infty}^{\infty} c_p e^{ip\theta} \tag{37}$$

with the condition

$$\sum_{p=-\infty}^{\infty} \mid p \mid c_p c_{-p} < \infty, \tag{38}$$

then the strong limit theorem of Szegö asserts that

$$\ln D_N(f) = (N+1)c_0(f) + \sum_{n=1}^{\infty} n c_n c_{-n} + o(1). \tag{39}$$

However, the formula becomes harder to determine when the symbol f has singularities on the circle; indeed the condition (38) does not hold when the symbol is of form

$$f(e^{i\theta}) = f_1(e^{i\theta}) \prod_{r=1}^{T} s_{\beta_r}(\theta - \theta_r) t_{\alpha_r}(\theta - \theta_r), \tag{40}$$

where

$$s_\beta(\theta) = e^{-i\beta(\pi-\theta)},\ 0 < \theta < 2\pi, \tag{41}$$

and where

$$t_\alpha(\theta) =\mid 1 - e^{i\theta} \mid^{2\alpha_r},\ \ \mid \alpha \mid < \frac{1}{2}. \tag{42}$$

The multiplicative function f_1 is supposed to be sufficiently smooth, non vanishing, with winding number zero and satisfying the condition (38). When $f_1 = 1$, the symbol f is a called a symbol with a pure Fisher–Hartwig singularity symbol. The Fisher–Hartwig conjecture asserts that for $N \to \infty$

$$D_N(f) = GM(f_1)^{N+1}(N+1)^{\sum_{r=1}^{R}(\alpha_r^2 - \beta_r^2)} E_{f_1} + o(1), \tag{43}$$

where

$$E_{f_1} = E[f_1] \prod_{r=1}^{R} \left((f_1)_+(\chi_r)^{-\alpha_r+\beta_r}(f_1)_+(\chi_r)^{-\alpha_r-\beta_r}\right)$$
$$\times \prod_{1\leq s\neq r\leq R} \left(1 - e^{i(\theta-\theta_r)})^{-(\alpha_r+\beta_r)(\alpha_s-\beta_s)}\right)$$
$$\times \frac{G(1+\alpha_r+\beta_r)(G(1+\alpha_r-\beta_r)}{G(1+2\alpha_r)},$$

where

$$E[f_1] = \exp\left(\sum_{k=1}^{\infty} k \ \widehat{\log f_1(\theta)}(k)\widehat{\log f_1(\theta)}(-k)\right),$$

the notation G stands for the Barnes G-function: an entire analytic function defined by

$$G(z) = (2\pi)^{\frac{z}{2}} e^{-\frac{z(z+1)}{2} - \frac{Cz^2}{2}} \prod_{n=1}^{\infty} \left\{\left(1+\frac{z}{n}\right)^n e^{-z+\frac{z^2}{n}}\right\},$$

where $C = 0.577$ is the Euler's constant. We have $G(z+1) = \Gamma(z)G(z)$, $G(0) = 0$ and $G(1) = 1$. The first result on the conjecture has been completely solved by Ehrarhdt and Silbermann (1997) in the case of one singularity ($R = 1$). Thanks to the pioneering works of Basor, Boettcher, Widom and their collaborators, the conjecture has been completely solved in the general case.

6 A Sketch of the Proof of the Main Result

We give only a sketch of the proof of i). The interested reader will find more details in Kateb *et al.* (2004).
To prove i), we begin to study the asymptotic behavior of the first element $(T_N^{-1}(f))_{1,1}$. By some elementary algebraic manipulations based on Cramer's rule, we get

$$(T_N(f))^{-1}_{1,1} = \frac{D_{N-1}(f)}{D_N(f)}, \tag{44}$$

the formula giving the asymptotic behavior of the determinants implies

$$(T_N(f))^{-1}_{1,1} = \frac{1}{G(f_1)} \frac{N^{\alpha^2}}{(N+1)^{\alpha^2}} (1+o(1))$$

and then

$$(T_N(f))^{-1}_{1,1} = \frac{1}{G(f_1)} \left(1 - \frac{\alpha^2}{N} + o\left(\frac{1}{N}\right)\right).$$

The formula contains a useful information; indeed after setting $k = 0$ in formula (38), we deduce after identification

$$\sum_{m=0}^{\infty} F_{2m,N}(0)\left(\frac{\sin(\pi\alpha)}{\pi}\right)^{2m+2} = \frac{\alpha^2}{N}(1+o(1)). \tag{45}$$

This relation is important to guess the asymptotic behavior of the other first elements. Indeed since the inverting formula learns us that

$$\begin{aligned}(T_N(f))^{-1}_{k+1,1} &= \beta_k^{(\alpha)}\beta_0^{(\alpha)} \\ &\quad -\beta_0^{(\alpha)}\sum_{u=0}^{k}\bar{\beta}_u^{(\alpha)}\sum_{m=0}^{\infty}F_{2m,N}(k-u)\left(\frac{\sin(\pi\alpha)}{\pi}\right)^{2m+2}(1+o(1)),\end{aligned}$$

and since for small values of k we have

$$\frac{k}{N} = o(1) \Rightarrow F_{2m,N}(k-u) = F_{2m,N}(0) + O\left(\frac{1}{N^2}\right),$$

we then obtain

$$\begin{aligned}(T_N(f))^{-1}_{k+1,1} &= \bar{\beta}_k^{(\alpha)}\beta_0^{(\alpha)} \\ &\quad -\beta_0^{(\alpha)}\sum_{u=0}^{k}\bar{\beta}_u^{(\alpha)}\left(\sum_{m=0}^{\infty}F_{2m,0}(k-u)\left(\frac{\sin(\pi\alpha)}{\pi}\right)^{2m+2}\right)(1+o(1)) \\ &= \beta_0^{(\alpha)}\left(\bar{\beta}_k^{(\alpha)} - \frac{\alpha^2}{N}\sum_{u=0}^{k}\bar{\beta}_u^{(\alpha)}\right) + o\left(\frac{1}{N}\right) \qquad (46)\\ &= \beta_0^{(\alpha)}\left(\bar{\beta}_k^{(\alpha)} - \frac{\alpha^2}{N}\bar{\beta}_k^{(\alpha+1)}\right) + o\left(\frac{1}{N}\right). \qquad (47)\end{aligned}$$

The detailed proof of *ii*) and *iii*) is out of the scope of this work. We refer the reader to the papers by Kateb *et al.* (2004) and Rambour and Seghier (2003).

7 Application to the Prediction of Long–Range Dependent Processes

Generating long–range dependent processes is of interest for statisticians who whish to analyze the finite sample properties of statistics and estimators related to this type of processes. Usually, for linear processes in the conditional mean, one uses the Durbin–Levinson, henceforth DL, algorithm, which consists in computing recursively the autoregressive coefficients for a fractional Gaussian noise process. Assume that we wish to generate samples of a fractional ARMA, FARIMA(0,d,0), process $\{\nu_t\}_{-\infty}^{\infty}$ defined as

$$(1-L)^d\nu_t = \xi_t, \quad \xi_t \sim N(0,\sigma_\xi^2). \tag{48}$$

We generate the sequence $\{Y_t\}$ with the same finite dimensional distribution as $\{\nu_t\}$ as follows:

$$Y_0 = v_0^{1/2}\xi_0 \tag{49}$$

$$Y_t = \sum_{j=1}^{t} \phi_{tj} Y_{t-j} + v_t^{1/2}\xi_t, \quad t \geq 1,$$

$$v_0 = \mathrm{Var}(\nu_0), \quad \phi_{tt} = \frac{d}{t-d}, \quad v_t = v_{t-1}(1 - \phi_{tt}^2), \tag{50}$$

$$\phi_{ti} = \phi_{t-1,i} - \phi_{tt}\phi_{t-1,t-i}, \quad 0 \leq i \leq t-1.$$

A FARIMA(p, d, q) sequence can be generated from this FARIMA$(0, d, 0)$ sequence.

As mentioned above, this algorithm is of order N^2, which is not computationally demanding for very large samples, e.g., 10,000 observations. For simulation purpose, the triangular matrix of coefficients ϕ_{ij} is computed once, stored in memory, which is largely feasible using the storage capacities of modern computers, and then used for each replication. Thus, for that particular purpose, there is no advantage in computing the elements of the triangular matrix $[\phi_{ij}]$ using the approximation method defined by equations (15), (16) and (18). Furthermore, Bardet *al.* (2003) compared several generators of long–range dependent processes using several tests, and concluded that the Durbin–Levinson, although it is a good algorithm, is not the best one. Later in this chapter, we will compare simulations obtained with both methods for computing the coefficients ϕ_{ij}, using in each case the same sequence of error terms $\{\xi_t\}$ for illustrating the accuracy of the approximation.

The point of view is completely different if we are observing a sample of observations $Y_0, \ldots, Y_T$, and we whish to forecast the realization of Y_t at time $t+1$, i.e., compute $\hat{Y}_{t+1}$:

$$\hat{Y}_{t+1} = \phi_{t+1,1}Y_t + \phi_{t+1,2}Y_{t-1} + \cdots + \phi_{t+1,t+1}Y_0. \tag{51}$$

In that case, there is no need to recursively compute the elements ϕ_{ij}, $i < t+1$ for computing the sequence $\{\phi_{t+1,j}\}$, $j = 1, \ldots, t+1$. We approximate it with the sequence $\{\psi_{t+1,j}\}$ computed using formulas (15), (16) and (18).

An open question is then the choice for the threshold k_{inf} i.e., the switch between approximations (15) and (16), and the threshold k_{sup}, for the choice between the approximations (16) and (18). We consider several sample sizes, $N = 100, 200, \ldots, 1000, 1500, \ldots, 18500$, several values for the degree of long–range dependence, $d = 0.05, 0.10, \ldots, 0.45$, and find for each configuration the optimal values k_{inf} and k_{sup} which minimize the distance:

$$\sum_{j=1}^{N} (\phi_{N,j} - \psi_{N,j})^2, \tag{52}$$

where $\{\phi_{N,j}\}$ and $\{\psi_{N,j}\}$ respectively denote the sequence of coefficients computed with the Durbin–Levinson algorithm and the approximation formulas.

Table 1 below reports the optimal values for k_{inf} and k_{sup} for $N = 100$ and various values for d, and the fraction of the sequence $\{\psi_{N,j}\}$ below each threshold, i.e., k_{inf}/N and k_{sup}/N.

Table 1. Optimal values for k_{inf} and k_{sup}, for $N = 100$ and various values for the memory parameter d

d	k_{inf}	k_{sup}	k_{inf}/N	k_{sup}/N
0.05	20	98	0.2	0.98
0.10	20	98	0.2	0.98
0.15	20	97	0.2	0.97
0.20	20	97	0.2	0.97
0.25	20	97	0.2	0.97
0.30	20	97	0.2	0.97
0.35	20	97	0.2	0.97
0.40	20	97	0.2	0.97
0.45	20	97	0.2	0.97

Figure 1 illustrates the difference between the coefficients $\psi_{100,i} - \phi_{100,i}$ for these optimal values for k_{inf} and k_{sup}. The magnitude of the difference $\psi_{100,i} - \phi_{100,i}$ is very small, and takes its highest values at the thresholds k_{inf} and k_{sup} between the approximation formulas.

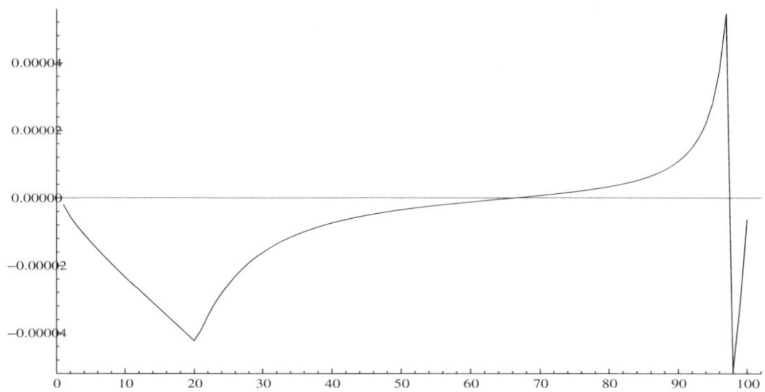

Fig. 1. Difference between the coefficients $\psi_{100,i}$ and $\phi_{100,i}$, $i = 1, \ldots, 100$ respectively given by the approximation formula and by the Durbin–Levinson algorithm. $N = 100$, $k_{\text{inf}} = 20$, $k_{\text{sup}} = 97$, $d = 0.35$

Table 2. Optimal values for k_{inf} and k_{sup}, for $d = 0.35$ and various values for the polynomial size N

N	k_{inf}	k_{sup}	k_{inf}/N	k_{sup}/N
200	32	196	0.1600	0.9800
300	42	295	0.1400	0.9833
400	52	394	0.1300	0.9850
500	60	493	0.1200	0.9860
600	68	593	0.1133	0.9883
700	76	692	0.1086	0.9886
800	83	791	0.1037	0.9887
900	90	891	0.1000	0.9900
1000	96	990	0.0960	0.9900
1500	127	1488	0.0847	0.9920
2000	154	1987	0.0770	0.9935
2500	180	2485	0.0720	0.9940
3000	203	2984	0.0677	0.9947
3500	225	3482	0.0643	0.9949
4000	247	3981	0.0617	0.9952
4500	267	4480	0.0593	0.9956
5000	287	4979	0.0574	0.9958
5500	306	5478	0.0556	0.9960
6000	324	5977	0.0540	0.9962
7000	360	6975	0.0514	0.9964
8000	393	7974	0.0491	0.9967
9000	426	8972	0.0473	0.9969
10000	457	9971	0.0457	0.9971
11000	487	10969	0.0443	0.9972
12000	517	11968	0.0431	0.9973
13000	545	12967	0.0419	0.9975
14000	573	13965	0.0409	0.9975
15000	600	14964	0.0400	0.9976
16000	627	15963	0.0392	0.9977
17000	653	16962	0.0384	0.9978
18000	678	17961	0.0377	0.9978
19000	703	18960	0.0370	0.9979
20000	728	19959	0.0364	0.9979
30000	955	29949	0.0318	0.9983
40000	1158	39942	0.0289	0.9985
50000	1345	49935	0.0269	0.9987
60000	1520	59929	0.0253	0.9989
70000	1685	69923	0.0241	0.9990
80000	1843	79918	0.0230	0.9990
90000	1994	89913	0.0221	0.9990
100000	2134	99908	0.0214	0.9991

Since the choice for k_{inf} and k_{sup} is not much affected by the value of the memory parameter d, we report on table 2 the optimal values for k_{inf} and k_{sup} for various polynomial sizes N, from $N = 200$ to $N = 100,000$, and $d = 0.35$. As N gets large, both k_{inf} and k_{sup} increases, but more slowly as $k_{\text{inf}}/N \to 0$ and $(N - k_{\text{sup}})/N \to 0$, i.e.,

$$1/k_{\text{inf}} + k_{\text{inf}}/N \to 0, \quad 1/(N-k_{\text{sup}}) + (N-k_{\text{sup}})/N \to 0 \quad \text{as} \quad N \to \infty. \tag{53}$$

The choice for k_{inf} and k_{sup} resembles a bandwidth selection problem.

Figure 2 below illustrates the difference between the coefficients $\psi_{15000,i} - \phi_{15000,i}$, $i = 1, \ldots, 15000$

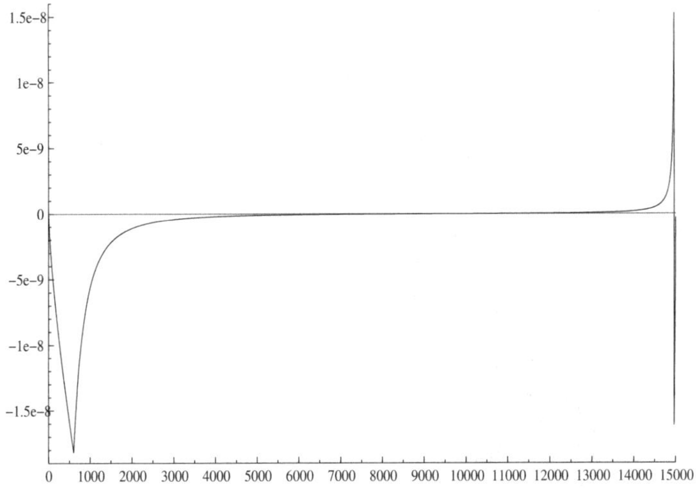

Fig. 2. Difference between the coefficients $\psi_{15000,i}$ and $\phi_{15000,i}$, $i = 1, \ldots, 15000$ respectively given by the approximation formula and by the Durbin–Levinson algorithm. $N = 15000$, $k_{\text{inf}} = 600$, $k_{\text{sup}} = 14964$, $d = 0.30$

7.1 Monte Carlo Analysis

As we can see from figures 1 and 2, even for small sample size, there is no significant difference between the coefficients generated by the DL algorithm, and the coefficients generated by the approximation formula. As a consequence, long range dependent series generated by both formula, should not differ too much, provided that the same sequence of pseudo–error terms $\{\xi_t\}$ is used for both series. We generate the two processes:

$$Y_0 = v_0^{1/2}\xi_0 \tag{54}$$

$$Y_t = \sum_{j=1}^{t} \phi_{tj} Y_{t-j} + v_t^{1/2}\xi_t, \quad t \geq 1, \tag{55}$$

$$Z_0 = v_0^{1/2}\xi_0 \tag{56}$$

$$Z_t = \sum_{j=1}^{t} \psi_{tj} Z_{t-j} + v_t^{1/2}\xi_t, \quad t \geq 1, \tag{57}$$

where the sequence $\{v_t\}$ is defined by equation (50). In fact, the 100 first observations of the process $\{Z_t\}$ are generated by the Durbin–Levinson algorithm, the next ones are generated using equation (57). We also do not use at each time t the optimal values for k_{inf} and k_{sup}. Using the results from table 2, we set $k_{\text{inf}} = [0.05t]$ and $k_{\text{sup}} = [0.99t]$, i.e., the lower and upper thresholds are respectively set to 5% and 99% of the polynomials. As we will see later, this rough choice works remarkably well.

We check the accuracy of the approximation by estimating the long memory parameter of the two processes with the wavelet estimator by Veitch and Abry (1999), the local Whittle (LW) estimator proposed by Künsch (1987) and further analyzed by Robinson (1995), and the estimator based on a log–periodogram (LP) regression in the spectral domain proposed by Geweke and Porter–Hudak (1983). See also the chapter by Teyssière and Abry (2005) in this volume which compares these three estimators for various processes.

Since we are using the wavelet estimator, we consider samples of size $N = 10,000$. For this estimator, we consider the ranges of lowest octaves $j_1 = 1, \ldots, 7$ and the number of vanishing moments $\mathcal{N} = 2, \ldots, 6$. For the LW estimator, we select the optimal bandwidth m using the iterative procedure proposed by Henry and Robinson (1996); see also Henry (2001). For the LP estimator, we set the optimal bandwidth $m = [N^{0.8}]$.[4] For the choice of the optimal bandwidth for both LW and LP, see the chapter by Henry (2005) in this volume.

We consider several values for $d = 0.05, 0.10, \ldots, 0.45$. Because of the large number of results, we only report some of them. Since the coefficients ϕ_{tj} and ψ_{tj} do not differ too much, and we are using the same sequence of pseudo error terms $\{\xi_t\}$, the estimation results between each simulated sequence are very similar. Table 3 below reports some estimates obtained from a fractional Gaussian noise with $d = 0.25$:

[4] We also used Henry's (2001) optimal bandwidth for the LP estimator. Since results do not differ too much from the LW estimator, with a slightly higher Root Mean Squared Error (RMSE), we do not report the results here.

Table 3. Estimates of the long memory parameter for the two series Y_t and Z_t respectively generated using equations (54)–(55) and equations (56)–(57). For the wavelet estimator, we set $j_1 = 5$ and $\mathcal{N} = 2$. $d = 0.25$, $N = 10000$

Replication	LW_Y	LP_Y	Wav_Y	LW_Z	LP_Z	Wav_Z
1	0.3366	0.2140	0.3069	0.3365	0.2140	0.3069
2	0.2473	0.2507	0.1933	0.2471	0.2507	0.1933
3	0.2314	0.2400	0.2565	0.2313	0.2400	0.2565
4	0.2092	0.2552	0.2576	0.2092	0.2552	0.2576
5	0.2334	0.2541	0.2737	0.2381	0.2541	0.2737

For the LP and wavelet estimators, the difference is of order 10^{-5}. The slight difference for the Local Whittle (LW) estimator is due to a difference in the selection of the optimal bandwidth m for the processes $\{Y_t\}$ and $\{Z_t\}$: as we will see in Figure 3 in section 7.2, there is a negligible discrepancy between the periodogram of both processes, which results in a small difference in the optimal bandwidth parameter.

Table 4 below reports the average estimates over 5000 replication of the memory parameter of both processes $\{Y_t\}$ and $\{Z_t\}$.

Table 4. Average estimates of the long memory parameter for the processes $\{Y_t\}$ and $\{Z_t\}$ respectively generated using equations (54)–(55) and equations (56)–(57). For the wavelet estimator, we set $j_1 = 5$ and $\mathcal{N} = 2$. $N = 10000$, 5000 simulations. The Root Mean Squared Errors are between parentheses

d	LW_Y	LP_Y	Wav_Y	LW_Z	LP_Z	Wav_Z
0.05	0.0486	0.0493	0.0508	0.0486	0.0493	0.0508
	(0.0220)	(0.0162)	(0.0372)	(0.0220)	(0.0162)	(0.0372)
0.10	0.0979	0.0992	0.1009	0.0979	0.0992	0.1008
	(0.0238)	(0.0162)	(0.0373)	(0.0237)	(0.0162)	(0.0373)
0.15	0.1471	0.1492	0.1508	0.1469	0.1492	0.1509
	(0.0271)	(0.0162)	(0.0367)	(0.0266)	(0.0162)	(0.0373)
0.20	0.1961	0.1993	0.2010	0.1961	0.1993	0.2010
	(0.0297)	(0.0162)	(0.0374)	(0.0297)	(0.0162)	(0.0374)
0.25	0.2458	0.2494	0.2511	0.2457	0.2494	0.2511
	(0.0335)	(0.0162)	(0.0375)	(0.0337)	(0.0162)	(0.0375)
0.30	0.2938	0.2996	0.3012	0.2939	0.2996	0.3012
	(0.0395)	(0.0162)	(0.0377)	(0.0389)	(0.0162)	(0.0377)
0.35	0.3420	0.3498	0.3512	0.3419	0.3498	0.3512
	(0.0458)	(0.0162)	(0.0378)	(0.0458)	(0.0162)	(0.0378)
0.40	0.3895	0.4001	0.4013	0.3894	0.4001	0.4013
	(0.0528)	(0.0162)	(0.0381)	(0.0529)	(0.0162)	(0.0381)
0.45	0.4371	0.4505	0.4514	0.4369	0.4505	0.4514
	(0.0615)	(0.0163)	(0.0383)	(0.0616)	(0.0163)	(0.0383)

Although the thresholds parameters $k_{\inf}$ and $k_{\sup}$ have not been optimally chosen, the memory properties for both processes look very similar, which emphasizes the accuracy of the approximation formulas (15), (16) and (18) even for generating long–range dependent processes. Thus, these formulas are reliable for 1–step ahead forecasting.

7.2 Goodness–of–Fit Test for the Spectral Density

Besides the comparison of the semiparametric estimation of the long–memory parameter for the series generated by the DL algorithm and the approximation formula, we conducted a parametric goodness–of–fit test based on the spectral density, and test whether the spectral densities of the processes generated by the DL algorithm and the approximation formula are the same. Figure 3 below displays an example of the periodogram $I_{DL}(\lambda)$ of a series generated with the DL algorithm with the periodogram of the series generated with the approximation formula $I_A(\lambda)$ using the same sequence of pseudo–error terms:

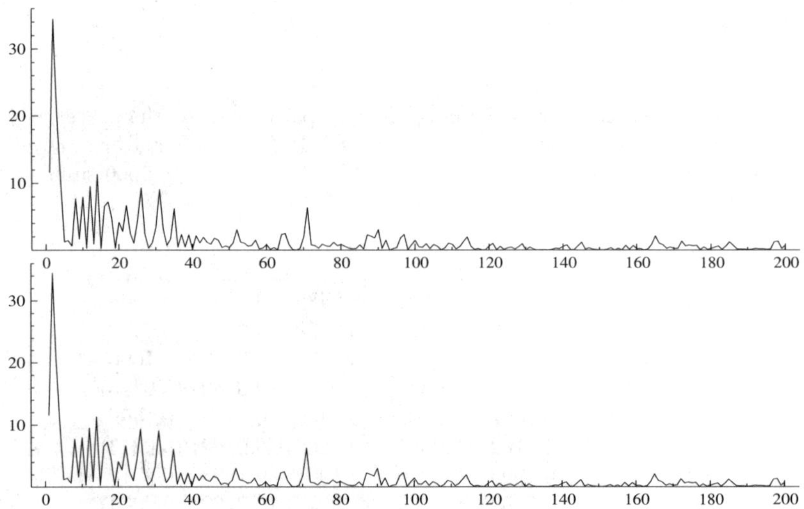

Fig. 3. Periodograms of two series generated by the DL algorithms (above) and the approximation formula (below)

The difference between the two estimated spectra is not noticeable, which explains the negligible difference between the LW estimators for the series generated with either processes; see discussion of table 3 above.

Bardet *et al.* (2003) use the goodness–of–fit test by Chen and Deo (2004) and concluded that the DL algorithm, although a good one, is not the best algorithm for generating long–range dependent data. We carried a little scale experiment using the goodness–of–fit test by Fay and Philippe (2002) based on the information measure, or logarithmic contrast, which compares the spectral densities f and g

$$S(f,g) = \log \int_0^{2\pi} \frac{f(\lambda}{g(\lambda)} \frac{d\lambda}{2\pi} - \int_0^{2\pi} \log \frac{f(\lambda}{g(\lambda)} \frac{d\lambda}{2\pi}. \tag{58}$$

The test by Fay and Philippe compares the (possibly) pooled and (possibly) tapered periodogram of the series with the "plug–in" estimated spectra, using the Whittle estimator, of a FARIMA$(0,d,0)$ process. According to these authors, this test has a better size than the test by Chen and Deo (2002). Both algorithms give close values for this test, which is not surprising as tapering the periodogram will reduce again the difference between the spectra of the series generated by the two algorithms.[5]

7.3 Forecasting Volatility

We consider here the issue of volatility forecasting, which is important in risk management, and present several volatility models which could rely on the forecasting property of the Durbin–Levinson algorithm as the best one–step ahead linear predictor. Since this algorithm is recursive, the approximation formula considered here is of interest as very large samples, e.g., samples of size over 10,000 observations, are common in finance. For instance, the chapter by Lavielle and Teyssière (2005) in this volume consider samples over 16,000 observations. The application, evaluation and comparison of the forecasting performance[6] of these different models will be carried in a subsequent research work that will deal with the issue of h–steps ahead prediction.

Standard volatility models are the Generalized ARCH processes and the stochastic volatility models. Robinson (1991) generalized the class of GARCH processes to the long–memory case, while Breidt *et al.* (1998) introduced the long–memory stochastic volatility process, henceforth LMSV, defined as:

$$r_t = \sigma_t \zeta_t, \quad \zeta_t \sim N(0,1), \tag{59}$$

$$\sigma_t = \sigma \exp(X_t/2), \quad X_t \sim \text{FARIMA}(p,d,q), \tag{60}$$

where σ is a scale parameter, and X_t are independent of ζ_t. The process $\{r_t^2\}$ is linearized as follows:

$$\log r_t^2 = \log \sigma^2 + E(\log \zeta_t^2) + u_t + \left(\log \zeta_t^2 - E(\log \zeta_t^2)\right) = \mu + u_t + \varepsilon_t, \tag{61}$$

where $\{\varepsilon_t\}$ is iid, $E(\varepsilon_t) = 0$ and $\text{Var}(\varepsilon_t) = \sigma_\varepsilon^2$. Since $\zeta \sim N(0,1)$, then $\log \zeta^2$ is distributed as the log of a χ_1^2 random variable, thus $E(\log \zeta^2) = 1.27$ and $\sigma_\varepsilon^2 = \pi^2/2$.

[5] As in Fay and Philippe (2002) we set the tapering order $p = 1$, and the size of the pooling block $m = 5$.

[6] See Christoffersen and Diebold (2000), Diebold, Gunter and Tay (1998).

The spectral density of the process $\{\log r_t^2\}$ is given by:

$$f(\lambda) = \frac{\sigma_e^2}{2\pi} \frac{|\theta(e^{-i\lambda})|^2}{|\phi(e^{-i\lambda})|^2} \left|1 - e^{-i\lambda}\right|^{-2d} + \frac{\sigma_\varepsilon^2}{2\pi}, \quad \lambda \in (-\pi, \pi) \tag{62}$$

which is the sum of the spectrum of the long–range dependent process $\{X_t\}$ + $\sigma_\varepsilon^2/2\pi$, and then has the same shape as the spectrum of a long–range dependent process.

Since the series $\{\log |r_t|^\delta\}$, with $\delta > 0$ displays long–range dependence, the intensity of which is the highest for $\delta = 1$, we can forecast $\log r_{T+1}^2$ having observed a sample $\log r_0^2, \dots, \log r_T^2$; see Chen, Hurvich and Lu (2004).[7] Since the DL algorithm provides the best linear forecast $\widehat{\log r_{T+1}^2}$, our fast approximation method can provide the same forecast more quickly.

Brockwell and Dalhaus (2004) generalized the DL algorithm for the case of step–h prediction, i.e., predicting X_{T+h} having observed only the sample $X_1, \dots, X_T$. Lux (2003) considered the application of this generalized DL algorithm for the prediction of the volatility process using the multifractal model of asset returns (MMAR) by Mandelbrot, Fisher and Calvet (1997), although the modeling of the volatility by a FARIMA$(0, d, 0)$ process is a rather rough approximation of the statistical properties of the MMAR process.

A third application of the (generalized) DL algorithm is the prediction of the series of absolute asset prices returns $|r_t|$. The periodogram and the autocorrelation function (ACF) of absolute (and squared) returns have the typical shape of a long–range dependent process, i.e., a pole near the zero frequency and a hyperbolic decay of the ACF. However, as emphasized by Granger (2000), absolute (and squared) returns share the properties of the second moments of strongly dependent processes, but not the properties of the first moments as unlike FARIMA processes, absolute (and squared) returns empirical processes do not display local trends. Thus, modeling the absolute returns by a simple FARIMA process is irrelevant.

It has been claimed that the best way to model the volatility of the returns r_t is by using a change–point GARCH(1,1) process, i.e., a GARCH(1,1) process with changing coefficients so that the unconditional variance of the process changes significantly, the magnitude of the jumps in the unconditional variance being positively correlated with the intensity of long–range dependence in the absolute (and squared) returns process; see e.g., Mikosch and Stărică (1999, 2003, 2004). Stărică and Granger (2001) even modeled the returns process by a simple change in variance process.

However, since absolute returns are used as a measure of risk, see Granger and Ding (1995), modeling this process is of interest. Granger and Hyung (2004) reconcile both approaches by modeling the absolute returns using both a FARIMA$(0, d, 0)$ process and a change–point (occasional break) process,

[7] Deo, Hurvich and Lu (2003) propose to forecast $|r_{T+1}|^\delta$ from $|r_0|^\delta, \dots, |r_T|^\delta$, where δ is a "scaling" factor which makes the distribution of $\{|r_t|^\delta\}$ close to the normal distribution.

$$(1-L)^d|r_t| = m_o + \varepsilon_t + \sum_{j=1}^{t} q_t\zeta_t, \quad \zeta_t \sim iid(0,\sigma_\zeta^2), \tag{63}$$

where $q_t \in \{0,1\}$ is a two–state process governed by a Markov process. Granger and Hyung (2004) found that this combination improves the forecasting of the absolute returns.

References

1. Abry, P., Flandrin, P., Taqqu, M. and Veitch, D. (2003). Self–similarity and long–range dependence through the wavelet lens. In: Doukhan, P., Oppenheim, G. and Taqqu, M.S. (Eds.), *Long–Range Dependence: Theory and Applications.* Birkhauser, Boston, pp. 527–556.
2. Bardet, J-M., Lang, G., Oppenheim, G., Philippe, A. and Taqqu, M.S. (2003). Generators of Long–range Dependent Processes: A Survey. In: Doukhan, P., Oppenheim, G. and Taqqu, M.S. (Eds.), *Long-Range Dependence: Theory and Applications*, 579–623. Birkhauser, Boston.
3. Beran, J. (1994). *Statistics for Long–Memory Processes.* Chapman & Hall, New York.
4. Bleher P.M. (1981). Inversion of Toeplitz matrices *Trans. Moscow Math. Soc.*, **2**, 201–224.
5. Bollerslev, T. and Mikkelsen, H.-O. (1996). Modelling and pricing long–memory in stock market volatility. *Journal of Econometrics*, **73,** 151–184.
6. Breidt, J., Crato, N. and de Lima, P. (1998). On the detection and estimation of long memory in stochastic volatility. *Journal of Econometrics*, **83,** 325–348.
7. Brockwell, P.J. and Dalhaus, R. (2004). Generalized Levinson–Durbin and Burg algorithms. *Journal of Econometrics*, **118,** 129–149.
8. Chen, W., Hurvich, C.M. and Lu, Y. (2004). On the correlation matrix of the discrete Fourier transform and the fast solution of large Toeplitz systems for long–memory time series. *Preprint.*
9. Chen, W. and Deo, R. (2004). A generalized portmanteau goodness–of–fit test for time series models. *Econometric Theory*, **20,** 382–416.
10. Christoffersen, P.F. and Diebold, F.X. (2000). How relevant is volatility forecasting for financial management. *Review of Economics and Statistics*, **82,** 12–22.
11. Deo, R., Hurvich, C.M. and Lu, Y. (2003). Forecasting realized volatility using a long memory stochastic volatility model: estimation, prediction and seasonal adjustment. *Preprint.*
12. Diebold, F.X., Gunter, T. and Tay, A. (1998). Evaluating density forecasts, with application to financial risk management. *International Economic Review*, **39,** 863–883.
13. Ehrhardt, T. and Silbermann, B. (1997). Toeplitz determinants with one Fisher–Hartwig singularity, *Journal of Functional Analysis*, **148,** 229-256.
14. Fay, G. and Philippe, A. (2002). Goodness–of–fit test for long range dependent processes. *ESAIM: Probability and Statistics*, **6,** 239–258.
15. Fisher, M.E. and Hartwig, R.E. (1968). Toeplitz determinants: some applications, theorems and conjectures. *Adv. Chem. Phys*, **15,** 333–353.

16. Geweke, J. and Porter–Hudak, S. (1983). The estimation and application of long memory time series models. *Journal of Time Series Analysis*, **4,** 221–238.
17. Granger, C.W.J. (2002). Long memory, volatility, risk and distribution. *Preprint.*
18. Granger, C.W.J. (2000). Current perspectives on long memory processes. *Academia Economic Papers*, **28,** 1–16.
19. Granger, C.W.J. and Hyung, N. (2004). Occasional structural breaks and long memory with an application to the S&P 500 absolute stock returns. *Journal of Empirical Finance*, **11,** 399–421.
20. Granger, C.W.J. and Ding, Z. (1995). Some properties of absolute returns, an alternative measure of risk. *Annales d'Économie et de Statistique*, **40,** 67–91.
21. Henry, M. (2005). Bandwidth choice, optimal rates and adaptivity in semiparametric estimation of long memory. In: Teyssière, G., Kirman, A. (Eds.), *Long–Memory in Economics.* Springer Verlag, Berlin. Appears in this volume.
22. Henry, M. (2001). Robust automatic bandwidth for long–memory. *Journal of Time Series Analysis*, **22,** 293–316.
23. Henry, M. and Robinson, P.M. (1996). Bandwidth choice in Gaussian semiparametric estimation of long range dependence. In: Robinson, P.M., Rosenblatt, M. (Eds.), *Athens Conference on Applied Probability and Time Series Analysis, Time Series Analysis, In Memory of E.J. Hannan.* Lecture Notes in Statistics, Vol. 115. Springer Verlag, New York, pp. 220–232.
24. Kateb, D., Rambour, P. and Seghier, A. (2004). The inversion of Toeplitz matrices : the singular case *Prépublications de l'Université Paris-Sud* 2004-10.
25. Künsch, H.R. (1987). Statistical Aspects of Self-Similar Processes In Yu. Prohorov and V. V. Sazanov editors, *Proceedings of the First World Congress of the Bernoulli Society*, **1**, 67–74. VNU Science Press, Utrecht.
26. Lavielle, M. and Teyssière, G. (2005). Semiparametric adaptive detection of multiple change–points in asset price volatility. In: Teyssière, G., Kirman, A. (Eds.), *Long–Memory in Economics.* Springer Verlag, Berlin. Appears in this volume.
27. Lux, T. (2003). The multifractal model of asset returns: its estimation via GMM and its use for volatility forecasting. *Preprint.*
28. Mandelbrot, B.B., Fisher, A. and Calvet, L. (1997). The Multifractal model of asset returns. *Cowles Foundation Discussion Paper* 1164.
29. Mikosch, T. and Stărică, C. (2004). Non–stationarities in financial time series: the long range dependence and the IGARCH effects. *Review of Economics and Statistics*, **86,** 278–290.
30. Mikosch, T. and Stărică, C. (2003). Long–range dependence effects and ARCH modeling. In: Doukhan, P., Oppenheim, G. and Taqqu, M.S. (Eds.), *Long–Range Dependence: Theory and Applications.* Birkhauser, Boston, pp. 439–459.
31. Mikosch, T. and Stărică, C. (1999). Change of structure in financial time series, long range dependence and the GARCH model. *Preprint.*
32. Rambour, P. and Seghier, A. (2003). Inversion des matrices de Toeplitz à symboles singuliers. Extension d'un résultat de Keysten. Submitted to *Annales de l'Institut de Fourier.*
33. Robinson, P.M. (1995). Gaussian semiparametric estimation of long–range dependence. *The Annals of Statistics*, **23,** 1630–1661.
34. Robinson, P.M. (1991). Testing for strong serial correlation and dynamic conditional heteroskedasticity in multiple regression. *Journal of Econometrics*, **47,** 67–84.

35. Stărică, C. and Granger, C.W.J. (2001). Non–stationarities in stock returns. *Review of Economics and Statistics*, forthcoming.
36. Teyssière, G. and Abry, P. (2005). Wavelet analysis of nonlinear long–range dependent processes. Applications to financial time series. In: Teyssière, G. and Kirman, A., (Eds.), *Long–Memory in Economics.* Springer Verlag, Berlin. Appears in this volume.
37. Veitch, D. and Abry, P. (1999). A wavelet based joint estimator of the parameters of long–range dependence. *IEEE Transactions on Information Theory*, **45,** 878–897.
38. Wu, T.T. (1966). Theory of Toeplitz determinants and the spin correlations of the two–dimensional Ising Model. I. *Physical Review*, **149,** 380–401.
39. Zygmund, A. (1959). *Trigonomic Series*, second edition. Cambridge University Press.

Part II

Economic Models

A Nonlinear Structural Model for Volatility Clustering*

Andrea Gaunersdorfer[1] and Cars Hommes[2]

[1] Department of Business Studies, University of Vienna. andrea.gaunersdorfer@univie.ac.at
[2] Center for Nonlinear Dynamics in Economics and Finance, University of Amsterdam. C.H.Hommes@uva.nl

Summary. A simple nonlinear structural model of endogenous belief heterogeneity is proposed. News about fundamentals is an IID random process, but nevertheless volatility clustering occurs as an endogenous phenomenon caused by the interaction between different types of traders, fundamentalists and technical analysts. The belief types are driven by adaptive, evolutionary dynamics according to the success of the prediction strategies as measured by accumulated realized profits, conditioned upon price deviations from the rational expectations fundamental price. Asset prices switch irregularly between two different regimes – periods of small price fluctuations and periods of large price changes triggered by random news and reinforced by technical trading – thus, creating time varying volatility similar to that observed in real financial data.

* Earlier versions of this paper were presented at the 27th annual meeting of the EFA, August 23–26, 2000, London Business School, the International Workshop on Financial Statistics, July 5–8, 1999, University of Hong Kong, the Workshop on Expectational and Learning Dynamics in Financial Markets, University of Technology, Sydney, December 13–14, 1999, the Workshop on Economic Dynamics, January 13–15, 2000, University of Amsterdam, and the workshop Beyond Efficiency and Equilibrium, May 18-20, 2000 at the Santa Fe Institute. We thank participants of all workshops for stimulating discussions. Special thanks are due to Arnoud Boot, Peter Boswijk, Buz Brock, Carl Chiarella, Dee Dechert, Engelbert Dockner, Doyne Farmer, Gerwin Griffioen, Alan Kirman, Blake LeBaron, Thomas Lux, Ulrich Müller, and Josef Zechner. Detailed comments by two anonymous referees have led to several improvements. We also would like to thank Roy van der Weide and Sebastiano Manzan for their assistance with the numerical simulations. This research was supported by the Austrian Science Foundation (FWF) under grant SFB#010 ('Adaptive Information Systems and Modelling in Economics and Management Science.') and by the Netherlands Organization for Scientific Research (NWO) under an NWO-MaG Pionier grant.

1 Introduction

Volatility clustering is one of the most important 'stylized facts' in financial time series data. Whereas price changes themselves appear to be unpredictable, the magnitude of those changes, as measured e.g., by the absolute or squared returns, appears to be predictable in the sense that large changes tend to be followed by large changes – of either sign – and small changes tend to be followed by small changes. Asset price fluctuations are thus characterized by episodes of low volatility, with small price changes, irregularly interchanged by episodes of high volatility, with large price changes. This phenomenon was first observed by Mandelbrot (1963) in commodity prices.[3] Since the pioneering papers by Engle (1982) and Bollerslev (1986) on autoregressive conditional heteroskedastic (ARCH) models and their generalization to GARCH models, volatility clustering has been shown to be present in a wide variety of financial assets including stocks, market indices, exchange rates, and interest rate securities.[4]

In empirical finance, volatility clustering is usually modeled by a *statistical model*, for example by a (G)ARCH model or one of its extensions, where the conditional variance of returns follows a low order autoregressive process. Other approaches to modeling volatility clustering and long memory by statistical models include fractionally integrated GARCH or similar long memory models, see e.g., Granger and Ding (1996), Baillie *et al.* (1996), Breidt *et al.* (1998), and multi-fractal models, see Mandelbrot (1997, 1999). Whereas all these models are extremely useful as a statistical description of the data, they do not offer a structural explanation of why volatility clustering is present in so many financial time series. Rather the statistical models postulate that the phenomenon has an exogenous source and is for example caused by the clustered arrival of random 'news' about economic fundamentals.

The volatility of financial assets is a key feature for measuring risk underlying many investment decisions in financial practice. It is therefore important to gain theoretical insight into economic forces that may contribute to or amplify volatility and cause, at least in part, its clustering. The need for an equilibrium theory and a possible relation with technical trading rules and overreaction was e.g., already suggested in Lo and MacKinlay (1990), p. 176: '*... 'the stock market overreaction' hypothesis, the notion that investors are subject to waves of optimism and pessimism and therefore create a kind of 'momentum' that causes prices to temporarily swing away from their fundamental values,*' and '*..., a well-articulated equilibrium theory of overreaction with sharp empirical implications has yet to be developed.*' More recent work in behavioral finance has also emphasized the role of 'market psychology' and 'investor sentiment'

[3] Mandelbrot (1963), pp. 418–419 notes that Houthakker stressed this fact for daily cotton prices, at several conferences and private conversation.

[4] See, for example, Pagan (1996) or Brock (1997) for further discussion of 'stylized facts' that are observed in financial data.

in financial markets; see e.g., Shleifer (2000), Shefrin (2000), and Hirshleifer (2001) for recent surveys.

In this paper we present a simple nonlinear *structural equilibrium model* where price changes are driven by a combination of exogenous random news about fundamentals and evolutionary forces underlying the trading process itself. Volatility clustering becomes an endogenous phenomenon caused by the interaction between *heterogeneous* traders, fundamentalists and technical analysts, having different trading strategies and expectations about future prices and dividends of a risky asset. Fundamentalists believe that prices will move towards its fundamental rational expectations (RE) value, as given by the expected discounted sum of future dividends.[5] In contrast, the technical analysts observe past prices and try to extrapolate historical patterns. The chartists are not completely unaware of the fundamental price however, and condition their technical trading rule upon the deviation of the actual price from its fundamental value. The fractions of the two different trader types change over time according to evolutionary fitness, as measured by accumulated realized profits or wealth, conditioned upon price deviations from the RE fundamental price.

The heterogeneous market is characterized by an irregular switching between phases of low volatility, where price changes are small, and phases of high volatility, where small price changes due to random news are reinforced and may become large due to trend following trading rules. Volatility clustering is thus driven by heterogeneity and conditional evolutionary learning. Although our model is very simple, it is able to generate autocorrelation patterns of returns, and absolute and squared returns similar to those observed in daily S&P 500 data.

Recently, closely related heterogeneous agent models generating volatility clustering have been introduced e.g., in LeBaron *et al.* (1999), Lux and Marchesi (1999, 2000), Kirman and Teyssière (2002), and DeGrauwe and Grimaldi (2004). An interesting feature of our model is that, due to heterogeneity in expectations and switching between strategies, the deterministic skeleton (i.e., the model with exogenous shocks shut off to zero) of our evolutionary model is a *nonlinear* dynamical system exhibiting (quasi)periodic and even chaotic fluctuations in asset prices and returns. Nonlinear dynamic models can generate a wide variety of irregular patterns. In particular, our nonlinear heterogeneous agent model exhibits an important feature naturally suited to describe volatility clustering, namely *coexistence of attractors*. This means that, depending upon the initial state, different types of long run dynamical behavior can occur. In particular, our evolutionary model exhibits coexistence of a stable

[5] As a special case we will discuss an example where traders do not believe that prices move towards a fundamental value, but believe that markets are efficient and (since the fundamental value is held constant in our model) that fluctuations are completely random. Thus, they believe that the last observed price is the best predictor for the future price.

steady state and a stable limit cycle. Hence, depending on initial conditions of the market, prices will either settle down to the locally stable fundamental steady state price, or converge to a stable cycle, fluctuating in a regular pattern around the fundamental steady state price. In the presence of dynamic noise, the market will then switch irregularly between close to the fundamental steady state fluctuations, with small price changes, and periodic fluctuations, triggered by technical trading, with large price changes. It is important to note that coexistence of attractors is a *structurally stable* phenomenon, which is by no means special for our conditionally evolutionary systems, but occurs naturally in nonlinear dynamic models, and moreover is robust with respect to small perturbations.

Whereas the fundamentalists have some 'rational valuation' of the risky asset, the technical analysts use a simple extrapolation rule to forecast asset prices. An important critique from 'rational expectations finance' upon heterogeneous agent models using simple habitual rule of thumb forecasting rules is that 'irrational' traders will *not* survive in the market. Brock and Hommes (1997a, 1998) have discussed this point extensively and stress the fact that in an evolutionary framework technical analysts are not 'irrational,' but they are in fact *boundedly rational*, since in periods when prices deviate from the RE fundamental price, chartists make better forecasts and earn higher profits than fundamentalists. See also the survey in Hommes (2001) or the interview with William Brock in Woodford (2000).

We would like to relate our work to some other recent literature. Agent based evolutionary modeling of financial markets is becoming quite popular and recent contributions include the computational oriented work on the Santa Fe artificial stock market, see Arthur *et al.* (1997), LeBaron *et al.* (1999), the stochastic multi-agent models of Lux and Marchesi (1999, 2000), genetic learning in Arifovic and Gencay (2000), the multi-agent model of Youssefmir and Huberman (1997), and the evolutionary markets based on out-of-equilibrium price formation rules by Farmer and Joshi (2002). [6] Another recent branch of work concerns adaptive learning in asset markets. Timmermann (1993, 1996) e.g., shows that excess volatility in stock returns can arise under learning processes that converge (slowly) to RE. Routledge (1999) investigates adaptive learning in the Grossman-Stiglitz (1980) model where traders can choose to acquire a costly signal about dividends, and derives conditions under which the learning process converges to RE.[7] An important characteristic that dis-

[6] An early example of a heterogeneous agent model is Zeeman (1974); other examples include Frankel and Froot (1986), Kirman (1991), Chiarella (1992), Brock (1993), and Lux (1995).

[7] In Routledge (1999), the fraction of informed traders is fixed over time. De Fontnouvelle (2000) investigates a Grossman-Stiglitz model where traders can choose to buy a costly signal about dividends, with fractions of informed and uninformed traders changing over time according to evolutionary fitness. de Fontnouvelle (2000) is in fact an application of the *Adaptive Rational Equilibrium Dynamics (ARED)* framework of Brock and Hommes (1997a), which is also underlying our

tinguishes our approach is the heterogeneity in expectation rules, with time varying fractions of trader types driven by evolutionary competition. These adaptive, evolutionary forces can lead to endogenous asset price fluctuations around the (stable or unstable) benchmark RE fundamental steady state, thus creating excess volatility and volatility clustering.

We view our model as a simple formalization of general ideas from behavioral finance, where markets are populated by different agents using trading strategies (partly) based on 'psychological heuristics.' In our framework the fractions of trading strategies change over time driven by evolutionary fitness, such as profits and wealth, conditioned upon market indicators.[8] In such a heterogeneous boundedly rational world simple trading strategies survive evolutionary competition. A convenient feature of our model is that the traditional benchmark rational expectations model is nested as a special case within the heterogeneous framework. Our model thus provides a link between the traditional theory and the new behavioral approach to finance.

The paper is organized as follows. Section 2 presents the conditional evolutionary asset pricing model with fundamentalists and technical analysts. The dynamics of the deterministic skeleton of the model is discussed in section 3. In section 4 we compare the time series properties of the model, in particular the autocorrelation patterns of returns, squared returns, and absolute return with those of daily S&P 500 data. Finally, section 5 presents some concluding remarks.

2 A Heterogeneous Agents Model

Our nonlinear model for volatility clustering will be a standard discounted value asset pricing model with two types of traders, fundamentalists and technical analysts. The model is closely related to the *Adaptive Belief Systems (ABS)*, that is, the present discounted value asset pricing model with *heterogeneous* beliefs and evolutionary learning introduced by Brock and Hommes (1998). However, our technical analysts condition their price forecasts upon the deviation of the actual price from the rational expectations fundamental price, similar to the approach taken in the Santa Fe artificial stock market in Arthur *et al.* (1997) and LeBaron *et al.* (1999).

Agents can either invest their money in a risk free asset, say a T-bill, that pays a fixed rate of return r, or they can invest their money in a risky asset, for example a large stock or a market index traded at price p_t (ex-dividend) at time t, that pays uncertain dividends y_t in future periods t, and therefore has an uncertain return. Wealth in period $t+1$ of trader type h is given by

heterogeneous agent asset pricing model, to the Grossman-Stiglitz model leading to unpredictable (chaotic) fluctuations in asset prices.

8 Goetzmann and Massa (2000) and Manzan (2003) provide recent empirical evidence for the existence of different trader types in the market and that their impact on the market changes over time.

$$W_{h,t+1} = (1+r)W_{h,t} + (p_{t+1} + y_{t+1} - (1+r)p_t)z_{ht}, \tag{1}$$

where z_{ht} is the demand of the risky asset for trader type h. Let E_{ht} and V_{ht} denote the 'beliefs' (forecasts) of trader type h about conditional expectation and conditional variance. Agents are assumed to be myopic mean-variance maximizers so that the demand z_{ht} for the risky asset by type h solves

$$\max_{z_{ht}}\{E_{ht}[W_{h,t+1}] - \frac{a}{2}V_{ht}[W_{h,t+1}]\}, \tag{2}$$

where a is the risk aversion parameter. The demand z_{ht} of type h for the risky asset is then given by

$$z_{ht} = \frac{E_{ht}[p_{t+1} + y_{t+1} - (1+r)p_t]}{aV_{ht}[p_{t+1} + y_{t+1} - (1+r)p_t]} = \frac{E_{ht}[p_{t+1} + y_{t+1} - (1+r)p_t]}{a\sigma^2}, \tag{3}$$

where the beliefs about conditional variance $V_{ht}[p_{t+1} + y_{t+1} - (1+r)p_t] = \sigma^2$ are assumed to be constant over time and equal for all types.[9] Let z^s denote the supply of outside risky shares per investor, assumed to be constant, and let n_{ht} denote the fraction of type h at date t. Equilibrium of demand and supply yields

$$\sum_{h=1}^{H} n_{ht}\frac{E_{ht}[p_{t+1} + y_{t+1} - (1+r)p_t]}{a\sigma^2} = z^s, \tag{4}$$

where H is the number of different trader types.

In the case of zero supply of outside risky assets, i.e., $z^s = 0$,[10] the market equilibrium equation may be rewritten as

$$(1+r)p_t = \sum_{h=1}^{H} n_{ht}E_{ht}(p_{t+1} + y_{t+1}). \tag{5}$$

In a world where all traders are identical and expectations are *homogeneous* the arbitrage market equilibrium equation (5) for the price p_t of the risky asset reduces to

$$(1+r)p_t = E_t(p_{t+1} + y_{t+1}), \tag{6}$$

where E_t denotes the common conditional expectation of all traders at the beginning of period t, based on a publically available information set $\mathcal{F}_t$ such as past prices and dividends. The arbitrage equation (6) states that today's price of the risky asset must be equal to the sum of tomorrow's expected price and expected dividend, discounted by the risk free interest rate. It is well known that in a world where expectations are *homogeneous*, where *all* traders

[9] Gaunersdorfer (2000) analyzes the case with time varying beliefs about variances and shows that, in the case of an IID dividend process, the results are quite similar to those with constant ones. Therefore we concentrate on this simple case.

[10] In the general case one can introduce a risk adjusted dividend $y^{\#}_{t+1} = y_{t+1} - a\sigma^2 z^s$ to obtain the market equilibrium equation (5), as in Brock (1997).

are rational, and where it is *common knowledge* that all traders are rational, the fundamental rational expectations equilibrium price, or the *fundamental price* is

$$p_t^* = \sum_{k=1}^{\infty} \frac{E_t(y_{t+k})}{(1+r)^k}, \tag{7}$$

given by the discounted sum of expected future dividends. We will focus on the simplest case of an IID dividend process y_t with mean $E_t(y_{t+1}) = \bar{y}$, so that the fundamental price is constant and given by[11]

$$p^* = \sum_{k=1}^{\infty} \frac{\bar{y}}{(1+r)^k} = \frac{\bar{y}}{r}. \tag{8}$$

It is important to note that so-called speculative bubble solutions, growing at a constant rate $1+r$, also satisfy the arbitrage equation (6) at each date. In a homogeneous, perfectly rational world the existence of these *speculative bubbles* is excluded by the transversality condition

$$\lim_{t\to\infty} \frac{E(p_t)}{(1+r)^t} = 0,$$

and the constant fundamental solution (8) is the only solution of (6) satisfying this condition. Along a speculative bubble solution traders would have perfect foresight, but prices would diverge to infinity. In a homogeneous, perfectly rational world traders realize that speculative bubbles cannot last forever and therefore, they will never get started.

In the asset pricing model with heterogeneous beliefs, market equilibrium in (5) states that the price p_t of the risky asset equals the discounted value of tomorrow's expected price plus tomorrow's expected dividend, averaged over all different trader types. In such a *heterogeneous* world, temporary bubbles with prices deviating from the fundamental, may arise, when the fractions of traders believing in those bubbles is large enough. Notice that, within our heterogeneous agents equilibrium model (5) the standard present discounted value model is nested as a special case. In the nested RE benchmark, asset prices are only driven by economic fundamentals. In contrast, the heterogeneous agent model generates excess volatility driven by evolutionary competition between different trading strategies, leading to unpredictability and volatility clustering in asset returns.

In order to complete the model, we have to be more precise about traders' expectations (forecasts) about future prices and dividends. For simplicity we focus on the case where expectations about future dividends are the same for all traders and given by

$$E_{ht}(y_{t+1}) = E_t(y_{t+1}) = \bar{y}, \tag{9}$$

[11] Notice that in our setup, the constant benchmark fundamental $p^* = \bar{y}/r$ could easily be replaced by another, more realistic time varying fundamental price p_t^*.

for each type h. All traders are thus able to derive the fundamental price $p^* = \bar{y}/r$ in (8) that would prevail in a perfectly rational world. Traders nevertheless believe that in a heterogeneous world prices will in general *deviate* from their fundamental value. We focus on a simple case with two types of traders, with expected prices given respectively by[12]

$$E_{1t}[p_{t+1}] \equiv p^e_{1,t+1} = p^* + v(p_{t-1} - p^*), \qquad 0 \le v \le 1, \tag{10}$$
$$E_{2t}[p_{t+1}] \equiv p^e_{2,t+1} = p_{t-1} + g(p_{t-1} - p_{t-2}), \qquad g \ge 0. \tag{11}$$

Traders of type 1 are *fundamentalists*, believing that tomorrow's price will move in the direction of the fundamental price p^* by a factor v. Of special interest is the case $v = 1$, for which

$$E_{1t}[p_{t+1}] \equiv p^e_{1,t+1} = p_{t-1}. \tag{12}$$

We call this type of traders *EMH-believers*, since the naive forecast of the last observed price as prediction for tomorrow's price is consistent with an efficient market where prices follow a random walk. Trader type 2 are simple *trend followers*, extrapolating the latest observed price change. The market equilibrium equation (5) in a heterogeneous world with fundamentalists and chartists as in (10)–(11), with common expectations on dividends as in (9), becomes

$$(1+r)p_t = n_{1t}(p^* + v(p_{t-1} - p^*)) + n_{2t}(p_{t-1} + g(p_{t-1} - p_{t-2})) + \bar{y}, \tag{13}$$

where n_{1t} and n_{2t} represent the fraction of fundamentalists and chartists, respectively, at date t. At this point we also would like to introduce (additive) dynamic noise into the system, to obtain

$$(1+r)p_t = n_{1t}(p^* + v(p_{t-1} - p^*)) + n_{2t}(p_{t-1} + g(p_{t-1} - p_{t-2})) + \bar{y} + \varepsilon_t, \tag{14}$$

where ε_t are IID random variables representing the fact that this deterministic model is in fact too simple to capture all dynamics of a financial market. Our model can at best be only an approximation of the real world. One can interprete this noise term also as coming from *noise traders*, i.e., traders, whose behavior is not explained by the model but considered as exogenously given; see, for example, Kyle (1985).

The market equilibrium equation (14) represents the first part of the model. The second, *conditionally evolutionary* part of the model describes how the fractions of fundamentalists and technical analysts change over time. The basic idea is that fractions are updated according to past performance, conditioned upon the deviation of actual prices from the fundamental price. Agents are *boundedly rational* in the sense that most of them will choose the forecasting rule that performed best in the recent past, conditioned upon deviations from the fundamental. Performance will be measured by accumulated

[12] For example, Frankel and Froot (1986) and Kirman (1998) have been using exactly the same fundamental and chartist trader types.

realized past profits. Note that realized excess returns per share over period t to period $t+1$, can be computed as

$$R_{t+1} = p_{t+1} + y_{t+1} - (1+r)p_t = p_{t+1} - p^* - (1+r)(p_t - p^*) + \delta_{t+1}, \quad (15)$$

where $\delta_{t+1} = y_{t+1} - \bar{y}$, $E_t(\delta_{t+1}) = 0$. In the general case where the dividend process y_t is not IID, δ_{t+1} is a martingale difference sequence w.r.t. $\mathcal{F}_t$. This term represents *intrinsic uncertainty* about economic fundamentals in a financial market, in our case unexpected random news about future dividends. Thus, realized excess returns (15) can be decomposed in an EMH-term δ_t and a *speculative* endogenous dynamic term explained by the theory represented here.

The first, evolutionary part of the updating of fractions of fundamentalists and technical analysts is described by the discrete choice probabilities[13]

$$\tilde{n}_{ht} = \exp[\beta U_{h,t-1}]/Z_{t-1}, \qquad h = 1, 2 \quad (16)$$

where $Z_{t-1} = \sum_{h=1}^{2} \exp[\beta U_{h,t-1}]$ is just a normalization factor such that the fractions add up to one. $U_{h,t-1}$ measures the evolutionary fitness of predictor h in period $t-1$, given by accumulated realized past profits as discussed below. The key feature of (16) is that strategies or forecasting rules are ranked according to their fitness and the higher the ranking, the more traders will follow that strategy. The parameter β is called the intensity of choice, measuring how fast the mass of traders will switch to the optimal prediction strategy. In the special case $\beta = 0$, both fractions $\tilde{n}_{ht}$ will be constant and equal to 1/2. In the other extreme case $\beta = \infty$, in each period *all* traders will use the same, optimal strategy.

We assume that traders use observed data for evaluating their prediction rules. Thus, a natural candidate for evolutionary fitness is *accumulated realized profits*,[14] as given by

$$\begin{aligned} U_{ht} &:= R_t z_{h,t-1} + \eta U_{h,t-1} \\ &= (p_t + y_t - (1+r)p_{t-1}) \frac{E_{h,t-1}[p_t + y_t - (1+r)p_{t-1})}{a\sigma^2} + \eta U_{h,t-1} \\ &= \frac{1}{a\sigma^2}(p_t + y_t - (1+r)p_{t-1})(p^e_{h,t} + \bar{y} - (1+r)p_{t-1}) + \eta U_{h,t-1}. \quad (17) \end{aligned}$$

[13] The discrete choice probabilities coincide with the well known 'Gibbs'-probabilities in interacting particle systems in physics. Discrete choice probabilities can be derived from a random utility model when the number of agents tends to infinity. See Manski and McFadden (1981) and Anderson, de Palma and Thisse (1993) for an extensive discussion of discrete choice models and applications in economics.

[14] Gaunersdorfer and Hommes (2000) analyze the model with *risk adjusted* realized profits or, equivalently, utilities derived from realized profits, as performance measure. The results are very similar to the model presented here, showing the robustness of the dynamic behavior of the model w.r.t. the evolutionary fitness measure.

The first term defines realized excess return of the risky asset over the risk free asset times the demand for the risky asset by trader type h. The parameter η, $0 \leq \eta \leq 1+r$, in the second term is a memory parameter measuring how fast past fitness is discounted for strategy selection. In the extreme case $\eta = 0$, fitness equals realized net profit in the previous period. In the case with infinite memory, i.e., $\eta = 1$, fitness equals accumulated realized net profits over the entire past. In the intermediate case $0 < \eta < 1$, the weight given to past realized profits decreases exponentially with time. Notice also that for $\eta = 1+r$ fitness (17) coincides exactly with the hypothetical accumulated wealth (1) of a trader who always would have used trading strategy h. It should be emphasized that the key feature of this evolutionary mechanism is that *traders switch to strategies that have earned more money in the recent past.* The memory parameter simply measures the weight given to past earnings for strategy selection.

In the second step of updating of fractions, the conditioning on deviations from the fundamental by the technical traders is modeled as

$$n_{2t} = \tilde{n}_{2t} \exp[-(p_{t-1} - p^*)^2/\alpha], \qquad \alpha > 0 \tag{18}$$

$$n_{1t} = 1 - n_{2t}. \tag{19}$$

According to (18) the fraction of technical traders decreases more, the further prices deviate from their fundamental value p^*. As long as prices are close to the fundamental, updating of fractions will almost completely be determined by evolutionary fitness, that is, by (16)–(17). But when prices move far away from the fundamental, the correction term $\exp[-(p_{t-1} - p^*)^2/\alpha]$ in (18) becomes small, representing the fact that more and more chartists start believing that a price correction towards the fundamental price is about to occur. Our conditional evolutionary framework thus models the fact that technical traders are conditioning their charts upon information about fundamentals, as is common practice in real markets. A similar approach is for example in DeGrauwe *et al.* (1993). The conditioning of their charts upon economic fundamentals may be seen as a 'transversality condition' in a heterogeneous agent world, allowing for temporary speculative bubbles but not for unbounded bubbles; see Hommes (2001) for a discussion of this point.

The timing of the coupling between the market equilibrium equation (14) and the conditional evolutionary selection of strategies in (16)–(19) is important. The market equilibrium price p_t in (14) depends upon the fractions n_{ht}. The notation in (16), (18) and (19) stresses the fact that these fractions depend upon *past* fitnesses $U_{h,t-1}$, which in turn depend upon past prices p_{t-1} and dividends y_{t-1} in periods $t-1$ and further in the past. After the equilibrium price p_t has been revealed by the market, it will be used in evolutionary updating of beliefs and determining the new fractions $n_{h,t+1}$. These new fractions $n_{h,t+1}$ will then determine a new equilibrium price p_{t+1}, etc. In the adaptive belief system, market equilibrium prices and fractions of different trading strategies thus coevolve over time.

3 Model Dynamics

The noisy conditional evolutionary asset pricing model with fundamentalists versus chartists is given by (14), (16)–(19). In this section, we briefly discuss the dynamical behaviour of the *deterministic skeleton* of the model, where the noise terms δ_t and ε_t are both set equal to zero. Understanding of the dynamics of the underlying deterministic skeleton will be useful when we discuss the time series properties of the stochastic model in section 4.

Using the pricing equation (14) it follows easily that the unique steady state price level is the fundamental price, i.e., $p = p^*$. Since both forecasting rules (10) and (11) yield the same forecast at the steady state, the steady state fractions must satisfy $n_1^* = n_2^* = 0.5$. The model thus has a unique steady state where price equals its fundamental value and fractions of the two types are equal.

In order to investigate the stability of the steady state it is useful to rewrite the model in terms of lagged prices. The actual market price p_t in (14) depends on lagged prices p_{t-1} and p_{t-2} and on fractions n_{1t} and n_{2t}. According to (16) these fractions depend on the fitness $U_{h,t-1}$, which by (17) depend on p_{t-1}, p_{t-2}, $U_{h,t-2}$ and the forecasts $p^e_{h,t-1}$. Finally, the forecasts $p^e_{h,t-1}$ depend on p_{t-3} and p_{t-4}. We thus conclude that the market price p_t in (14) depends upon four lagged prices p_{t-j}, $1 \leq j \leq 4$, and the fitnesses $U_{h,t-2}$, so that the system is equivalent to a six dimensional (first order) dynamical system. A straightforward computation shows that the characteristic equation for the stability of the steady state is given by (see Gaunersdorfer (2001) for details)[15]

$$\lambda^2(\eta - \lambda)^2 \left(\lambda^2 - \frac{1+g+v}{2(1+r)}\lambda + \frac{g}{2(1+r)}\right) = 0. \tag{20}$$

Thus, the eigenvalues of the Jacobian are 0, η (both of multiplicity 2) and the roots λ_1, λ_2 of the quadratic polynomial in the last bracket. Note that these roots satisfy the relations

$$\lambda_1 + \lambda_2 = \frac{1+g+v}{2(1+r)} \qquad \text{and} \qquad \lambda_1 \lambda_2 = \frac{g}{2(1+r)}. \tag{21}$$

Also note that the eigenvalues 0 and η always lie inside the unit circle. Thus, the stability of the steady state is determined by the absolute values of λ_1 and λ_2.

The fundamental value p^* is a unique steady state, which is locally stable if the trend chasing parameter $g < 2(1+r)$. That is, if price does not differ too much from the fundamental value, it will converge towards it. As g is increased,

[15] Gaunersdorfer *et al.* (2003) present a detailed mathematical analysis of the *deterministic skeleton* of a slightly different version of the model, where the fitness measure is defined by risk adjusted past realized profits. The dynamics of the model presented here is very similar and in particular, the local stability analysis of the steady state is exactly the same.

the steady state is destabilized by a *Hopf bifurcation*[16] at $g = 2(1 + r)$ and a stable invariant 'circle' with periodic or quasiperiodic dynamics (stable limit cycle) emerges. The invariant circle may undergo bifurcations as well, turning into a strange (chaotic) attractor. This means, if trend chasing parameter g is large enough price will not settle down at the fundamental value but will fluctuate around it.

But even when trend chasing is weak (i.e., $g < 2(1 + r)$) price needs not converge to the fundamental value. There exists a region in parameter space for which two attractors, a stable steady state and a stable (quasi)periodic cycle or even a chaotic attractor, coexist (see figure 1, in this example the dynamics on the 'circle' which surrounds the fundamental steady state is quasiperiodic). Thus, it depends on the initial price if price will converge to the fundamental value or not. [17] [18]

So our nonlinear evolutionary system exhibits coexistence of a locally stable fundamental steady state and (quasi)periodic as well as chaotic fluctuations of asset prices and returns. When buffeted with dynamic noise, in such a case irregular switching occurs between close to the fundamental steady state fluctuations and (quasi)periodic fluctuations triggered by technical trading.

In the next section we analyze time series properties of the model buffeted with noise and present an example where the endogenous fluctuations in returns is characterized by volatility clustering.

4 Time Series Properties

We are interested in the statistical properties of time series generated by our model and how they compare with those of real data. In particular, we are interested in the autocorrelation structure of the returns, and absolute and squared returns generated from the heterogeneous agents market equilibrium model (14), (16), (17)–(19). Returns are defined as relative price changes,

[16] A bifurcation is a qualitative change in the dynamics when parameters change. See, for example, Kuznetsov (1998) for an extensive mathematical treatment of bifurcation theory.

[17] As a technical remark, Gaunersdorfer *et al.* (2003) show that the mathematical generating mechanism for these coexisting attractors is a so-called Chenciner or degenerate Hopf bifurcation (see Kuznetsov (1998), pp. 404–408). Any (noisy) model with two coexisting attractors produces some form of volatility clustering. We emphasize that the Chenciner bifurcation is not special, but it is a generic phenomenon in nonlinear dynamic models with at least two parameters.

[18] Coexistence of attractors is a *generic*, structurally stable phenomenon, occurring for an open set of parameter values. When the stable cycle disappears and the system has a strange (chaotic) attractor intermittency occurs. Recent mathematical results on homoclinic bifurcations have shown that strange attractors are *persistent* in the sense that they typically occur for a positive Lebesgue measure set of parameter values, see e.g., Palis and Takens (1993) for a mathematical treatment.

$$r_t = \frac{p_{t+1} - p_t}{p_t}. \tag{22}$$

We focus on a typical example in which strong volatility clustering occurs, with 'EMH-believers' ($v = 1$ in (10)) and technical traders. In the absence of random shocks ($\varepsilon_t \equiv \delta_t \equiv 0$), there are two coexisting attractors in the example, a locally stable fundamental steady state and an attracting quasiperiodic cycle, as illustrated in figure 1.[19] Depending upon the initial state, the system will settle down either to the stable fundamental steady state or to the stable cycle.

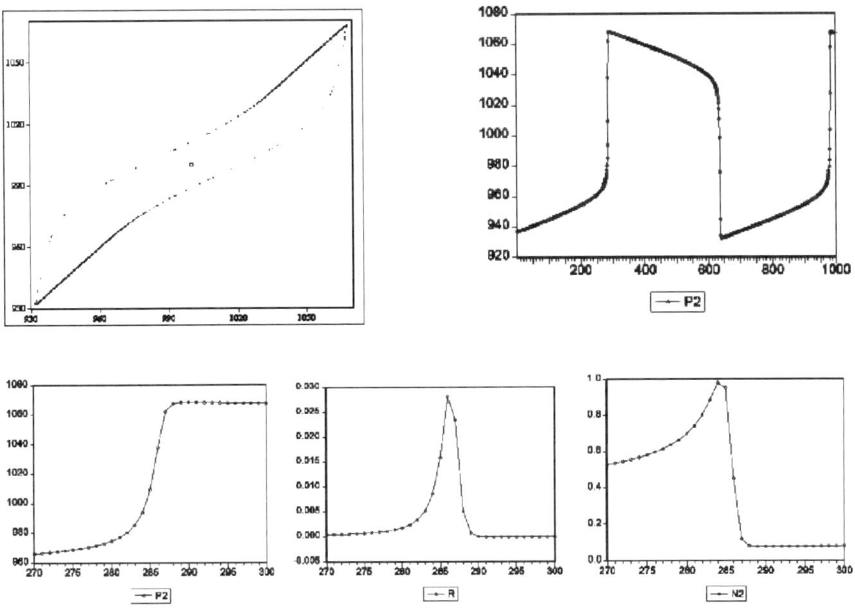

Fig. 1. Top panel: Left figure: phase space projection of prices p_t for deterministic skeleton without noise, where p_t is plotted against p_{t-1}: coexisting limit cycle and stable fundamental steady state $p^* = 1000$ (marked as a square). Right figure: corresponding time series along the limit cycle. Bottom panel: Time series of prices, returns, and fractions of trend followers. Parameter values: $\beta = 2$, $r = 0.001$, $v = 1$, $g = 1.9$, $\bar{y} = 1$, $\alpha = 1800$, $\eta = 0.99$, $a\sigma^2 = 1$, $\delta_t \equiv 0$ and $\varepsilon_t \equiv 0$

The time series of the deterministic skeleton of prices, returns, and fractions of EMH believers along the cycle, as shown in the bottom pannel of figure 1, yield important insight into the economic mechanism driving the price

[19] The memory parameter for all simulations in this paper is $\eta = 0.99$, so that for the strategy selection decision past realized profits are slowly discounted. Simulations with other memory parameters yield similar results.

movements. Prices start far below the fundamental price $p^* = 1000$. Since the trend followers condition their trading rules upon the deviation from the fundamental price, the market will be dominated by EMH believers. Prices will slowly increase in the direction of the fundamental and the fraction of trend followers starts increasing. As the fraction of trend followers increases, the increase in prices is reinforced and trend followers earn a lot of money, which in turn causes the fraction of trend followers to increase even more, etc. At some critical phase from periods 283–288 prices rapidly move to a higher level. During this phase returns increase and volatility jumps to a high value, with a peak around period 286. As the price level moves to a high level of about 1070 far above the fundamental price p^* in period 288, the fraction of trend followers drops to a low level of about 0.08, so that the market becomes dominated by EMH believers again. Prices decrease and move slowly in the direction of the fundamental price again[20] with small negative returns close to zero and with low volatility. Thereafter, the fraction of trend followers slowly increases again, finally causing a rapid decrease in prices to a value of about 930, far below the fundamental, in period 640. Prices slowly move into the direction of the fundamental again to complete a full price (quasi)cycle of about 700 periods. The price cycle is thus characterized by a period of small changes and low volatility when EMH-believers dominate the market, and periods of rapid increase or decrease of prices with high volatility. The periods of rapid change and high volatility are triggered by technical trading; the conditioning of their charts upon the fundamental prevents the price to move too far away from the fundamental and leads to a new period of low volatility.

Adding dynamic noise to the system destroys the regularity of prices and returns along the cycle and leads to an irregular switching between phases of low volatility, with returns close to zero, and phases of high volatility, initiated by technical trading. Figure 2 compares time series observations of the same example buffeted with dynamic noise with daily S&P 500 data.[21]

[20] In the case where all agents are EMH believers, the market equilibrium equation without noise (13) reduces to $p_t = (p_{t-1} + rp^*)/(1+r)$, which is a linear difference equation with fixed point p^* and stable eigenvalue $1/(1+r)$, so that prices always move slowly into the direction of the fundamental. Notice also that when all agents are EMH believers, the market equilibrium equation with noise (14) becomes $p_t = (p_{t-1} + rp^* + \varepsilon_t)/(1+r)$, which is a stationary AR(1) process with mean p^* and root $1/(1+r)$ close to 1, for r small. Hence, in the case when all traders believe in a random walk, the implied actual law of motion is very close to a random walk and EMH-believers only make small forecasting errors which may be hard to detect in the presence of noise.

[21] The noise level was chosen high enough to destroy the regularity in the price series such that autocorrelations in returns become insignificant for lags higher than one. But the noise should also not be too high in order not to destroy the structure imposed by the deterministic part of the model.

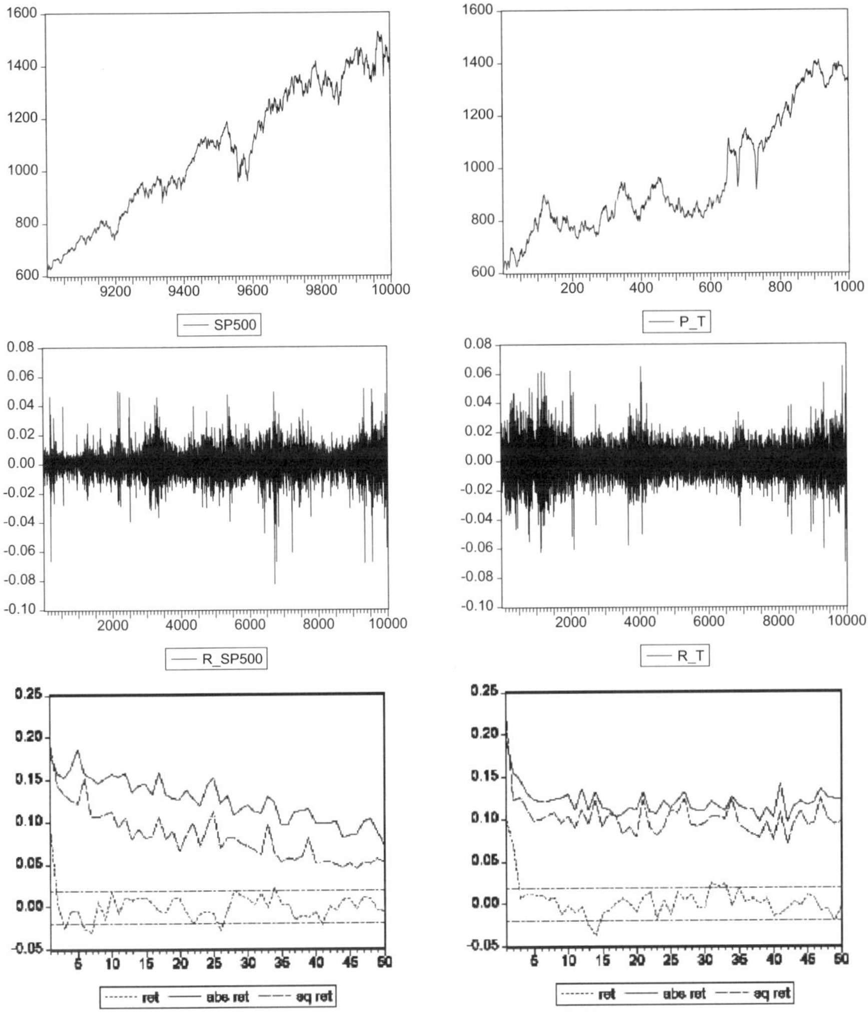

Fig. 2. Daily S&P 500 data (left panel; prices: 07/11/1996–05/10/2000, returns: 08/17/1961–05/10/2000) compared with data generated by our model (right panel), with dynamic noise $\varepsilon_t \sim N(0, 10^2)$ and other parameters as in figure 1: price series (top panel) return series (middle panel), and autocorrelation functions of returns, absolute returns, and squared returns (bottom panel)

Prices in our evolutionary model are highly persistent and *close* to having a *unit root*.[22] In fact, simulated price series including only a sample size of 1000 observations look 'similar' to real price series and the null hypothesis

[22] For $v = 1$ and $r = 0$ the characteristic polynomial of the Jacobian at the steady state has an eigenvalue equal to 1. Note that the Jacobian of a linear difference

of a unit root is not rejected, though the series is generated by a stationary model.[23] [24] The model price series exhibits sudden large movements, which are triggered by random shocks and amplified by technical trading. (Notice the big price changes between periods 650 and 750 in the right top panel when prices are close to the fundamental $p^* = 1000$, similar to the big changes in the deterministic model, cf. figure 1.) When prices move too far away from the fundamental value 1000, technical traders condition their rule upon the fundamental and switch to the EMH-belief. With many EMH believers in the market, prices have a (weak) tendency to return to the fundamental value. As prices get closer to the fundamental, trend following behavior may become dominating again and trigger another fast price movement. However, in contrast to the deterministic version of the model these big price movements do not occur regularly.

The middle panel of figure 2 compares return series of the S&P 500 over 40 years (where the October 1987 crash and the two days thereafter have been excluded[25]) with return series including 10000 observations generated by our model. The simulated return series is qualitatively similar to the S&P 500 daily return series and exhibits clustered volatility.

Table 1 shows some descriptive statistics for both return series. The means and medians of both return series are close to 0 and the range and standard deviations are comparable in size. The S&P 500 returns have negative skewness, which is not the case in our example.[26] This should not come as a surprise, because our simple stylized model is in fact symmetric around the fundamental steady state equilibrium, since both type of traders behave symmetrically with respect to high or low prices and with respect to positive or negative changes in prices. Finally, both return series show excess kurtosis, though the kurtosis coefficient of our example is smaller than the coefficients for the

---

equation $y_t = \alpha_0 + \sum_{k=1}^{L} \alpha_k y_{t-k}$ has an eigenvalue 1 if and only if the time series $y_t = \alpha_0 + \sum_{k=1}^{L} \alpha_k y_{t-k} + \varepsilon_t$ has a unit root equal to 1.

[23] For the price series presented in figure 2 test statistics for the simulated series are: Augmented Dickey-Fuller: −0.8237 (S&P 500: −1.1858), Phillips-Perron: −0.8403 (S&P 500: −1.2238). The MacKinnon critical values for rejection of the hypothesis of a unit root are: 1%: −3.4396, 5%: −2.8648, 10%: −2.5685.

[24] Notice that for price series with only 1000 observations the assumption of a stationary fundamental value seems quite reasonable. Whereas, if we would like to compare longer price series with real data we have to replace our IID dividend process by a non-stationary dividend process, e.g., by a geometric random walk. We intend to study such non-stationary evolutionary systems in future work.

[25] The returns for these days were about −0.20, +0.05, and +0.09. In particular, the crash affects the autocorrelations of squared S&P 500 returns, which drop to small values of 0.03 or less for all lags $k \geq 10$ when the crash is included.

[26] Skewness statistics are not significant nor of the same sign for all markets. Nevertheless, some authors examine the skewness in addition to excess kurtosis. Harvey and Siddique (2000) argue that skewness may be important in investment decisions because of induced asymmetries in realized returns.

S&P 500 returns. This may be due to the fact that in our simple evolutionary system chartists' price expectations are always conditioned upon the same distance function of price deviations from the fundamental price, i.e., upon the weighted distance $(p_{t-1}^2 - p^*)^2/\alpha$ as described by (18). Nevertheless, our simple stylized evolutionary model clearly exhibits excess kurtosis.

Table 1. Descriptive statistics for returns shown in figure 2. (**) null hypothesis of normality rejected at the 1% level

	S&P 500	Simulation
Mean	0.000348	0.000076
Median	0.000214	0.000028
Maximum	0.051152	0.065194
Minimum	−0.082789	−0.070051
Std. Dev.	0.008658	0.011038
Skewness	−0.187095 (**)	0.044317
Kurtosis	8.512094 (**)	5.579890 (**)

We next turn to the time series patterns of returns fluctuations and the phenomenon of volatility clustering. In real financial data autocorrelation functions (ACF) of returns are roughly zero at all lags. For high frequencies they are slightly negative for individual securities and slightly positive for stock indices. Autocorrelations functions of volatility measures such as absolute or squared returns are positive for all lags with slow decay for stock indices and a faster decay for individual stocks. This is the well-known stylized fact known as volatility clustering.

Figure 2 (bottom panel) shows autocorrelation plots of the first 50 lags of the return series and the series of absolute and squared returns. Both return series have significant, but small autocorrelations at the first lag ($\rho_1 = 0.092$ for the S&P 500 and $\rho_1 = 0.099$ for our example). For the S&P 500 the autocorrelation coefficient at the second lag is insignificant and at the third lag slightly negative significant ($\rho_2 = 0.005$, $\rho_3 = -0.025$), whereas in our simulation the autocorrelation coefficient is small but significant at the second lag ($\rho_2 = 0.070$) and insignificant for the third lag ($\rho_3 = 0.007$). For all higher order lags autocorrelations coefficients are close to zero and almost always insignificant. Our noisy conditional evolutionary model thus has almost no linear dependence in the return series. [27]

[27] Brock and Hommes (1997b) calibrate their evolutionary asset pricing model to ten years of monthly IBM prices and returns. They present (noisy) chaotic time series with autocorrelations of prices and returns similar to the autocorrelation structure in IBM prices and returns. In particular, the noisy chaotic return series have (almost) no significant autocorrelations. However, these series do not exhibit volatility clustering, since there are no significant autocorrelations in squared returns.

The bottom panel in figure 2 also shows that for the absolute and squared returns the autocorrelations coefficients of the first 50 lags are strongly significant and positive. Although our model is only six dimensional it is able to generate apparent long memory effects. Table 2 reports the numerical values of the autocorrelation coefficients at the first 5 lags, which are comparable in size for both series.

Table 2. Autocorrelations of the absolute and squared returns shown in figure 2

	S&P 500		Simulation	
lag n	$\|r_t\|$	r_t^2	$\|r_t\|$	r_t^2
1	0.179	0.190	0.193	0.219
2	0.158	0.144	0.156	0.123
3	0.153	0.133	0.147	0.127
4	0.164	0.126	0.131	0.112
5	0.186	0.122	0.124	0.098

Finally, we estimate a simple GARCH(1,1) model on the return series. [28] As is well known, for many financial return series the sum of the ARCH(1) coefficient γ_1 and the GARCH(1) coefficient γ_2 is smaller than but close to unity, representing the fact that the squared error term in the return equation follows a stationary, but highly persistent process. The estimated parameters are given in table 3.

Table 3. GARCH(1,1) estimations for the returns shown in figure 2

	γ_1	γ_2	$\gamma_1 + \gamma_2$
S&P 500	0.069	0.929	0.998
Simulation	0.034	0.963	0.997

Our conditional evolutionary model thus exhibits long memory with long range autocorrelations and captures the phenomenon of volatility clustering.

Let us finally briefly discuss the generality of the presented example. In order to get strong volatility clustering, the parameter $v = 1$ (or v very close to 1) is important, but the results are fairly robust with respect to the choices of the other parameter values.[29] We find volatility clustering also for parameters

[28] The estimations are done with EViews.

[29] As mentioned above, for $v = 1$ the system is close to having a unit root and prices are highly persistent (cf. footnotes 22 and 23). Gaunersdorfer (2001) presents an example of strong volatility clustering for $v = 0.9$ for shorter time series. However, the null hypothesis of a unit root is clearly rejected.

where the fundamental value is a globally stable steady state. For parameter values close to the region where a stable steady state and a stable limit cycle coexist, price paths only converge slowly towards the fundamental value and look similar to price paths converging to a limit cycle. Especially, when the system is buffeted with dynamic noise it is difficult to decide if parameters are chosen in or out of the global stability region.

In general, when $0 \leq v < 1$ volatility clustering becomes weaker, and sometimes also significant autocorrelations in returns may arise. The fact that $v = 1$ or v very close to 1 (so that type 1 are EMH-believers or fundamentalists adapting only slowly into the direction of the fundamental), yields the strongest volatility clustering results may be understood as follows. When EMH-believers dominate the market asset prices are highly persistent and mean reversion is weak, since the evolutionary system is close to having a unit root (see footnote 20). Apparently, the interaction between unit root behavior far from the fundamental steady state with relatively small price changes driven only by exogenous news, and larger price changes due to amplification by trend following rules in some neighborhood around the fundamental price yields the strongest form of volatility clustering. We emphasize that all these results have been obtained for an IID dividend process and a corresponding constant fundamental price p^*. Including a non-stationary dividend process and accordingly a non-stationary time varying fundamental process p_t^* may lead to stronger volatility clustering also in the case $0 \leq v < 1$. We leave this conjecture for future work.

5 Concluding Remarks

We have presented a nonlinear structural model for volatility clustering. Fluctuations in asset prices and returns are caused by a combination of random news about economic fundamentals and evolutionary forces. Two typical trader types have been distinguished. Traders of the first type are *fundamentalists* ('*smart money*' traders), believing that the price of an asset returns to its fundamental value given by the discounted sum of future dividends or '*EMH-believers*,' believing that prices follow a random walk. Traders of the second type are *chartists* or *technical analysts*, believing that asset prices are not solely determined by fundamentals, but that they may be predicted in the short run by simple technical trading rules based upon patterns in past prices, such as trends or cycles. The fraction of each of the two types is determined by an evolutionary fitness measure, given by accumulated profits, conditioned upon how far prices deviate from their fundamental value. This leads to a highly nonlinear, conditionally evolutionary learning model buffeted with noise.

The time series properties of our model are similar to important stylized facts observed in many real financial series. In particular, the autocorrelation structure of the returns and absolute and squared return series of our noisy nonlinear evolutionary system are similar to those observed in daily S&P 500 data, with little or no linear dependence in returns and high persistence and long memory in absolute and squared returns. Although the model is simple, it captures the first two moments of the distribution of real asset returns. Our model thus might serve as a good starting point for a structural explanation – by a tractable model – of further stylized facts in finance, such as cross correlation between volatility and volume.

The *generic* mathematical mechanism generating volatility clustering is the coexistence of a stable fundamental steady state and a stable (quasi)periodic cycle. But there is also a strikingly simple economic intuition of why the phenomenon of volatility clustering should in fact be expected in our conditionally evolutionary system. When EMH-believers dominate the market prices are highly persistent, changes in asset prices are small and only driven by news, returns are close to zero and volatility is low. As prices move towards the fundamental, the influence of trend followers gradually increases, which reinforces the price trend. When trend followers start dominating the market, a rapid change in asset prices occurs with large (positive or negative) returns and high volatility. The price trend cannot persist forever, since prices cannot move away too far from the fundamental because technical traders condition their charts upon the fundamental. In the noisy conditionally evolutionary system both, the low and the high volatility phases, are persistent and the interaction between the two phases is highly irregular. The nonlinear interaction between heterogeneous trading rules in a noisy environment thus causes unpredictable asset returns and at the same time volatility clustering and the associated predictability in absolute and squared returns.

Our model is also able to explain empirical facts like 'fat tails,' i.e., it generates excess kurtosis in the returns. This is due to the fact that the model implies a decomposition of returns into two terms, one martingale difference sequence part according to the conventional EMH theory, and an extra *speculative* term added by the evolutionary theory. The heterogeneity in the model thus creates *excess volatility*.

However, because of the simplicity of the model there are also some shortcomings compared to real financial data, which we would like to discuss briefly. Our model does not generate return series which exhibit strong skewness. This is due to the fact that our agents use trading rules which are exactly symmetric with respect to the constant fundamental value of the risky asset. As a consequence, the evolutionary model is also symmetric with respect to the fundamental price. Another shortcoming is that our model is stationary and therefore it is not able to generate long growing price series. By replacing our IID dividend process by a non-stationary dividend process, e.g., by a geometric random walk, prices will also rapidly increase, similar to real series. We intend to study such non-stationary models within the presented framework in future work.[30]

Other important topics for future work are concerned with the welfare implications and the wealth distribution of our heterogeneous agents economy. What can be said about the total wealth in a multi-agent financial market where prices may (temporary) deviate from their fundamental compared to the RE benchmark? How will wealth be distributed among traders? What would be an optimal investment strategy, a fundamentalists strategy, a trend following strategy or a switching strategy, in such a heterogeneous world? Notice that these are nontrivial questions, because although the number of trading types or strategies is only two in our setup, an underlying assumption of the discrete choice model for strategy selection is that the number of traders in the population is large, in fact infinite. To investigate wealth dynamics one thus has to keep track of the wealth distribution over an infinite population of traders. One could of course consider hypothetical wealth generated by a trader always sticking to the same type or strategy, but in our evolutionary world the majority of traders switch strategy in each time period based upon accumulated realized profits in the recent past. Addressing these important issues is beyond the scope of the present paper, but we plan to study welfare implications and wealth dynamics in future work.

In our model excess volatility and volatility clustering are created or reinforced by the trading process itself, which seems to be in line with common financial practice. If the evolutionary interaction of boundedly rational, speculative trading strategies amplifies volatility, this has important consequences for risk management and regulatory policy issues in real financial markets. Our model predicts that 'good' or 'bad' news about economic fundamentals may be *amplified* by evolutionary forces. Small fundamental causes may thus occasionally have big consequences and trigger large changes in asset prices. In the time of globalization of international financial markets, small shocks in fundamentals in one part of the world may thus cause large changes of asset prices in another part of the world. Our simple structural model shows that a stylized version of this theory already fits real financial data surprisingly well. Our results thus call for more financial research in this area to build more realistic models to asses investors' risk to speculative trading and evolutionary amplification of changes in underlying fundamentals.

[30] Hommes (2002) contains some simulations of the model with a non-stationary fundamental.

References

1. Anderson, S., de Palma, A. and Thisse, J. (1993). *Discrete Choice Theory of Product Differentiation.* MIT Press, Cambridge.
2. Arifovic, J. and Gencay, R. (2000). Statistical properties of genetic learning in a model of exchange rate. *Journal of Economic Dynamics and Control,* **24,** 981–1005.
3. Arthur, W.B., Holland, J.H., LeBaron, B., Palmer, R. and Taylor, P. (1997). Asset pricing under endogenous expectations in an artificial stock market, in: Arthur, W.B., Durlauf, S.N., and Lane, D.A. (Eds.), *The Economy as an Evolving Complex System II.* Redwood City, Addison-Wesley, pp. 15–44.
4. Baillie, R.T., Bollerslev, T. and Mikkelsen, H.O. (1996). Fractionally integrated generalized autoregressive conditional heteroskedasticity. *Journal of Econometrics,* **74,** 3–30.
5. Bollerslev, T. (1986). Generalized autoregressive conditional heteroscedasticity. *Journal of Econometrics,* **31,** 307–327.
6. Breidt, F.J., Crato, N. and de Lima, P. (1998). The detection and estimation of long memory in stochastic volatility. *Journal of Econometrics,* **83,** 325–348 .
7. Brock, W.A. (1993). Pathways to randomness in the economy: Emergent nonlinearity and chaos in economics and finance. *Estudios Económicos,* **8,** 3–55.
8. Brock, W.A. (1997). Asset price behavior in complex environments, in: Arthur, W.B., Durlauf, S.N., and Lane, D.A. (Eds.), *The Economy as an Evolving Complex System II.* Addison-Wesley, Reading, MA, pp. 385–423.
9. Brock, W.A. and Hommes, C.H. (1997a). A rational route to randomness. *Econometrica,* **65,** 1059–1095.
10. Brock, W.A. and Hommes, C.H. (1997b). Models of complexity in economics and finance, in: Hey, C., Schumacher, J.M., Hanzon, B., and Praagman, C. (Eds.), *System Dynamics in Economic and Financial Models.* John Wiley & Sons, Chichester, pp. 3–41.
11. Brock, W.A. and Hommes, C.H. (1998). Heterogeneous beliefs and routes to chaos in a simple asset pricing model. *Journal of Economic Dynamics and Control,* **22,** 1235–74.
12. Chiarella, C. (1992). The dynamics of speculative behaviour. *Annals of Operations Research,* **37,** 101–123.
13. DeGrauwe, P., DeWachter, H. and Embrechts, M. (1993). *Exchange rate theory. Chaotic models of foreign exchange markets.* Blackwell, Oxford.
14. DeGrauwe, P. and Grimaldi, M. (2004). Exchange rate puzzles: a tale of switching attractors, *Working paper,* University of Leuven.
15. Engle, R.F. (1982). Autoregressive conditional heteroscedasticity with estimates of the variance of United Kingdom inflation. *Econometrica,* **50,** 987–1007.
16. Farmer, J.D. and Joshi, S. (2002). The price dynamics of common trading strategies. *Journal of Economic Behavior & Organization,* **49,** 149–171.
17. de Fontnouvelle, P. (2000). Information dynamics in financial markets. *Macroeconomic Dynamics,* **4,** 139–169.
18. Frankel, J.A. and Froot, K.A. (1986). Understanding the US Dollar in the Eighties: the Expectations of Chartists and Fundamentalists. *Economic Record,* Special Issue 1986, 24–38.
19. Gaunersdorfer, A. (2000). Endogenous fluctuations in a simple asset pricing model with heterogeneous beliefs. *Journal of Economic Dynamics and Control,* **24,** 799–831.

20. Gaunersdorfer, A. (2001). Adaptive beliefs and the volatility of asset prices. *Central European Journal of Operations Research*, **9,** 5–30.
21. Gaunersdorfer A. and Hommes C.H. (2000). A Nonlinear Structural Model for Volatility Clustering – Working Paper Version. *SFB Working Paper* No. 63, University of Vienna and Vienna University of Economics and Business Administration, and *CeNDEF Working Paper* WP 00-02, University of Amsterdam.
22. Gaunersdorfer, A., Hommes, C.H. and Wagener, F.O.J. (2003). Bifurcation routes to volatility clustering under evolutionary learning. *CeNDEF Working Paper* 03-03, University of Amsterdam.
23. Goetzmann, W. and Massa, M. (2000). Daily momentum and contrarian behavior of index fund investors. *Journal of Financial and Quantitative Analysis*, forthcoming.
24. Granger, C.W.J. and Ding, Z. (1996). Varieties of long memory models. *Journal of Econometrics*, **73,** 61–77.
25. Grossman, S. and Stiglitz, J. (1980). On the impossibility of informationally efficient markets. *American Economic Review*, **70,** 393–408.
26. Harvey C.R. and Siddique A. (2000). Conditional skewness in asset pricing tests. *Journal of Finance*, **55,** 1263–1295.
27. Hirshleifer, D. (2001). Investor psychology and asset pricing. *Journal of Finance*, **56,** 1533–1597.
28. Hommes, C.H. (2001). Financial markets as nonlinear adaptive evolutionary systems. *Quantitative Finance*, **1,** 149–167.
29. Hommes, C.H. (2002). Modeling the stylized facts in finance through simple nonlinear adaptive systems. *Proc. Nat. Ac. Sc.*, **99,** 7221–7228.
30. Kirman, A. (1991). Epidemics of opinion and speculative bubbles in financial markets, in: Taylor, M. (Ed.), *Money and Financial Markets.* Macmillan, London, pp. 354–368.
31. Kirman, A. (1998). On the nature of gurus. *Preprint.*
32. Kirman, A. and Teyssière, G. (2002). Microeconomic models for long memory in the volatility of financial time series. *Studies in Nonlinear Dynamics & Econometrics*, **5,** 281–302.
33. Kuznetsov, Y.A. (1998). *Elements of Applied Bifurcation Theory* (2nd ed.). Springer Verlag, New York.
34. Kyle, A.S. (1985). Continuous auctions and insider trading. *Econometrica*, **53,** 1315–1335.
35. LeBaron, B., Arthur, W.B. and Palmer, R. (1999). Time series properties of an artificial stock market. *Journal of Economic Dynamics and Control*, **23,** 1487–1516.
36. Lo, A.W. and MacKinlay, A.C. (1990). When are contrarian profits due to stock market overreaction? *The Review of Financial Studies*, **3,** 175–205.
37. Lux, T. (1995). Herd behavior, bubbles and crashes. *The Economic Journal*, **105,** 881–896.
38. Lux, T. and Marchesi, M. (1999). Scaling and criticality in a stochastic multiagent model of a financial market. *Nature*, **397,** 498–500.
39. Lux, T. and Marchesi, M. (2000). Volatility clustering in financial markets: A microsimulation of interacting agents. *International Journal of Theoretical and Applied Finance*, **3,** 675–702 .
40. Mandelbrot, B. (1963). The variation of certain speculative prices. *Journal of Business*, **36,** 394–419.

41. Mandelbrot, B. (1997). *Fractals and Scaling in Finance: Discontinuity, Concentration, Risk.* Springer Verlag, New York.
42. Mandelbrot, B. (1999). A multi-fractal walk down Wall Street. *Scientific American*, **280, (2)** 50–53.
43. Manski, C. and McFadden, D. (1981). *Structural Analysis of Discrete Data with Econometric Applications.* MIT Press, Cambridge.
44. Manzan, S. (2003). Essays in Nonlinear Economic Dynamics. *Thesis*, University of Amsterdam.
45. Pagan, A. (1996). The econometrics of financial markets, *Journal of Empirical Finance*, **3,** 15–102.
46. Palis, J. and Takens, F. (1993). *Hyperbolicity and Sensitive Chaotic Dynamics at Homoclinic Bifurcations.* Cambridge University Press, New York.
47. Routledge B.R. (1999). Adaptive learning in financial markets. *The Review of Financial Studies*, **12,** 1165–1202.
48. Shefrin, H. (2000). *Beyond Greed and Fear. Understanding Behavioral Finance and the Psychology of Investing.* Harvard Business School Press, Boston.
49. Shleifer, A. (2000). *Inefficient Markets. An Introduction to Behavioral Finance.* Clarendon Lectures in Economics, Oxford University Press.
50. Timmermann, A. (1993). How learning in financial markets generates excess volatility and predictability in stock prices. *Quarterly Journal of Economics*, **108,** 1135–1145.
51. Timmermann, A. (1996). Excess volatility and predictability of stock prices in autoregressive dividend models with learning. *Review of Economic Studies*, **63,** 523–557.
52. Woodford, M. (2000). An interview with William A. Brock. *Macroeconomic Dynamics*, **4,** 108–138.
53. Youssefmir, M. and Huberman, B.A. (1997). Clustered volatility in multiagent dynamics. *Journal of Economic Behavior & Organization*, **32,** 101–118.
54. Zeeman, E.C. (1974). On the unstable behaviour of stock exchanges. *Journal of Mathematical Economics*, **1,** 39–49.

Volatility Clustering in Financial Markets: Empirical Facts and Agent–Based Models

Rama Cont*

Centre de Mathématiques Appliquées, Ecole Polytechnique,
F-91128 Palaiseau, France. Rama.Cont@polytechnique.fr

Summary. Time series of financial asset returns often exhibit the *volatility clustering* property: large changes in prices tend to cluster together, resulting in persistence of the amplitudes of price changes. After recalling various methods for quantifying and modeling this phenomenon, we discuss several economic mechanisms which have been proposed to explain the origin of this volatility clustering in terms of behavior of market participants and the news arrival process. A common feature of these models seems to be a switching between low and high activity regimes with heavy-tailed durations of regimes. Finally, we discuss a simple agent-based model which links such variations in market activity to threshold behavior of market participants and suggests a link between volatility clustering and investor inertia.

1 Introduction

The study of statistical properties of financial time series has revealed a wealth of interesting stylized facts which seem to be common to a wide variety of markets, instruments and periods (Ding *et al.*, 1993, Guillaume *et al.*, 1997, Pagan, 1996 , Cont, 2001):

- **Excess volatility:** many empirical studies point out to the fact that it is difficult to justify the observed level of variability in asset returns by variations in "fundamental" economic variables. In particular, the occurrence of large (negative or positive) returns is not always explainable by the arrival of new information on the market (Cutler *et al.*, 1989).
- **Heavy tails:** the (unconditional) distribution of returns displays a heavy tail with positive excess kurtosis.

* The author thanks Alan Kirman and Gilles Teyssière for their infinite patience and participants in the CNRS Summer School on Complex Systems in the Social Sciences (ENS Lyon, 2004) for their stimulating feedback. The last section of this paper is based on joint work with F. Ghoulmie and J.P. Nadal.

- **Absence of autocorrelations in returns:** (linear) autocorrelations of asset returns are often insignificant, except for very small intraday time scales ($\simeq$ 20 minutes) where microstructure effects come into play.
- **Volatility clustering:** as noted by Mandelbrot (1963), "large changes tend to be followed by large changes, of either sign, and small changes tend to be followed by small changes." A quantitative manifestation of this fact is that, while returns themselves are uncorrelated, absolute returns $|r_t|$ or their squares display a positive, significant and slowly decaying autocorrelation function: $\text{corr}(|r_t|, |r_{t+\tau}|) > 0$ for τ ranging from a few minutes to a several weeks.
- **Volume/volatility correlation:** trading volume is positively correlated with market volatility. Moreover, trading volume and volatility show the same type of "long memory" behavior (Lobato and Velasco, 2000).

Among these properties, the phenomenon of volatility clustering has intrigued many researchers and oriented in a major way the development of stochastic models in finance –GARCH models and stochastic volatility models are intended primarily to model this phenomenon. Also, it has inspired much debate as to whether there is long-range dependence in volatility. We review some of these issues in Section 2. As noted by the participants of this econometric debate (Willinger *et al.*, 1999, Mikosch and Stărică, 2003), statistical analysis alone is not likely to provide a definite answer for the presence or absence of long-range dependence phenomenon in stock returns or volatility, unless economic mechanisms are proposed to understand the origin of such phenomena.

Some insights into these economic mechanisms are given by agent-based models of financial markets. Agent-based market models attempt to explain the origin of the observed behavior of market prices in terms of simple, stylized, behavioral rules of market participants, (Chiarella *et al.*, 2003, Lux, 1998, Lux and Marchesi, 2000, LeBaron 2001a): in this approach a financial market is modeled as a system of heterogeneous, interacting agents and several examples of such models have been shown to generate price behavior similar to those observed in real markets. We review some of these approached in Section 3 and discuss how they lead to volatility clustering.

Most of these agent-based models are complex in structure and have been studied using Monte Carlo simulations. As noted also by LeBaron (2000), due to the complexity of such models it is often not clear *which* aspect of the model is responsible for generating the stylized facts and whether all the ingredients of the model are indeed required for explaining empirical observations. In Section 4 we present an agent-based model capable of generating time series of asset returns with properties similar to the stylized facts above, but which is simple enough in structure so the origins of volatility clustering can be traced back to agents behavior. This model points to a link between investor inertia and volatility clustering and provide an economic explanation for the switching mechanism proposed in the econometrics literature as an origin of volatility clustering.

2 Volatility Clustering in Financial Time Series

Denote by S_t the price of a financial asset — a stock, an exchange rate or a market index — and $X_t = \ln S_t$ its logarithm. Given a *time scale* Δ, the log return at scale Δ is defined as:

$$r_t = X_{t+\Delta} - X_t = \ln(\frac{S_{t+\Delta}}{S_t}). \tag{1}$$

Δ may vary between a minute (or even seconds) for tick data to several days. Observations are sampled at discrete times $t_n = n\Delta$. Time lags will be denoted by the Greek letter τ; typically, τ will be a multiple of Δ in estimations. For example, if Δ =1 day, $\mathrm{corr}[r_{t+\tau}, r_t]$ denotes the correlation between the daily return at period t and the daily return τ periods later.

2.1 Empirical Behavior of Autocorrelation Functions

A typical display of daily log-returns is shown in figure 1: the volatility clustering feature is seen graphically from the presence of sustained periods of high or low volatility. As noted above, the autocorrelation of returns is typically insignificant at lags between a few minutes and a month. An example is shown in figure 2 (left). This "spectral whiteness" of returns can be attributed to the activity of arbitrageurs who exploit linear correlations in returns via trend following strategies, see Mandelbrot (1971). By contrast, the autocorrelation function of absolute returns remains positive over lags of several weeks and decays slowly to zero: figure 2 (right) shows this decay for SLM stock (NYSE). This observation is remarkably stable across asset classes and time periods and is regarded as a typical manifestation of volatility clustering (Bollerslev *et al.*, 1992, Ding *et al.*, 1993, Cont *et al.*, 1997, Guillaume *et al.*, 1997). Similar behavior is observed for the autocorrelation of squared returns, see Bollerslev *et al.* (1992), and more generally for $|r_t|^\alpha$ (Ding *et al.*, 1993, Ding and Granger, 1996, Cont *et al.*, 1997), but it seems to be most significant for $\alpha = 1$, i.e. absolute returns (Ding *et al.*, 1993).

GARCH models (Bollerslev *et al.*, 1992, Engle, 1995) were among the first models to take into account the volatility clustering phenomenon. In a GARCH(1,1) model the (squared) volatility depends on last periods volatility:

$$r_t = \sigma_t \varepsilon_t, \quad \sigma_t^2 = a_0 + a\sigma_{t-1}^2 + b\varepsilon_t^2, \quad 0 < a + b < 1, \tag{2}$$

leading to positive autocorrelation in the volatility process σ_t, with a rate of decay governed by $a + b$: the closer $a + b$ is to 1, the slower the decay of the autocorrelation of σ_t. The constraint $a + b < 1$ allows for the existence of a

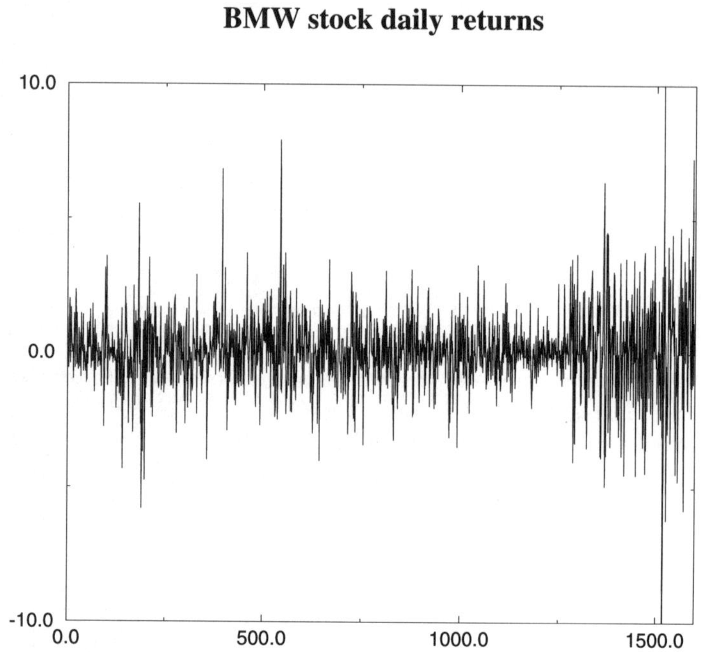

Fig. 1. Large changes cluster together: BMW daily log-returns. $\Delta = 1$ day

stationary solution, while the upper limit $a + b = 1$ corresponds to the case of an integrated process. Estimations of GARCH(1,1) on stock and index returns usually yield $a + b$ very close to 1 (Bollerslev *et al.*, 1992). For this reason the volatility clustering phenomenon is sometimes called a "GARCH effect"; one should keep in mind however that volatility clustering is a "non-parametric" property and is not intrinsically linked to a GARCH specification.

While GARCH models give rise to exponential decay in autocorrelations of absolute or squared returns, the empirical autocorrelations are similar to a power law; see Cont *et al.* (1997), Guillaume *et al.* (1997):

$$C_{|r|}(\tau) = \mathrm{corr}(|r_t|, |r_{t+\tau}|) \simeq \frac{c}{\tau^{\beta}},$$

with an exponent $\beta \leq 0.5$, (Cont *et al.*, 1997, Breidt *et al.*, 1998), which suggests the presence of "long-range" dependence in amplitudes of returns, discussed below.

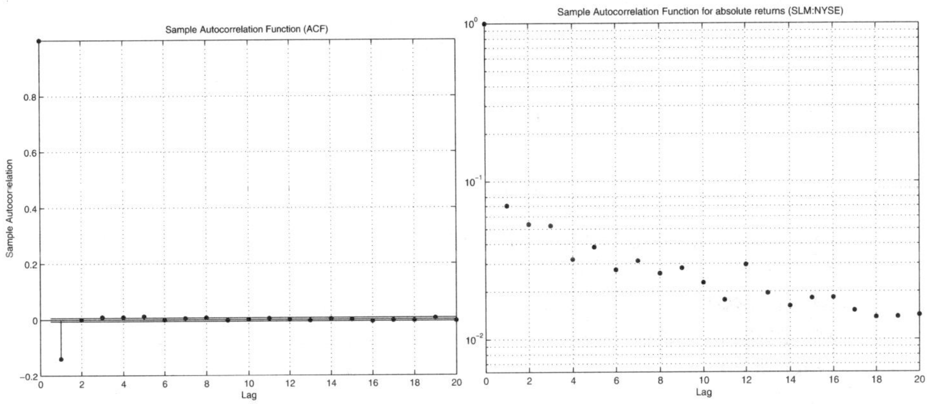

Fig. 2. SLM stock, NYSE, $\Delta = 5$ minutes. Left: autocorrelation function of log-returns. Right: autocorrelation of absolute log-returns

2.2 Long Range Dependence

Let us recall briefly the commonly used definitions of long range dependence, based on the autocorrelation function of a process:

Definition 1 (Long range dependence). *A stationary process Y_t (with finite variance) is said to have* long range dependence *if its autocorrelation function $C(\tau) = \mathrm{corr}(Y_t, Y_{t+\tau})$ decays as a power of the lag τ:*

$$C(\tau) = \mathrm{corr}(Y_t, Y_{t+\tau}) \underset{\tau\to\infty}{\sim} \frac{L(\tau)}{\tau^{1-2d}}, \quad 0 < d < \frac{1}{2}, \tag{3}$$

where L is slowly varying at infinity, i.e. verifies $\forall a > 0$, $\frac{L(at)}{L(t)} \to 1$ as $t \to \infty$.

By contrast, one speaks of "short range dependence" if the autocorrelation function decreases at a geometric rate:

$$\exists K > 0, c \in]0,1[, \quad |C(\tau)| \le K c^{\tau}. \tag{4}$$

Obviously, (3) and (4) are not the only possibilities for the behavior of the autocorrelation function at large lags: there are many other possible decays rates, intermediate between a power decay and a geometric decay. However, it is noteworthy that in all stochastic models used in the financial modeling literature, the behavior of returns and their absolute values fall within one of the two categories.

The long range dependence property (3) hinges upon the behavior of the autocorrelation function at *large* lags, a quantity which may be difficult to estimate empirically, see Beran (1994). For this reason, models with long-range dependence are often formulated in terms of self-similar processes, which allow to extrapolate across time scales and deduce long time behavior from short time behavior, which is more readily observed. A stochastic process $(X_t)_{t\geq 0}$ is said to be self-similar if there exists $H > 0$ such that for any scaling factor $c > 0$, the processes $(X_{ct})_{t\geq 0}$ and $(c^H X_t)_{t\geq 0}$ have the same law:

$$(X_{ct})_{t\geq 0} \stackrel{d}{=} (c^H X_t)_{t\geq 0}. \tag{5}$$

H is called the self-similarity exponent of the process X. Note that a self-similar process cannot be stationary, so the above definition of long-range dependence cannot hold for a self-similar process, but eventually for its increments (if they are stationary). The typical example of self-similar process whose increments exhibit long range dependence is fractional Brownian motion (Mandelbrot and Van Ness, 1968).

But self-similarity does not imply long-range dependence in any way: α-stable Lévy processes provide examples of self-similar processes with *independent* increments. Nor is self-similarity implied by long range dependence: Cheridito (2004) gives several examples of Gaussian semimartingales with the same long range dependence features as fractional Brownian noise but with no self-similarity (thus very different "short range" properties and sample path behavior). The example of fractional Brownian motion is thus misleading in this regard, since it conveys the idea that these two properties are associated. When testing for long range dependence in a model based on fractional Brownian motion, we thus test the joint hypothesis of self-similarity *and* long-range dependence and strict self-similarity is not observed to hold in asset returns (Cont *et al.*, 1997, Cont, 2001).

A fallacy often encountered in the literature is that long range dependence in returns is incompatible with absence of (continuous-time) arbitrage. Again, the origin of this idea can be traced back to models based on fractional Brownian motion: since fractional Brownian motion is not a semimartingale, a model in which the (log)-price are described by a fractional Brownian motion is not arbitrage-free (in the continuous-time sense); see Rogers (1997). This result (and the fact that fractional Brownian motions fails to be a semimartingale) crucially depends on the *local* behavior of its sample paths, not on its long range dependence property. Cheridito (2004) gives several examples of Gaussian processes with the same long range dependence features as fractional Brownian motion, but which are semimartingales and lead to arbitrage-free models.

2.3 Dependence in Stock Returns

The volatility clustering feature indicates that asset returns are not independent across time; on the other hand the absence of linear autocorrelation shows that their dependence is nonlinear. Whether this dependence is "short range" or "long range" has been the object of many empirical studies.

The idea that stock returns could exhibit long range dependence was first suggested by Mandelbrot (1971) and subsequently observed in many empirical studies using R/S analysis (Mandelbrot and Taqqu, 1979). Such tests have been criticized by Lo (1991) who pointed out that, after accounting for short range dependence, they might yield a different result and proposed a modified test statistic. Lo's statistic highly depends on the way "short range" dependence is accounted for and shows a bias towards rejecting long range dependence, see Teverovsky *et al.* (1999). The final empirical conclusions are therefore less clear (Willinger *et al.*, 1999).

However, the absence of long range dependence in returns may be compatible with its presence in absolute returns or "volatility". As noted by Heyde (2002), one should distinguish long range dependence in signs of increments, when $\text{sign}(r_t)$ verifies (3), from long range dependence in amplitudes, when $|r_t|$ verifies (3). Asset returns do not seem to possess long range dependence in signs (Heyde, 2002). Many authors have thus suggested models, such as FIGARCH (Baillie *et al.*, 1996), in which returns have no autocorrelation but their amplitudes have long range dependence; see Doukhan *et al.* (2003).

It has been argued (LeBaron, 2001b, Barndorff-Nielsen and Shephard, 2001) that the decay of $C_{|r|}(\tau)$ can also be reproduced by a superposition of several exponentials, indicating that the dependence is characterized by multiple time scales. In fact, an operational definition of long range dependence is that the time scale of dependence in a sample of length T is found to be of the order of T: dependence extends over the whole sample. Interestingly, the largest time scale in LeBaron (2001b) is found to be of the order of ... the sample size, a prediction which would be compatible with long-range dependence!

Many of these studies test for long range dependence in returns, volatility,etc. by examining sample autocorrelations, Hurst exponents etc. but if time series of asset returns indeed possess the two features of heavy tails *and* long range dependence, then many of the standard estimation procedures for these quantities may fail to work (Resnick, 1998). For example, sample autocorrelation functions may fail to be consistent estimators of the true autocorrelation of returns in the price generating process: Resnick *et al.* (1999) give examples of such processes where sample autocorrelations converge to *random* values as sample size grows! Also, in cases where the sample ACF is consistent, its estimation error can have a heavy-tailed asymptotic distribution, leading to large errors. The situation is even worse for autocorrelations of squared returns, see Mikosch and Stărică (2000). Thus, one must be cautious in identifying behavior of *sample* autocorrelation with the autocorrelations of the return process.

Slow decay of sample autocorrelation functions may possibly arise from other mechanism than long-range dependence. For example, note that nonstationarity of the returns may also generate spurious effects which can be mistaken for long-range dependence in the volatility. However, we will not go to the extreme of suggesting, as in Mikosch and Stărică (2003), that the slow decay of sample autocorrelations of absolute returns is a pure artefact due to non-stationarity. "Non-stationarity" does not suggest a modeling approach and it seems highly unlikely that unstructured non-stationarity would lead to such a robust, stylized behavior for the sample autocorrelations of absolute returns, stable across asset classes and time periods. The robustness of these empirical facts call for an explanation, which "non-stationarity" does not provide. Of course, these mechanisms are not mutually exclusive: a recent study by Granger and Hyung (2004) illustrates the interplay of these two effects by combining an underlying long memory process with occasional structural breaks.

Independently of the econometric debate on the "true nature" of the return generating process, one can take into account such empirical observations without pinpointing a specific stochastic model by testing for similar behavior of sample autocorrelations in agent-based models (described below), and using sample autocorrelations for indirect inference (Gourieroux *et al.*, 1993), of the parameters of such models.

3 Mechanisms for Volatility Clustering

While GARCH, FIGARCH and stochastic volatility models propose statistical constructions which mimick volatility clustering in financial time series, they do not provide any economic explanation for it. We discuss here possible mechanisms which have been proposed for the origin of volatility clustering.

3.1 Heterogeneous Arrival Rates of Information

Heterogeneity in agent's time scale has been considered as a possible origin for various stylized facts (Guillaume *et al.*, 1997). Long term investors naturally focus on long-term behavior of prices, whereas traders aim to exploit short-term fluctuations.

Granger (1980) suggested that long memory in economic time series can be due to the aggregation of a cross section of time series with different persistence levels. This argument was proposed by Andersen and Bollerslev (1997) as a possible explanation for volatility clustering in terms of aggregation of different information flows.

The effects of the diversity in time horizons on price dynamics have also been studied by Lebaron (2001a) in an artificial stock market, showing that the presence of heterogeneity in horizons may lead to an increase in return variability, as well as volatility-volume relationships similar to those of actual markets.

3.2 Evolutionary Models

Several studies have considered modeling financial markets by analogy with ecological systems where various trading strategies co-exist and evolve via a "natural selection" mechanism, according to their relative profitability, (Arthur *et al.*, 1997, Arifovic and Gencay, 2000, LeBaron *et al.*, 1999, LeBaron, 2001a). The idea of these models, the prototype of which is the Santa Fe artificial stock market, (Arthur *et al.*, 1997, LeBaron *et al.*, 1999), is that a financial market can be viewed as a population of agents, identified by their (set of) decision rules. A decision rule is defined as a mapping from an agents information set (price history, trading volume, other economic indicators) to the set of actions (buy, sell, no trade). The evolution of agents decision rule is often modeled using a genetic algorithm (Holland, 1992). The specification and simulation of such evolutionary models can be quite involved and specialized simulation platforms have been developed to allow the user to specify variants of agents strategies and evolution rules. Other evolutionary models represent the evolution by a deterministic dynamical system which, through the complex price dynamics it generate, are able to mimick some "statistical" properties of the returns process, including volatility clustering; see Hommes *et al.* (2003).

Though the Santa Fe market model is capable of qualitatively replicating some of the stylized facts (LeBaron *et al.*, 1999) precise comparisons with empirical observations are still lacking. Indeed, given the large number of parameters, it is not possible to calibrate the parameters in order to interpret the time periods in the simulations as "days" or "minutes" etc. thereby leading to a lack of reference for empirical comparisons.

More importantly, the competition between numerous strategies in such complex simulation models does not allow to pinpoint a single mechanism as being responsible for volatility clustering or other stylized properties. Models in which a dominant mechanism is at work are more helpful in this respect; we will now discuss some instances of such models.

3.3 Behavioral Switching

The economic literature contains examples where switching of economic agents between two behavioral patterns leads to large aggregate fluctuations, Kirman (1993): in the context of financial markets, these behavioral patterns can be seen as trading rules and the resulting aggregate fluctuations as large movements in the market price i.e. heavy tails in returns. Recently, models based on this idea have also been shown to generate volatility clustering (Kirman and Teyssière, 2002, Lux and Marchesi, 2000).

Lux and Marchesi (2000) study an agent-based model in which heavy tails of asset returns and volatility clustering arise from behavioral switching of market participants between fundamentalist and chartist behavior. Fundamentalists expect that the price follows the fundamental value in the long run. Noise traders try to identify price trends, which results in a tendency to herding. Agents are allowed to switch between these two behaviors according to the performance of the various strategies. Noise traders evaluate their performance according to realized gains, whereas for the fundamentalists, performance is measured according to the difference between the price and the fundamental value, which represents the anticipated gain of a "convergence trade". This decision-making process is driven by an exogenous fundamental value, which follows a Gaussian random walk. Price changes are brought about by a market maker reacting to imbalances between demand and supply. Most of the time, a stable and efficient market results. However, its usual tranquil performance is interspersed by sudden transient phases of destabilization. An outbreak of volatility occurs if the fraction of agents using chartist techniques surpasses a certain threshold value, but such phases are quickly brought to an end by stabilizing tendencies. This behavioral switching is believed be the cause of volatility clustering, long memory and heavy tails in the Lux-Marchesi (2000) model.

Kirman and Teyssière (2002) have proposed a variant of Kirman (1993) in which the proportion $\alpha(t)$ of fundamentalists in the market follows a Markov chain, of the type used in epidemiological models, describing herding of opinions. Simulation of this model exihibit autocorrelation patterns in absolute returns with a behavior similar to that described in Section 2.

3.4 The Role of Investor Inertia

As argued by Liu (2000), the presence of a Markovian regime switching mechanism in volatility can lead to volatility clustering, is not sufficient to generate long-range dependence in absolute returns. More important than the switching is the fact the time spent in each regime –the duration of regimes– should have a heavy-tailed distribution (Pourahmadi, 1988, Taqqu and Levy, 1986). By contrast with Markov switching, which leads to short range correlations, this mechanism has been called "renewal switching".[2]

Bayraktar *et al.* (2003) study a model where an order flow with random, heavy-tailed, durations between trades leads to long range dependence in returns. When the durations τ_n of the inactivity periods have a distribution of the form $\mathbb{P}(\tau_n \geq t) = t^{-\alpha}L(t)$, conditions are given under which, in the limit of a large number of agents randomly submitting orders, the price process in this models converges to a a process with Hurst exponent $H = (3-\alpha)/2 > 1/2$. In this model the randomness (and the heavy tailed nature) of the durations between trades are both exogenous ingredients, chosen in a way that generates long range dependence in the returns. However, as noted above, empirical observations point to clustering and persistence in *volatility* rather than in returns so such a result does not seem to be consistent with the stylized facts.

By contrast, as noted above, regime switching in *volatility* with heavy-tailed durations could lead to volatility clustering. Although in the agent-based models discussed above, it may not be easy to speak of well-defined

[2] See the chapter by Giraitis, Leipus and Surgailis in this volume for a review on renewal switching models.

"regimes" of activity, but Giardina and Bouchaud (2003) argue that this is indeed the mechanism which generates volatility clustering in the Lux-Marchesi (2000) and other models discussed above. In these models, agents switch between strategies based on their relative performance; Giardina and Bouchaud argue that this (cumulative) relative performance index actually behaves in time like a random walk, so the switching times can be interpreted as times when the random walk crosses zero: the interval between successive zero-crossings is then known to be heavy-tailed, with a power-law decay of exponent 3/2.

4 Volatility Clustering and Threshold Behavior

While switching between high and low volatility states is probably the mechanism leading to volatility clustering in many of the agent-based models discussed above, this explanation is not easy to trace back to the level of agent behavior, partly because the models described above contain various other ingredients whose contribution to the overall behavior is thus blurred. We now discuss a simple model (Cont *et al.*, 2004) reproducing several stylized empirical facts, where the origin of volatility clustering can be clearly traced back to investor inertia, caused by threshold response of investors to news arrivals.

4.1 An Agent–Based Model for Volatility Clustering

Our model describes a market where a single asset, whose price is denoted by S_t, is traded by N agents. Trading takes place at discrete periods $t = 0, 1, 2, \ldots$. We will see that, provided the parameters of the model are chosen in a certain range, we will be able to interpret these periods as "trading days". At each period, agents have the possibility to send an order to the market for buying or selling a unit of asset: denoting by $\phi_i(t)$ the demand of the agent, we have $\phi_i(t) = 1$ for a buy order and $\phi_i(t) = -1$. We allow the value $\phi_i(t)$ to be zero; the agent is then inactive at period t. The inflow of public information is modeled by a sequence of IID Gaussian random variables $(\epsilon_t, t = 0, 1, 2, \ldots)$ with $\epsilon_t \sim N(0, D^2)$. ϵ_t represents the value of a common signal received by all agents at date $t-1$. The signal ϵ_t is a forecast of the future return r_t and each agent has to decide whether the information conveyed by ϵ_t is significant, in which case she will place a buy or sell order according to the sign of ϵ_t.

The trading rule of each agent $i = 1, \ldots, N$ is represented by a (time–varying) decision threshold $\theta_i(t)$. The threshold $\theta_i(t)$ can be viewed as the agents (subjective) view on volatility. The trading rule we study may be seen as a stylized example of threshold behavior: without sufficient external stimulus ($|\epsilon_t| \leq \theta_i(t)$), an agent remains inactive $\phi_i(t) = 0$ and if the external signal is above a certain threshold, the agent will act: if $\epsilon_t > \theta_i(t)$, $\phi_i(t) = 1$, if $\epsilon_t < -\theta_i(t)$, $\phi_i(t) = -1$. The corresponding demand generated by the agent is therefore given by:

$$\phi_i(t) = 1_{\epsilon_t > \theta_i(t)} - 1_{\epsilon_t < -\theta_i(t)}. \tag{6}$$

The excess demand is then given by $Z_t = \sum_{i=1}^{N} \phi_i(t)$. A non-zero value of Z produces a change in the price given by

$$r_t = \ln \frac{S_t}{S_{t-1}} = g(\frac{Z_t}{N}), \tag{7}$$

where the price impact function $g : \mathbb{R} \mapsto \mathbb{R}$ is an increasing function with $g(0) = 0$. We define the (normalized) market depth λ by : $g'(0) = \frac{1}{\lambda}$. Examples are a linear price impact $g(z) = z/\lambda$ or $g(z) = \arctan(z/\lambda)$, both having been used in various disequilibrium models.

Initially, we start from a population distribution F_0 of thresholds: $\theta_i(0), i = 1, \ldots, N$ are positive IID variables drawn from F_0. Updating of strategies is *asynchronous*: at each time step, any agent i has a probability $0 \leq s \leq 1$ of updating her threshold $\theta_i(t)$. Thus, in a large population, q represents the fraction of agents updating their views at any period; $1/q$ represents the typical time period during which an agent will hold a given view $\theta_i(t)$. If periods are to be interpreted as days, q is typically a small number $s \simeq 10^{-1} - 10^{-3}$. When an agent updates her threshold, she sets it to be equal to the recently observed absolute return, which is an indicator of recent volatility $|r_t| = |\ln \frac{S_t}{S_{t-1}}|$. Introducing IID random variables $u_i(t), i = 1, \ldots, N, t \geq 0$ uniformly distributed on $[0, 1]$, which indicate whether agent i updates her threshold or not:

$$\theta_i(t) = 1_{u_i(t) < s} |r_t| + 1_{u_i(t) \geq s} \theta_i(t-1). \tag{8}$$

This way of updating can be seen as a stylized version of various estimators of volatility based on moving averages of absolute or squared returns. It is also corroborated by a recent empirical study by Zovko and Farmer (2002), who show that traders use recent volatility as a signal when placing orders.

The asynchronous updating scheme proposed here avoids introducing an artificial ordering of agents as in sequential choice models. As noted above, the heterogeneity of time scales of intervention of agents is a feature believed to be important for generating persistence in volatility (Andersen and Bollerslev, 1997, Granger, 1980, LeBaron, 2000). The random nature of updating in this model is a parsimonious way to introduce heterogeneity in time scales without introducing extra parameters. Given this random updating scheme, even if we start from an initially homogeneous population $\theta_i(0) = \theta_0$, heterogeneity creeps into the population through the updating process and evolves in a random manner, leading to a history-dependent disordered system.

Let us recall the main ingredients of the model. At each time period:

1. agents receive a common signal $\epsilon(t) \sim N(0, D^2)$

2. each agent i compares the signal to her threshold $\theta_i(t)$
3. if $|\epsilon(t)| > \theta_i(t)$ the agent considers the signal as significant and generates an order $\phi_i(t)$ according to (6).
4. The market price is impacted by the excess demand and moves according to (7).
5. Each agent updates, with probability q, her threshold according to (8).

Compared to most agent–based models considered in the literature, there is no exogenous "fundamental price" process and we do not distinguish between "fundamentalist" and "chartist" traders. Also, the same information is available to all agents but they differ in the way they *process* the information. We do not introduce any "social interaction" among agents: no notion of locality, lattice or graph structure is introduced. The model has very few parameters: q describes the average updating frequency, D the standard deviation of the noise representing the news arrival process, the market depth λ and the number of agents N which is typically large. We will observe nevertheless that this simple model generates time series of returns with interesting dynamics and properties similar to empirically observed properties of asset returns.

4.2 Simulation Results

In order for a direct comparison with empirical stylized facts to be meaningful, we compute sample moments as in the case of empirical data, by averaging over the (single) sample path. After simulating a sample path of the price S_t for $T = 10^4$ periods, we compute the time series of returns $r_t = \ln(S_t/S_{t-1}), t = 1,\ldots,T$, their histogram, a moving average estimator of the standard deviation of returns ("volatility"), the sample autocorrelation function of returns and the sample autocorrelation function of absolute returns. In order to decrease the sensitivity of results to initial conditions, we allow for a transitory regime and discard the first 10^3 periods before averaging.

In order to interpret the trading periods as "days" and compare the results with properties of daily returns, we note that when g is linear $|r_t| \leq \frac{1}{\lambda}$ and choose $5 \leq \lambda \leq 20$ which allows a (maximal) range of daily returns between 5% and 20%. Also, the amplitude D of the input noise can be chosen such as to reproduce a realistic range of values for the (annualized) volatility: this leads to choosing D in the range $10^{-3} - 10^{-2}$. Let us emphasize that we are discussing the calibration of the *order of magnitude* of parameters, not fine–tuning them to a set of critical values. The results discussed in the sequel are generic within this range of parameters. Figures 3 and 4 illustrate typical sample paths obtained with different parameter values: they all generate series of returns with realistic ranges and realistic values of annualized volatility. For each series, we represent the histogram of returns both in linear and logarithmic scales, the ACF of returns C_r, the ACF of absolute returns $C_{|r|}$. The return series obtained possess regularities which match the properties outlined in the introduction (Cont *et al.*, 2004):

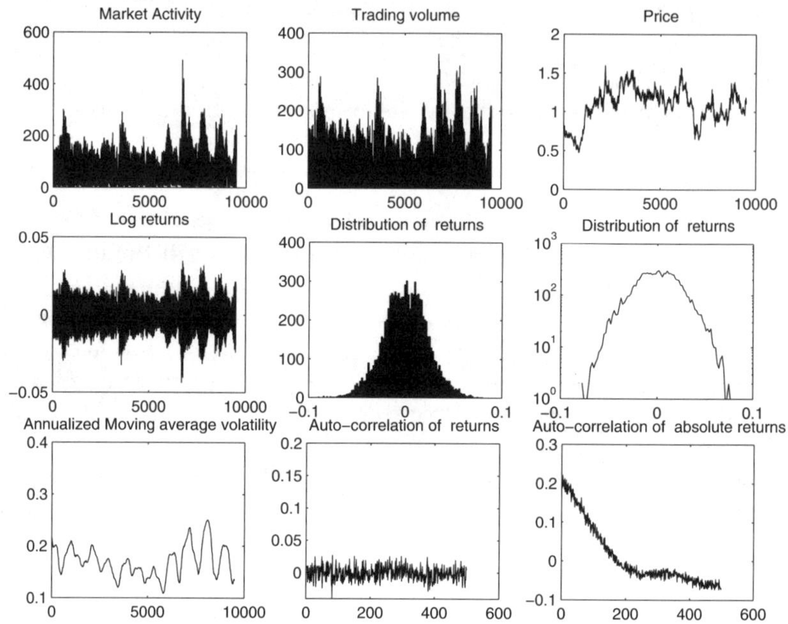

Fig. 3. Numerical simulation of the model with updating frequency $q = 0.01$ (average updating period: 100 "days") $N = 1000$ agents, $D = 0.001$ and $\lambda = 10$

1. Excess volatility: the sample standard deviation of returns can be much larger than the standard deviation of the input noise representing news arrivals $\hat{\sigma}(t) \gg D$.
2. Mean-reverting volatility: the market price fluctuates endlessly and the volatility, as measured by the moving average estimator $\hat{\sigma}(t)$, does neither to zero nor to infinity and displays a mean-reverting behavior.
3. The simulated process generates a leptokurtic distribution of returns with (semi)heavy tails, with an excess kurtosis around $\kappa \simeq 7$. As shown in the logarithmic histogram plots in figures 3–4, the tails exhibit an approximately exponential decay, as observed in various studies of daily returns (Ding *et al.*, 1993).
4. The returns are uncorrelated: the sample autocorrelation function of the returns exhibits an insignificant value (very similar to that of asset returns) at all lags, indicating the absence of linear serial dependence in the returns.
5. Volatility clustering: the autocorrelation function of absolute returns remains significantly positive over many time lags, corresponding to persistence of the amplitude of returns a time scale $\simeq 1/q$.

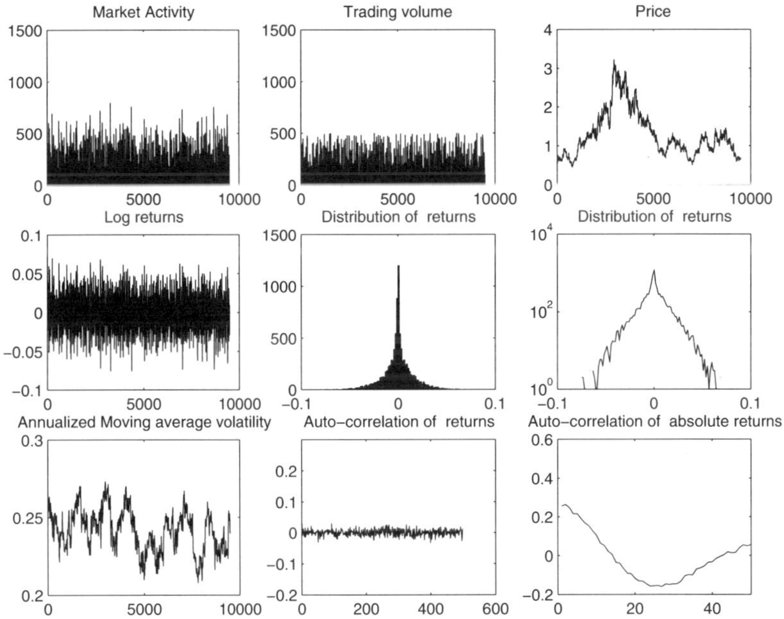

Fig. 4. Numerical simulation of the model with updating frequency $q = 0.1$ (average updating period: 10 "days") $N = 1500$ agents, $D = 0.001$ and $\lambda = 10$

4.3 Theoretical Analysis

Contrarily to some of the models discussed above, this model is simple enough to allow for a theoretical study of its qualitative studies (Cont *et al.*, 2004). Let us being by examining two limiting cases:

1. **Feedback without heterogeneity:** In the case where $q = 1$, all agents synchronously update their threshold at each period. Consequently, the agents have the same thresholds, given by the last periods absolute return: $\theta_i(t) = |r_{t-1}|$ and will therefore generate the same order: $Z_t = N\phi_1(t) \in \{0, N, -N\}$. So, the return r_t depends on the past only through the absolute return $|r_{t-1}|$:

$$r_t = f(|r_{t-1}, \epsilon_t|) = g(N)1_{\epsilon_t > |r_{t-1}|} + g(-N)1_{\epsilon_t < -|r_{t-1}|},$$

a dependence structure typical of ARCH models (Engle, 1995), leading to uncorrelated returns and volatility clustering. In this case, the distribution of r_t conditional on $|r_{t-1}|$ is actually a trinomial distribution: $r_t \in \{0, g(N), g(-N)\}$, which is not realistic. Simulation studies show that a similar behavior persists for $1 - q \ll 1$, leading to tri-modal distributions. This confirms our intuition that the updating probability q should be chosen small.

2. **Heterogeneity without feedback:** In the case where $q = 0$, no updating takes places: the trading strategies, given by the thresholds θ_i, are unaffected by the price behavior and the *feedback* effect is not present anymore. Heterogeneity is still present: the distribution of the thresholds remains identical to what it was at $t = 0$. The return r_t depends only on ϵ_t :

$$r_t = g(\frac{1}{N}\sum_{i=1}^{N} 1_{\epsilon_t > \theta_i} - 1_{\epsilon_t < -\theta_i}) = F(\epsilon_t).$$

We conclude therefore that the returns are IID random variables, obtained by transforming the Gaussian IID sequence (ϵ_t) by the nonlinear function F given in (9), whose properties depend on the (initial) distribution of thresholds $(\theta_i, i = 1, \ldots, N)$. The log–price then follows a (non–Gaussian) random walk and the model does not exhibit volatility clustering.

The two limiting cases above show that, in order to obtain the interesting statistical properties observed in the simulated examples shown above, it is necessary to have $0 < q \ll 1$: both feedback and heterogeneity are essential ingredients. In the general case we have the following properties:

- **Markovian dynamics**: the thresholds $[\theta_i(t), i = 1, \ldots, N]$ follow a Markov chain in $\{g(k), k = 0, \ldots, N\}$. We have $\theta_i(t+1) = \theta_i(t)$ with probability $1 - q$ and

$$\theta_i(t+1) = |r_t| = |g(\frac{1}{N}\sum_{i=1}^{N}[1_{\epsilon_t > \theta_i} - 1_{\epsilon_t < -\theta_i}])| \qquad \text{with probability } q. \quad (9)$$

  In fact given that agents are indistinguishable and only the empirical distribution of threshold values affects the returns, defining $N_k(t) = \sum_{i=1}^{N} 1_{[0,a_k[}(\theta_i(t))$ then $(N_k(t), k = 0, \ldots, N-1)_{t=0,1,\ldots}$ evolves as a Markov chain in $\{0, \ldots, N\}^N$. $N(t) = (N_k(t), k = 0, \ldots, N-1)$ is none other than the (cumulative) population distribution of the thresholds. The fact that $N(t)$ itself follows a Markov chain means that the population distribution of thresholds is a *random measure* on $\{0, \ldots, N\}$, which is characteristic of disordered systems (Mézard *et al.*, 1984), even if we start from a deterministic set of values for the initial thresholds (even identical ones). Here the disorder is endogenous and is generated by the random updating mechanism.
- **Excess volatility**: In this model, the volatility of the news arrival process is quantified by D which is the standard deviation of the external noise ϵ_t, whereas the volatility of the returns can be measured a posteriori as the (conditional or unconditional) standard deviation of r_t. As seen from the nonlinear relation between ϵ_t and r_t,

$$r_t = g(\frac{\sum_{i=1}^{N} 1_{\epsilon_t > \theta_i(t)} - 1_{\epsilon_t < -\theta_i(t)}}{\lambda N}), \qquad (10)$$

even after conditioning on the current states of agents $\theta_i(t), i = 1, \dots, N$, equation (10) yields a nonlinear relation between the input noise ϵ_t and the returns which can have the effect of amplifying the noise by an order of magnitude or more. In the simulation example shown in figure 3, $D = 10^{-3}$ which corresponds to an annualized volatility of 1.6%, while the annualized volatility of returns is in the range of 20%, an order of magnitude larger: the order of magnitude of the volatility of returns may be quite different from that of the input noise.

- **Absence of autocorrelation**
  From the dynamic equations of the model

$$Z_t = \frac{1}{N}\sum_{i=1}^{N}\phi_i(t) = \frac{1}{N}\sum_{i=1}^{N}[1_{\epsilon_t>\theta_i} - 1_{\epsilon_t<-\theta_i}], \tag{11}$$

$$r_t = g(Z_t) = g(\frac{1}{N}\sum_{i=1}^{N}[1_{\epsilon_t>\theta_i} - 1_{\epsilon_t<-\theta_i}]), \tag{12}$$

  one can deduce that, if g is an odd function (in particular if g is linear) then asset returns $(r_t)_{t\geq 0}$ are uncorrelated: $\text{cov}(r_t, r_{t+1})$=0. This is due to the fact that the trading/ nontrading decision is based only on the amplitude of the signal, not its sign. The sign of the return is determined by the sign of the common signal, which is independent across periods.

- **Investor inertia**
  Except in times of crisis or market crash, at a given point in time only a small proportion of stockholders are actually trading in the market. As a result, the (daily) order flow for a typical stock can be much smaller than the market capitalization. This phenomenon, sometimes referred to as *investor inertia*, is a generic outcome in our model due to threshold behavior of agents. Starting from an initial holding of $\pi_i(0)$, the quantity of asset held by agent i is given by $\pi_i(t) = \sum_{\tau=0}^{t}\phi_i(\tau)$. Figure 4.3 displays the evolution of the portfolio $\pi_i(t)$ of a typical agent: short periods of activity (trading) are separated by long periods of inertia, where the portfolio remains constant. This "inertia" increases in periods of high volatility, an effect similar to the behavior of risk-averse agent.

- **Mean reversion and clustering of volatility**
  Many market microstructure models –especially those with learning or evolution– converge over large time intervals to an equilibrium where prices and other aggregate quantities cease to fluctuate randomly. By contrast, in the present model, prices fluctuate endlessly and the volatility exhibits mean-reverting behavior. Suppose we are in a period of "low volatility"; the amplitude $|r_t|$ of returns is small. Agents who update their thresholds will therefore update them to small values, become more sensitive to news arrivals, thus generating higher excess demand and thus increasing the amplitude of returns. Conversely, in a period of high volatility, agents will update their threshold values to high values and become less reactive to

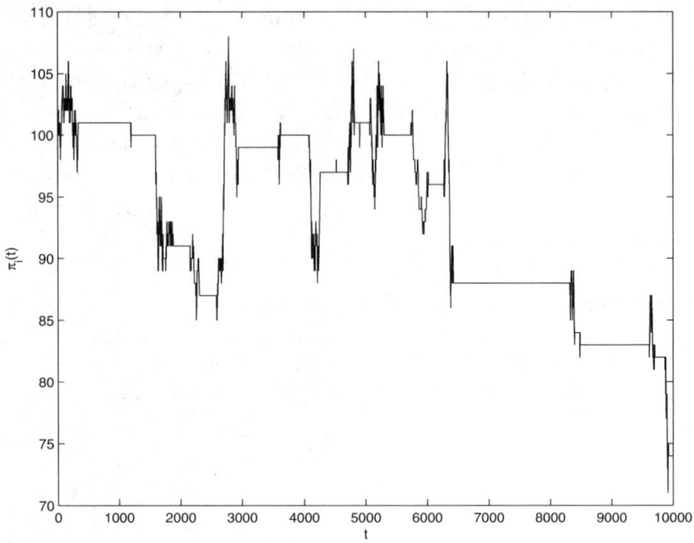

Fig. 5. Evolution of the portfolio of a typical agent, with long periods of inactivity punctuated by bursts of activity

the incoming signal: this increase in investor inertia will thus decrease the amplitude of returns. The mean reversion time in the volatility corresponds here to the time it takes for agents to adjust their thresholds to current market conditions, which is of order $\tau_c = 1/q$.

When the amplitude of the noise is small it can be shown (Cont *et al.*, 2004) that volatility decays exponentially in time and increases through upward "jumps". This behavior is actually similar to that of a class of stochastic volatility models, introduced by Barndorff-Nielsen and Shephard (2001) and successfully used to describe various econometric properties of returns.

5 Conclusion

Volatility clustering is recognized as a stylized property present in most financial time series. Agent-based models seek to explain volatility clustering in terms of behavior of market participants, described in terms of simple rules. We have discussed several agent-based models capable of generating volatility clustering. A common feature of these models seems to be the "switching" of the market between periods of high and low activity, with long durations of periods. Models differ in the mechanism which leadsz to this switching at the level of agents.

While the econometric debate on the short range or long range nature of dependence in volatility still goes on (and may probably never be resolved), agent-based models can provide motivation for choosing between alternative econometric specifications which are otherwise equally plausible in statistical terms, thus providing a useful complement to econometric analysis.

References

1. Andersen, T. and Bollerslev, T. (1997). Heterogeneous information arrivals and returns volatility dynamics. *Journal of finance*, **52,** 975–1005.
2. Arifovic, J. and Gencay, R. (2000). Statistical properties of genetic learning in a model of exchange rate. *Journal of Economic Dynamics and Control*, **24,** 981–1005.
3. Arthur, W., Holland, J., LeBaron, B., Palmer, J. and Tayler, P. (1997). Asset pricing under heterogeneous expectations in an artificial stock market, in: *The Economy as an Evolving Complex System.* Perseus Books, Reading MA, pp. 15–44.
4. Baillie, R.T., Bollerslev, T. and Mikkelsen, H.O. (1996). Fractionally integrated generalized autoregressive conditional heteroskedasticity. *Journal of Econometrics*, **74,** 3–30.
5. Barndorff-Nielsen, O.E. and Shephard, N. (2001). Non-Gaussian Ornstein-Uhlenbeck based models and some of their uses in financial econometrics. *Journal of the Royal Statistical Society* **B 63,** 167–241.
6. Bayraktar, E., Horst, U. and Sircar, K.R. (2003). A limit theorem for financial markets with inert investors. *Working paper* Princeton University,
7. Beran, J. (1994). *Statistics for Long-Memory Processes*, Chapman and Hall, New York.
8. Bollerslev, T., Chou, R. and Kroner, K. (1992). ARCH modeling in finance. *Journal of Econometrics*, **52,** 5–59.
9. Breidt, F.J., Crato, N. and De Lima, P. (1998). The detection and estimation of long memory in stochastic volatility. *Journal of Econometrics*, **83** , 325–348.
10. Cheridito, P. (2004). Gaussian moving averages, semimartingales and option pricing. *Stochastic Processes and Their Applications*, **109,** 47–68.
11. Chiarella, C., Gallegatti, M., Leombruni, R. and Palestrini, A. (2003). Asset price dynamics among heterogeneous interacting agents. *Computational Economics*, **22,** 213–223.
12. Cont, R. (2001). Empirical properties of asset returns: Stylized facts and statistical issues. *Quantitative Finance*, **1,** 1–14.
13. Cont, R., Bouchaud, J.-P. and Potters, M. (1997). Scaling in financial data: Stable laws and beyond, in: B. Dubrulle, F. Graner, and D. Sornette (eds.) *Scale Invariance and Beyond*, Springer, Berlin.
14. Cont, R., Ghoulmie, F. and Nadal, J.-P. (2004). Heterogeneity and feedback in an agent-based market model. *Working paper, Ecole Polytechnique.*
15. Cutler, D., Poterba, J. and Summers, L. (1989). What moves stock prices? *Journal of Portfolio Management*, **15,** 4–12.
16. Ding, Z., Granger, C.W.J. and Engle, R.F. (1993). A long memory property of stock market returns and a new model. *Journal of empirical finance*, **1,** 83–106.
17. Ding, Z. and Granger, C.W.J. (1996). Modeling volatility persistence of speculative returns: a new approach. *Journal of Econometrics*, **73,** 185–215.

18. Doukhan, P., Oppenheim, G. and Taqqu, M. S. (2003). *Theory and Applications of Long-Range Dependence*. Birkhäuser Boston Inc., Boston, MA.
19. Engle, R.F. (1995). *ARCH Models*. Oxford University Press, Oxford.
20. Giardina, I. and Bouchaud, J.-P. (2003). Bubbles, crashes and intermittency in agent based market models. *European Physical Journal*, **B. 31,** 421–437.
21. Giraitis, L., Leipus, R. and Surgailis, D. (2004). Recent advances in ARCH modelling. This volume.
22. Gourieroux, C., Monfort, A. and Renault, E. (1993). Indirect inference. *Journal of Applied Econometrics*, **8 ,** S85–S118.
23. Granger, C.W.J. (1980). Long memory relationships and the aggregation of dynamic models. *Journal of Econometrics*, **14,** 227–238.
24. Granger, C.W.J. and Hyung, N. (2004). Occasional structural breaks and long memory with an application to the S&P 500 absolute stock returns. *Journal of Empirical Finance*, **11,** 399-421.
25. Guillaume, D., Dacorogna, M.M., Davé, R., Müller, U., Olsen, R. and Pictet, O. (1997). From the birds eye view to the microscope: A survey of new stylized facts of the intraday foreign exchange markets. *Finance and Stochastics*, **1,** 95–131.
26. Heyde, C.C. (2002). On modes of long-range dependence. *Journal of Applied Probability*, **39,** 882–888.
27. Holland, J. (1992). *Adaptation in Natural and Artificial Systems*, Bradford Books.
28. Hommes, C.H., Gaunersdorfer, A. and Wagener, F.O. (2003). Bifurcation routes to volatility clustering under evolutionary learning. *Working paper, CenDEF.*
29. Kirman, A. (1993). Ants, rationality, and recruitment. *Quarterly Journal of Economics*, **108,** 137–156.
30. Kirman, A. and Teyssière, G. (2002). Microeconomic models for long memory in the volatility of financial time series. *Studies in Nonlinear Dynamics & Econometrics*, **5,** 281–302.
31. LeBaron, B. (2000). Agent-based computational finance : suggested readings and early research. *Journal of Economic Dynamics and Control*, **24,** 679–702.
32. LeBaron, B. (2001a). Evolution and time horizons in an agent-based stock market. *Macroeconomic Dynamics*, **5,** 225–254.
33. LeBaron, B. (2001b). Stochastic volatility as a simple generator of apparent financial power laws and long memory. *Quantitative Finance*, **1,** 621–631.
34. LeBaron, B., Arthur, B. and Palmer, R. (1999). Time series properties of an artificial stock market. *Journal of Economic Dynamics and Control*, **23,** 1487–1516.
35. Liu, M. (2000). Modeling long memory in stock market volatility. *Journal of Econometrics*, **99,** 139–171.
36. Lobato, I. and Velasco, C. (2000). Long memory in stock market trading volume. *Journal of Business and Economic Statistics*, **18,** 410-427.
37. Lo, A. (1991). Long-term memory in stock market prices. *Econometrica*, **59,** 1279–1313.
38. Lux, T. (1998). The socio-economic dynamics of speculative markets: interacting agents, chaos, and the fat tail of return distributions. *Journal of Economic Behavior and Organization*, **33,** 143–65.
39. Lux, T. and Marchesi, M. (2000). Volatility clustering in financial markets: a micro simulation of interacting agents. *International Journal of Theoretical and Applied Finance*, **3,** 675–702.

40. Mandelbrot, B.B. (1963). The variation of certain speculative prices. *Journal of Business*, **36,** 392–417.
41. Mandelbrot, B.B. (1971). When can price be arbitraged efficiently? A limit to the validity of the random walk and martingale models. *Review of Economics and Statistics*, **53,** 225–236.
42. Mandelbrot, B.B. and Taqqu M.S. (1979). Robust R/S analysis of long-run serial correlation, in *Proceedings of the 42nd session of the International Statistical Institute,* Vol. 2 (Manila, 1979), **48,** 69–99.
43. Mandelbrot, B.B. and Van Ness, J. (1968). Fractional Brownian motion, fractional noises and applications. *SIAM Review*, **10,** 422–437.
44. Mézard, M., Parisi, G. and Virasoro, M. (1984). *Spin Glass Theory and Beyond*, World Scientific.
45. Mikosch, T. and Stărică, C. (2000). Limit theory for the sample autocorrelations and extremes of a GARCH $(1, 1)$ process. *The Annals of Statististics*, **28,** 1427–1451.
46. Mikosch, T. and Stărică, C. (2003) Long–range dependence effects and ARCH modeling, in: Doukhan, P., Oppenheim, G. and Taqqu, M.S. (Eds.), *Long–Range Dependence: Theory and Applications.* Birkhauser, Boston, pp. 439–459.
47. Pagan, A. (1986). The econometrics of financial markets. *Journal of Empirical Finance*, **3,** 15–102.
48. Pourahmadi, M. (1988). Stationarity of the solution of $X_t = A_t X_t + 1$ and analysis of non-Gaussian dependent variables. *Journal of Time Series Analysis,* **9,** 225–239.
49. Resnick, S.I., Samorodnitsky, G. and Xue, F. (1999). How misleading can sample ACFs of stable MAs be? (Very!). *The Annals of Applied Probability,* **9,** 797–817.
50. Resnick, S.I. (1998). Why non-linearities can ruin the heavy-tailed modeler's day, in: Adler, P., Feldman, R. and Taqqu, M.S. (Eds.), *A Practical Guide to Heavy Tails: Statistical Techniques and Applications*, Birkhäuser Boston, Boston, MA, pp. 219–239.
51. Rogers, L.C.G. (1997). Arbitrage with fractional Brownian motion. *Mathematical Finance*, **7,** 95–105.
52. Taqqu, M.S. and Levy, J. (1986). Using renewal processes to generate long range dependence and high variability, in: Eberlein, E. and Taqqu, M.S. (Eds.), *Dependence in Probability and Statistics*, Birkhauser, Boston, pp. 73–89.
53. Teverovsky, V., Taqqu, M.S. and Willinger, W. (1999). A critical look at Lo's modified R/S statistic. *Journal of Statistical Planning and Inference,* **80,** 211–227.
54. Willinger, W., Taqqu, M.S. and Teverovsky, V. (1999). Long range dependence and stock returns. *Finance and Stochastics*, **3,** 1–13.
55. Zovko, I. and Farmer, J. D. (2002). The power of patience: a behavioral regularity in limit order placement. *Quantitative Finance*, **2,** 387–392.

The Microeconomic Foundations of Instability in Financial Markets*

Alan Kirman

School of Social Science. Institute for Advanced Study. kirman@ias.edu

1 Introduction

There are a number of salient features of financial time series, which are hard to explain with standard analysis. The efficient markets hypothesis as well as the idea of rational expectations, seem to be incompatible with the presence of "bubbles" and "herding" behaviour in financial markets, The existence of long memory in financial price series is well documented also, and is not easy to generate with standard financial market models. Yet the existence of all these phenomena seems to be widely accepted. This has led economists to try to develop theoretical models, which generate such phenomena. There is a real need to explain these phenomena since they are inextricably linked with the instability of foreign exchange markets for example. There has been growing concern over what is perceived to be the increasing volatility of financial markets in general and this is widely regarded as a recent phenomenon.

The purpose of this chapter is to present the building blocks of a class of models which have been proposed as an alternative to the more standard models associated with the CAPM paradigm. These models generate data, which, in some configurations, reproduce many of the stylised facts of the empirical series and in particular are capable of generating "long memory". The essential feature of the models is that they involve interaction between the various agents in the market. Agents forecast changes in prices and the actions that they take will be based on their forecasts. Their forecasts and their actions may be highly interdependent and it is this that generates the properties we are interested in. Perhaps the most characteristic feature of this model is that agents tend to herd on one type of behaviour and then to move together to another. This sort of "herding" behaviour is by no means

* This chapter draws heavily on joint work with Hans Foellmer, Ulrich Horst, Roman Riccioti, Gilles Teyssière, and Richard Topol. I would like to thank all of them and to insist on the fact that they bear no responsibility for any of the defects in what is presented here.

irrational, since the actions of agents reveal something about the information that the individuals possess; for a detailed account of how rational herding can occur see Chamley (2002).

In fact, bubbles and herding behaviour are far from new. Although we now tend to associate, "price bubbles" with financial markets such bubbles have been documented for a wide variety of markets over a considerable period of time. One of the earliest bubbles was that in the price of red mullet in the first century A.D. Cicero, Horace, Juvenal and Martial describe the red mullet fever. Interestingly, from our point of view, all of these authors attributed the high price of this Mediterranean fish to some sort of collective fad and regarded the behaviour of those who paid high prices for the fish as irrational. Although it is difficult to assess the real price now, good specimens could fetch the equivalent of the average monthly salary at the height of the craze. The bubble was ended by the imposition by the emperor of a sumptuary tax, which radically reduced the demand.

There have been many other historical bubbles, such as the Tulip, South Sea and Mississippi bubbles, and an excellent survey of these may be found in Garber (2000). However, Garber takes a particular stance and argues that these so-called bubbles do not correspond to what is usually meant by a bubble. It would be fair to say that, for most economists, a bubble occurs when there is a prolonged departure of the price of an asset from any value that might have been derived from some reasonable fundamentals associated with the asset in question. Garber's suggestion is that one can find fundamentals, which justify the prices that were observed but many contest his position, of course.

If we accept the idea that prolonged departures form fundamentals may occur then there are two possible approaches. We can attribute them to what Alan Greenspan referred to as "irrational exuberance" and simply argue that the market loses its head from time to time. Alternatively we might proceed in two steps. First give a careful definition of what we mean by "bubbles" and then devise tests for detecting them in empirical series and then, seek models, which would provide a theoretical basis for the existence of bubbles without a simple appeal to irrationality. There has indeed been a substantial literature on both of these aspects of the subject, (see for example, Blanchard and Watson (1982), Flood and Garber (1980), Meese (1986), Tirole (1985), West (1988), Woo (1987), Stiglitz (1990), Flood and Hodrick,(1990), Donaldson and Kamstra (1996), Avery and Zemsky (1998), Shiller (2000), and Brooks and Katsaris (2003).

A possibility suggested by several authors, see for example Lux and Sornette (2002), is that there are self-reinforcing swings of opinion which can cause departures from fundamentals, Shiller (1981) observes that attention seems to become focused on one share to another in financial markets without any particular change in the fundamentals associated with the share in question. Yet when he pursues this analysis further (Shiller (2000)) as the title of his book indicates, he takes up the phrase used by Greenspan, "Irrational

Exuberance" and he suggests, that such behaviour is indeed irrational. My basic purpose here is to argue that this may not the case.

Before going any further it is worth reflecting on what standard models would predict for financial market prices and then to see what sort of deviations from the predicted behaviour occur. A first observation is that a standard view of the evolution of the prices of financial assets is that they behave as a random walk or more precisely as geometric Brownian motion. This is what is at the basis of the Black-Scholes model. In such a case prices at time t would be characterised by the following equation:

$$dS_t = S_t(\mu dt + \sigma dWt). \tag{1}$$

Such an equation has the advantage that it is useful for pricing derivatives but misses some essential features of financial time series that we would like to capture. Consider the following example illustrated in Figure 1. The time series, which is that of the German Dax stock market, index from May 1994 to May 1999 seems to behave very much as predicted by (1).

Fig. 1. Dax 30, May 1994–May 1999

However, if we now complete the series by the observations from 1999 onwards the picture changes and we now have to explain the sudden turning point as seen clearly in Figure 2.

We need some explanation for the sudden turn-around of the market. One idea is to suggest that there was some major exogenous shock. Another is to try to build models, which will produce such turning points as endogenous phenomena.

My aim in this chapter is to examine the building blocks of models, which have as their goal to explain the sort of evolution illustrated in this example. I will then analyse in more detail a situation in which the participants in

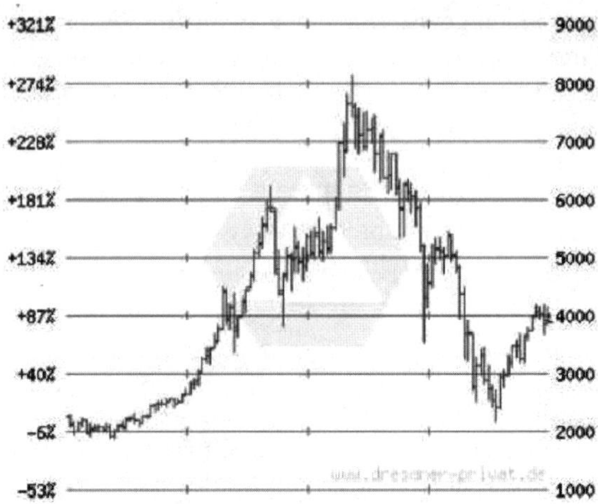

Fig. 2. Dax 30, May 1994–May 2004

the market can choose between several forecasting rules. The nature of these speculative and self-reinforcing rules will determine the demands of the various agents and determine the evolution of the equilibrium prices. I will give a simple example in which people have a prospect of investing at home or abroad and they are influenced in their choices of rules and hence in their decisions by the yields obtained by their past choices and by, of course, the movements of the exchange rate. In this model self-reinforcing changes in the exchange rate can occur, since as the number of individuals following a rule increases the success of that rule increases and more people tend to follow it. If the rule is an extrapolative one then the exchange rate will leave its "fundamental" value and a "bubble" will occur. Switches in, and transmission of, expectation formation cause this sort of bubble. For the reason mentioned this is self-reinforcing and causes people to herd on one particular alternative type of forecast and eventually to switch back to another rule. In this case, what is important is that there will be a substantial demand for the asset in question even if the underlying fundamentals do not seem to justify this.

In switching in this way, market participants are not being irrational. They will have good reason to focus on one opinion; one share or one currency for a period and then shift to another and a model of a stochastic process that results from such behaviour is proposed. Thus, it is the shifting composition of expectations that drives asset movements, in our case the exchange rate, and this is of course, at variance with the standard model in which expectations are homogeneous and rational and, of course, where no trade takes place. This is at variance with simple empirical observation of the spot market for foreign exchange where approximately $1.2 trillion per day was traded in 2001, for example.

As Bachetta and Van Wincoop (2003) point out, the obvious explanation lies in the heterogeneity of the agents on the market and, in particular, in the heterogeneity of their expectations. In the standard "representative agent" model there is no place for such heterogeneity and many authors have suggested that this is the reason for the poor predictive power of such models, (evidence for the latter is given by Meese and Rogoff (1983), Frankel and Rose (1995) and Cheung et al. (2002)). Furthermore, empirical observations suggest that expectations of actors on financial markets are indeed heterogeneous, see Chionis and MacDonald (2002).

A number of authors have introduced heterogeneous expectations into markets in different ways. One idea is simply to introduce agents who systematically have "wrong" expectations but who may survive nevertheless, (such models were pioneered by De Long et al. (1989, 1990) who introduced the now well known "noise traders". Such a solution to the problem is not very appealing and, if one takes account of the idea that agents may learn, it is difficult to accept that certain actors will persist in their error. Another alternative is to introduce dispersed information into the model and one approach suggested by Townsend (1983) is to have symmetrically dispersed information and to analyse the consequences of "higher order expectations", expectations about others expectations. The idea here is that a small amount of non-fundamental trade may generate considerable volatility since traders perceive movements in asset prices as conveying information about future values of fundamentals, (see Allen et al. (2003)). Again, despite the more sophisticated reasoning attributed to agents, in these models, a certain degree of irrational behaviour is needed to generate the results.

The building blocks of micro models, which may explain the empirical phenomena

I will focus in this chapter on models of single markets, so the analysis remains partial. Nevertheless, the link between the micro behaviour and aggregate phenomena will come out clearly. The micro models that we use to explain empirical phenomena use certain basic building blocks. Each of these is chosen, by those who analyse these problems, to create a coherent whole, which will be capable of explaining the observed empirical phenomena. Perhaps wrongly, we typically try to build models, which keep as much of the standard model as possible. There is no compelling reason to do this other than to answer the criticism that, a more conventional model could have made similar predictions. Indeed, in what follows I will stray from the conventional path but could no doubt, as I will suggest, stray even further.

The building blocks that we will need are: Firstly a specification of the objectives of the agents in the market. Secondly, the definition of the rules by which the individuals make their forecasts and how they make their choices between different rules. Thirdly, how from the previous two features demand is derived. Fourthly what equilibrium notion is used to make demands consistent? Fifthly, how the feedback from prices to forecasts works. Lastly, is there some appropriate notion of long-term equilibrium in such economies and how will its features reflect those found in observed empirical data? I shall take these one by one and with the aid of examples attempt to show the importance of these assumptions.

2 Agents Objective Function

The natural first approach for an economist, is to assume that agents have a utility function and furthermore to assume that the current utility of the agents on the markets in question depends on their expectations of future utility. A number of basic questions have to be answered if one takes this line. What sort of horizon should these agents have, and should their utility be influenced by the uncertainty of the outcomes on the market, for example? Then there is the question as to what form the utility function should take. One of the most commonly used utility functions is that based on the mean and variance of future wealth. This captures in a rudimentary way the trade off between risk and gain. It has obvious advantages from an analytical point of view, and well-known disadvantages as a description of human behaviour. One of its advantages is that the choices of economic agents who have this type of utility do not depend on their current wealth. Whilst this seems strange from an intuitive point of view, it is very attractive for the economist. If at any point in time the agents' choices are only influenced by current prices and anticipated prices there is no need to keep track of the evolution of each individual's wealth over time. Thus the system can be described by a very simple state space. Many other utility functions have been suggested in the literature, but my emphasis here is on the simplest interaction between agents as a motor for generating prices and not on the specific type of utility function that might produce plausible price dynamics. So in what follows it will be useful to keep this example in mind, for it can serve as a basis for the sort of model which will illustrate the way in which micro-behaviour leads to such phenomena as fat tails, volatility clustering and long memory in financial time series. Notice one important thing here which is that the objective function is typically the expectation of utility today. To simplify matters it is usually assumed that utility is separable and that what is maximised is the expected sum of future utilities discounted appropriately.

The horizon over which agents operate depends on many things. The traders on financial markets typically have very short horizons. Those who operate on the foreign exchange market are frequently constrained to clear their positions at the end of the day. This means that exchange rates for horizons beyond that are of no relevance to them. Those who place orders with brokers and with traders may well have longer horizons and the distribution of horizons over the participants in financial activity is undoubtedly one of the explanations of the enormous volume of trade observed on financial markets. Once again to simplify matters I will assume, in the two models used as examples, as do many others, that agents maximise tomorrow's expected utility. This sort of myopia can be thought of as corresponding to a very high discount rate and is worth looking at. Not so long ago it would have been considered unreasonable to make the assumption of myopia, but with the advent of a serious consideration of the notion of time preference and its origins, this assumption now seems no more unreasonable than many others. Strotz was probably the first to point out that exponential discounting was a very special form of time preference and has nothing particular to recommend it. Apart form hyperbolic discounting which attaches more weight to the present than would standard discounting, many alternative structures have been proposed. (For a good survey see Frederick et al. (2002)). The myopic case is just a special case of the β, δ preferences introduced for individual decision making by Elster (1989), which capture many of the qualitative features of hyperbolic discounting. Varying the assumptions on time preferences has also been suggested as a way of explaining some the anomalies observed in behaviour on financial markets.

In particular relaxing the standard assumption of exponential discounting provides a simple explanation for the volume of trade. If individuals attach more importance to the immediate future than to more distant horizons they may well, as is the case with hyperbolic discounting, take time inconsistent decisions. This means that they will constantly have to adjust their positions at each point in time and this is one explanation for the trade observed on financial markets.

For those economists who feel uncomfortable with this move away from standard rationality it is worth noting that we can always get back to one period ahead maximisation by introducing an "overlapping generations" model. In that setting agents only live for two periods and their assets are inherited by the next generation. The defects of this approach are too obvious to explain here and it is clearly a way to get around the horizon difficulty without weakening the standard assumptions.

The other side of the coin is the length of individuals' memory. This will be important when we examine how agents make their forecasts. In a sense, we have to reconcile the idea of how far ahead an agent can predict with how far back he can remember. We have also to consider the rate at which an agent discounts the future as opposed to how he discounts past gains.

3 Agents' Forecasts

Investment and speculation clearly depend, on the forecasts that people make as to future prices. The obvious question is then, how do agents make their forecasts? In keeping with everything I have said up to now it seems reasonable to assume that agents have different possible forecasting rules and must somehow choose between them. This is very different from the rational expectations approach in economics, which assumes that agents all have the correct vision of the stochastic process that governs prices in the future. In the latter case we take away any possibility that agents might learn over time about how prices evolve and simply assume, to close the model, that all of them understand the "true process" and forecast appropriately. There is no place in that equilibrium view for learning about the price process.

The rational expectations approach seems to be just a convenient construct for making equilibrium consistent but does not tell us anything about how agents actually form their forecasts and modify them. Furthermore, whilst it is true that rational expectations are consistent, it is not at all obvious that learning will lead to them. To be a little clearer, if we take the position that agents have to learn how to form their expectations, then as they learn, they will modify the latter. This, in turn will modify their demand and this will then change prices. Now, using the updated forecasting rule they will make a new forecast given the prices that have been realised and so forth. Why, as everybody learns in this way, the process should converge and why, in particular, it should converge to rational expectations, is not clear.

If we admit this then we should try to model the learning process and incorporate it directly into our model of a financial market. In the examples that I will use I suggest a rather simple approach. Agents have a number of forecasting rules amongst which they must choose and the way in which they choose amongst them has to be specified.

Let me look at two problems posed by this restriction, in turn. What we need to know is how many forecasting rules there are and what form they take. Provided that they are well defined, we can have any finite number of such rules. It is often the case that authors choose two rules as an example but there is no specific reason for this.

What form the rules should take poses an important question. Should they be based on past prices alone or should they be based on the values of certain economic variables, which are considered to be "fundamental" to the economy. In the first case future prices are forecast by extrapolating from past prices and, however sophisticated, such rules are referred to as "chartist". On the other hand, rules, which use the values of other economic variables, are frequently referred to as "fundamentalist".[2]

In the chartist case we have to specify the sort of extrapolation that the agent does. It could be a very simple moving average or some rather sophisticated time series construct. In either case we have to specify how many observations from the past the agent uses. That is, how long a memory he has. This choice is not innocent and the stability of the price process can be seriously affected by the presence of agents forecasting on the basis of very limited memory.

For the fundamentalist forecasters, there is one thing that is important and that is the choice of the variables that are considered as being fundamental in determining the price of the asset in question. Typically, the fundamentals are

[2] These terms were first coined in the foreign exchange literature by Frankel and Froot (1986).

considered as evolving exogenously. In fact, what are the fundamentals and how related they are to the price of the asset in question is a question that remains largely unanswered. What is worse, the view of the relationships linking economic variables is not fixed and if agents are persuaded of the relevance of some new variable it may actually become relevant. Michael Woodford (1990) showed how people could learn rationally to believe that some exogenous phenomenon has an effect on prices. In his case the exogenous variable was referred to as "sunspots" which, by their nature, have no intrinsic impact on the prices of assets. Yet, as people come to believe that sunspots affect prices they modify their demand for assets and this, in turn, causes them to be correlated with the sunspots. Thus, what are initially irrational beliefs become rational. This illustrates the very basic idea that what is fundamental depends to a large extent on what is perceived to be fundamental.

Once the set of rules, whether "fundamentalist", extrapolative, or some combination of the two, is established, we have to specify how agents choose amongst the rules.

4 The Choice Among Forecasting Rules: An Epidemiological Rule

There are various views that one could have as to how agents might reasonably choose their rule. A first example, which I will call an "epidemiological" rule, would be based on the influence of other market agents. The simplest idea is that, if I, as a chartist, meet a fundamentalist, there is a certain probability that I will be converted to using his rule. A stochastic process to describe this sort of stochastic process was developed with Hans Foellmer, (see Kirman (1993)).[3] Consider the following example.

Agents are faced with a price process s_t for a financial asset and form expectations about tomorrow's prices. There are two different ways[4] of forming expectations and each agent uses one of them. However, random meetings influence the expectations of the individual agents with other agents. Call the two methods of forming expectations the two "opinions" in the model and then if there are N agents define the state of the system by the number k of agents holding opinion one, i.e.

$$k \in \{0, \ldots, N\}.$$

The stochastic process governing the state evolves as follows. Two agents meet at random and the first is converted to the second's view with probability $1 - \delta$. There is also a small probability ϵ that the first agent will change her

[3] The original model was built to explain the asymmetric behaviour of foraging ants.

[4] As observed previously, there could be any finite number of ways of forming expectations and this would not change the nature of the results.

opinion independently of whom she meets. This is a technical necessity to prevent the process from being "absorbed" into one of the two states 0 or N, but can be allowed to go to zero as N becomes large.[5] Indeed, it is important in what follows, that ϵ be small.

The process then evolves as follows:

$$\begin{aligned} p(k,k+1) &= \left(1-\frac{k}{N}\right)\left(\epsilon+(1-\delta)\frac{k}{N-1}\right), \\ p(k,k) &= 1-p(k,k+1)-p(k,k-1), \\ p(k,k-1) &= \frac{k}{N}\left(\epsilon+(1-\delta)\frac{N-k}{N-1}\right). \end{aligned} \qquad (2)$$

The first problem is to look at the equilibrium distribution $\mu(k)$, $k = 0,1,\dots,N$, of the Markov Chain defined by (2).

This is important in the economic model since it describes the proportion of time that the system will spend in each state in the long run. Now the form of $\mu(k)$ will depend, naturally, on the values of δ and ϵ. The case of particular interest here is that in which $\mu(k)$ has the form indicated in figure 3:

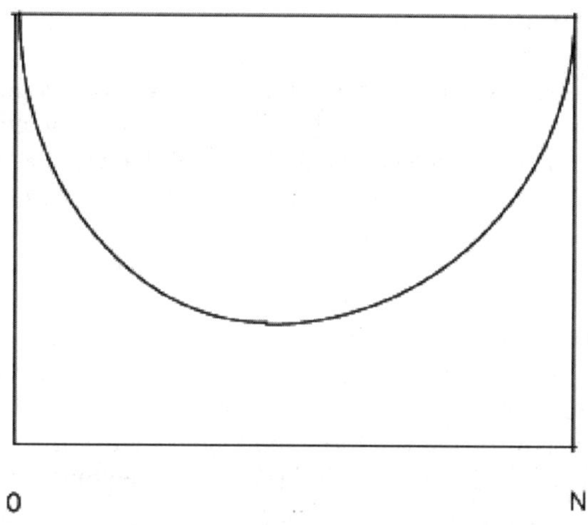

Fig. 3. $\mu(k)$, with $\epsilon = 0.005$ and $\delta = 0.01$

It is easy to see that if $\epsilon \leq \frac{1-\delta}{N-1}$ then $\mu(k)$ will indeed be convex. Thus this case, in which the process spends most of its time in the extremes, corresponds to the case in which the probability ϵ of "self conversion" is small relative to the probability of being converted by the person one meets. Although this

[5] This ϵ can be thought of as the replacement of some old agents in each new period by agents who may hold either opinion (see for example the evolutionary model of Young and Foster, 1991) or by some external shock which influences some people's expectations.

probability of conversion is independent of the numbers in each group, which type will actually meet which type depends on the relative numbers in each type at any moment, i.e. on the state of the system. Thus when one type is in the minority, conversion of any individual to that type is much less likely than when the numbers of the two types are fairly equal. It is this that yields a distribution which is concentrated on the extreme values.

Using this type of rule generates swings from a majority using one sort of rule to a majority using the other. This is a basis for models that yield time series with some of the characteristics found in empirical financial time series, (see Kirman (1991) and Kirman and Teyssiere (2002)). This line of research was pursued in Kirman and Teyssiere (2005) and different criteria were examined there for determining the probabilities of making forecasts. One important addition in that model was to add a second step, as in Kirman (1991) in which, once people had formed their opinion, they received a signal, which gave them an indication of the current majority opinion. This can lead them then to revise their opinion. In particular, this addition of a signal of market sentiment can make the swings between different opinions much more extreme when the signal is fairly accurate. Another idea is that the frequency with which agents have been met in the past should influence the probability of being converted by them. The latter idea was also used in Kirman and Teyssiere (2005) and they analysed the long range dependence properties of the resultant time series. They came to the conclusion that the behaviour of the time series of the volatility generated by these models was a combination of long-range dependence and of switching regimes.

Yet there is an objection to the type of probabilistic choice used in the epidemiological type of model. It depends on whom one has met, but not on the success of the rule adopted. It would seem reasonable to base the adoption of rules by individuals on some measure of how much was earned in the past b the various rules, or alternatively on how accurate those rules have proved to be.

5 A Choice Rule Based on Experience

Consider then an alternative type of rule in which the choice of a rule does depend on its success in the past. Now, in this case, to establish the transition probabilities for the agents we must first define their profit from following each of the possible forecasts. Suppose, to simplify matters that the cumulated wealth of each agent that he reinvests on the market for the asset at each point in time is given by W^i_{t+1} and the profit at time $t+1$ by,

$$\Pi^i_{t+1} = W^i_{t+1} - W^i_t . \tag{3}$$

What we have to do then, is to separate the wealth gained by following each of the forecasting rules in the past. Here we will make the assumption that the agent discounts, at each period, the gains he has made with a given forecast. This reflects the fact that he attaches less importance to information he gained in the past than to more current information. Making a further simplification, assume that there are two rules, a "fundamentalist" and a "chartist" rule. The updated profit from following the fundamentalist point of view is then given by, adding the current profit if the agent is following that rule and adding zero otherwise to the discounted wealth he has obtained in the past from following the same rule. More formally

$$\Pi_{t+1}^{if} = \Pi_{t+1}^{i}, \tag{4}$$

if, at time t agent i held the fundamentalist view and $\Pi_{t+1}^{if} = 0$ otherwise.[6]

Then we have the accumulated fundamentalist profit that is given by:

$$W_{t+1}^{if} = W_t^{if}(1-\omega) + \Pi_{t+1}^{if}. \tag{5}$$

Similarly for the chartist profit we have,

$$\Pi_{t+1}^{ic} = \Pi_{t+1}^{i}, \tag{6}$$

if the individual was following the chartist rule at t and

$$\Pi_{t+1}^{ic} = 0 \quad \text{otherwise}. \tag{7}$$

This gives accumulated profit from following the chartist rule of

$$W_{t+1}^{ic} = W_t^{ic}(1-\omega) + \Pi_{t+1}^{ic}. \tag{8}$$

Now, we have to explain how accumulated profits are mapped into the probabilities of choosing the forecasting rules. Here we assume that the probability of following the fundamentalist rule is given by,

$$P_t^{if} = \frac{e^{\beta W_t^{if}}}{e^{\beta W_t^{if}} + e^{\beta W_t^{ic}}}, \tag{9}$$

where β is a constant that reflects the importance the individual attaches to previous experience and obviously, since we only have two rules, the probability of becoming a chartist is given by:

$$P_t^{ic} = 1 - P_t^{if}. \tag{10}$$

[6] We could have assumed that the agents know, ex post, how the two rules have performed. In this case everybody would have the same experience and the probability of switching would be the same for everyone. Which approach is more acceptable depends on one's view as to the information available to the agents on the market, for a discussion of these two types of learning individual or social, see Vriend (2000).

The particular rule chosen here for the probability for the transition from one rule to another will be familiar from a number of economic applications (see McFadden (1974), Anderson et al. (1990), Brock (1993), Brock and Durlauf (2001a, 2001b). It has been widely used in the psychological literature on reinforcement learning, and in game theory, where it is referred to as the "quantal response rule". One way of deriving it is to consider the problem of the trade-off between obtaining information about sources of profit and exploiting those which have proved profitable in the past. This "exploration" versus "exploitation" arbitrage can be analysed by maximising a linear combination of the gain to be had from trying new alternatives and the expected gain given the experience in the past with the different rules, (see Brock (1993) and Weisbuch et al. (1998)).

It is clear that if β tends to infinity the individual tends toward being a "best responder" in the sense that he just chooses with certainty the opinion which has been most profitable in the past (see Blume (1993)). However, as β tends towards 0, the individual chooses each alternative with equal probability. In other words the individual learns nothing from his experience in using the alternatives. Other rules might be used to define the probabilities of switching but the logit rule is often used since it has rather plausible behavioural characteristics and it is analytically easy to use. A full discussion of the properties of the alternative ways of mapping previous profits into probabilities can be found in Weisbuch et al. (1998). One might also want to change the performance criterion on which the probabilities are based and Teyssiere (2003) has suggested the accuracy of the forecasting rule as an alternative. The problem with this is that, while it provides a sensible way of evaluating forecasting rules, it may lead to less profitable behaviour. Consider the case in which a rule or guru gives a forecast of a price rise which turns out to be inaccurate but in the right direction. It could then happen that ex post the trader will make more money as the result of the mistake, than he would have, had he had the correct forecast. Yet, if this is the case, perhaps one should choose a simpler rule based on the percentage of the time that the rule or guru gets the sign of the change right. There are many possibilities and no good reason to exclude most of them.

As observed earlier, the memory involved in evaluating the cumulated profit recursively in (8) could be limited to some finite number of periods. Alternatively, one might use the equivalent of hyperbolic discounting or quasi-hyperbolic discounting for discounting the past. This would mean giving much more weight to recent profits. The notion of limiting memory is akin to the idea that agents only forecast for a finite number of periods, and if this is the case, it does not seem reasonable to give them an unlimited memory. The statistician's response to this is to remark that one does not, with conventional discounting, need to keep the whole past history since the last cumulative observation is sufficient.

Again, the forecasting horizon of an agent may be closely linked to his objective function. As I have said, a trader who is judged on day to day performance has little interest in long term forecasting. However, no such stricture applies to his memory.

As has been mentioned, in expression (9) the constant β plays an important role and it would be of some interest to perform simulations with varying values of β and then comparing this to varying the discount factor. In any event, whatever the way in which the probabilities are constructed, once the forecasting rule is determined, agents will form their demand.

6 Agents' Demand

If agents have objective functions and forecasts we can determine their demand. This will also depend on their wealth, which will be a result of their behaviour in the past. Certain objective functions, as I have mentioned, lead to demands that do not depend on wealth. However, in general, this will not be the case and, fortunately, the formal analysis does not depend on this very arbitrary assumption. A reasonable alternative to the maximisation of some rather arbitrary objective function is to assume from the outset that we work with demand as a primitive concept and do not try to derive it. This is difficult for economists to accept, but given the unfounded nature of the standard assumptions on preferences, it is difficult to quarrel with. Such an approach pushes the problem off one step further since, if we do this, we have to decide on the nature of demand. However, this approach has one particular merit. It starts with something that is, at least in principle, observable.

A further step would be to start with aggregate demand, which is in an old tradition in economics. Doing this, however, would undermine the basic aim of explaining aggregate behaviour from the interaction of individuals. So, in the first of the examples that I will give I will stick to specified objective functions and in the other, specify individual demands directly.

Before I continue a note of caution is in order. Although the choice of rule by agents has been specified as stochastic, there are two ways of handling this. One way is to reduce the stochastic process to a deterministic one by using the "mean field" approach widely used in physics. This is what was done in Brock and Hommes (1997) and in Weisbuch et al. (2000). This is always open to the criticism that this is nothing more than an approximation and how good that approximation is has to be justified. An alternative approach is to analyse the stochastic process itself and to try to obtain results directly and I will come back to this in the last section.

7 The Type of Equilibrium Envisaged

Once demands are determined, then the standard approach is to define an equilibrium notion. Here again, there are various possibilities. A first idea would be to define what is called a "temporary equilibrium" for each period.

In this case today's demands are set equal to today's supply but markets will reopen in subsequent periods. This implies two things. Firstly that agents do not have a "correct" view of the future at each point in time and, secondly, that we have a series of temporary equilibria but no overall equilibrium. At each point in time there is a price but these prices do not necessarily converge to a steady state, though much of the early work on this theme did focus on the characterisation of such steady states.

Notice also that there is a hole in the literature dealing with the sort of model developed here. In general, it is not shown that there exists a temporary equilibrium at each point in time nor that it is unique. This is done in Kirman et al.(2006) but does not solve the problem of the analysis of the dynamic evolution of the sequence of temporary equilibria.

More ambitiously one might seek an equilibrium in which agents expectations are consistent with the stochastic price process. Such so-called "rational expectations" equilibria, have the advantage of inter-temporal consistency, but do not explain how agents arrive at the appropriate expectations. Furthermore, as mentioned, Woodford (1990) for example, has shown that if one tries to introduce a reasonable learning process for individuals, the economy or market can learn to have consistent expectations but which have no connection with the fundamental variables of the economy. People can learn to believe that "sunspots" are important in determining economic outcomes and, what is more, their beliefs will be self-fulfilling.

An alternative and more appealing approach would be to try to model the process of price formation in a more realistic fashion. On many financial markets, the prices that are announced at each point in time are those for each transaction and some system such as an "order book" is used to match bids and offers. In this case one would model the supplies and demands of individuals as resulting form current and expected prices and would use the type of algorithm actually used on markets such as the Paris stock exchange to match the bids and offers. The evolution of prices in an electronic order book has been studied extensively and particular attention has been paid to limit orders. To use this analysis in our context would require establishing how agents place their orders, and a specification of the mechanism through which orders are met and a price established. The particular protocol used can have considerable consequences for the evolution of prices as Domowitz (1993) has shown.

8 The Feedback from Equilibrium Prices to Forecasts and hence Demand

An important feature of financial market models is the feedback from past prices to current demand. Excess demands in one period determine the market price in that period but this price is then added to the price history and it is the latter, which will condition agents' forecasts. However, as soon as an agent modifies his forecast he modifies his demand for the financial asset in question. This in turn will mean that the equilibrium price in that period will be changed. In many cases, economists look for a steady state of this process, (see e.g. Grandmont (1983)). Yet, if agents choose their forecasting rules probabilistically, or if there is some other randomness in the model such a notion would not be appropriate. Depending on the source of randomness in the model, which might be in terms of the agents' incomes, or in terms of some exogenous demand for the asset, prices might follow a mean reverting random walk, geometric Brownian motion or some other such process. Perhaps most interesting is the case where agents choose their forecasts randomly and then with certain restrictions it can be shown that the price process is ergodic and has a unique limit distribution, (see Foellmer et al (2005)). This gives an alternative notion of equilibrium and I will come back to this. In particular, in the presence of chartists, the limit distribution exhibits fat tails and long memory.

9 An Example of a Model Exhibiting the Properties Described

This example is a simplified version of that used in Kirman et al (2006) and is closely related to the model developed in Kirman and Teyssiere (2005) There is an exchange market on which are traded two assets each of which is denominated in one of the two currencies domestic d and foreign f whose values are linked by the exchange rate s. s is the price of one unit of foreign currency in units of domestic currency, so that, from the point of view of a domestic investor a devaluation of her currency means an increase in the value of s. There are n domestic investors, and they measure their wealth in units of d. They have the same risk aversion and the same utility functions but they may hold one of two views as to the evolution of the exchange rate. Thus the driving force behind the dynamics in this model will be the variations in the proportions of those following each of the forecasting rules. In Kirman et al. (2006) we consider the investors in both countries, but for the example here consider all the investors as being domestic.

The first building block of this model is the choice of the set of forecasting rules. Here, we shall choose two types of rule, fundamentalist and chartist but, as I observed in my earlier remarks, any finite number of rules would fit into the same framework.

Consider the case of the investor who wishes to forecast the value of s_{t+1}. If she is a fundamentalist she believes there is an equilibrium value $\bar{s}_t$ to which the exchange rate will revert. We attribute the following form to these beliefs:

$$E^f(s_{t+1}|I_t) = \bar{s}_t + \sum_{j=0}^{M_f} \nu_j(s_{t-j+1} - \bar{s}_{t-j}), \quad \sum_{j=0}^{M_f} \nu_j = 1, \tag{11}$$

where M_f is the finite memory length of the fundamentalists.

If she is a chartist her forecast as to the future value of the exchange rate will be an extrapolation of its past values. We give these extrapolations the following form:

$$E^c(s_{t+1}|I_t) = \sum_{j=0}^{M_c} h_j s_{t-j}, \tag{12}$$

where the h_j are positive constants and M_c is the finite memory length of the chartists.

Now define the following variables at time t:

- ρ is the dividend in foreign currency paid on one unit of foreign currency,
- s_t is the exchange rate,
- f_t^i is the demand by individual i for foreign currency,
- d_t^i is the demand by individual i for domestic currency,
- r is the interest rate on domestic assets.

The individual's wealth at time t is determined by her investments in foreign and domestic assets and what she earned on each of them respectively. That is,

$$W_t^i = (1+r)d_{t-1}^i + s_t(1+\rho)f_{t-1}^i. \tag{13}$$

At each point in time the individual's demands for foreign and domestic assets must satisfy the budget constraint:

$$W_t^i = d_{t-1}^i + s_t f_t^i. \tag{14}$$

The gain for an individual in period t is given by:

$$g_t^i = W_t^i - (1-\omega)W_{t-1}^i, \tag{15}$$

where ω is the discount factor and the cumulative gain is is given by:

$$G_t^i = W_t^i - (1-\omega)^{t-1}W_1^i, \tag{16}$$

where W_1^i is individual i's wealth at the beginning of period one (before she chooses the rule to be used).

However, the gains at each period are determined by the demands for domestic and foreign currency in the previous period and these are in turn determined by the forecast that the individual made as to the exchange rate in the next period. The latter depends on whether she was following the fundamentalist or chartist rule.

Now we move to the next building block, how the rule is chosen. The particular assumption made here is that the choice of rule is probabilistic and the probabilities of choosing a rule are an individual behavioural learning process. As I explained earlier, in this case, the choice will depend on the success of the rules in the past in terms of the profit that was obtained when using them. Once again the probabilities could depend on other measures of success, such as the accuracy of the forecasting rules in the past, but in any particular model we have to decide on what governs the choice of rule. If it is to depend on the profitability of the rules in the past, we need to keep a total of the gains obtained by individual i up to period t by following the fundamentalist rule and similarly for the chartist rule. Define a random variable θ_t^i which will take on two values F and C, that is:

$$\begin{aligned} \theta_t^i &= F \quad \text{with probability} \quad p_t^i(F) \\ \theta_t^i &= C \quad \text{with probability} \quad 1 - p_t^i(F) = p_t^i(C). \end{aligned}$$

Now we can define an indicator function for the random variable and this is simply:

$$I_t(F) = 1 \quad \text{if} \quad \theta_t^i = F \quad \text{and0} \quad \text{if} \quad \theta_t^i = C, \tag{17}$$

$$I_t(C) = 1 \quad \text{if} \quad \theta_t^i = C \quad \text{and0} \quad \text{if} \quad \theta_t^i = F. \tag{18}$$

This leads to the gains for an individual when she was using each forecasting rule as:

$$G_t^i(F) = \sum_{r=1}^{t-1} I_r(F)(G_r^i - G_{r-1}^i), \tag{19}$$

$$G_t^i(C) = \sum_{r=1}^{t-1} I_r(C)(G_r^i - G_{r-1}^i), \tag{20}$$

with $G_0^i(F) = G_0^i(F) = 0$.

The investor's first piece of information is the rule θ_t^i she has drawn. What is the remaining information I_t^i of a domestic investor i at time t? She has her observations of the past values of her demands for foreign and domestic assets in the previous periods, the vector of observed exchange rates up to period $t-1$ and the cumulated gains that she has realised from using each of the two forecasting rules, fundamentalist and chartist. Thus we have:

$$I_t^i = \left\{ d_{t-1}^i, f_{t-1}^i, S_{t-1} = (s_1, \ldots, s_{t-1}), G_t^i(F), G_t^i(C) \right\}. \tag{21}$$

The total information available to the individual i once her forecasting rule has been determined is given by (I_t^i, θ_t^i).

Up to this point I have not specified how the demand function is determined and for this we will need to choose the next building block, the investor's objective function. Once we have made this choice we can turn our attention to the derivation of the ith individuals' demand for foreign assets and, given her budget constraint this will also determine her demand for domestic assets.

10 The Demand of an Investor for Foreign Currency

To make things tractable it is useful to make a compromise here. Consider the utility function from which the demand function is derived but, to simplify matters choose the very simple Mean Variance utility function which exhibits constant absolute risk aversion (CARA), and therefore its maximisation does not depend on the agent's wealth in the current period. For any investor i the expected utility function $E\left(\bar{W}_{t+1}^i | I_t^i, \theta_t^i\right)$ is defined by:

$$E\left(U\left(\bar{W}_{t+1}^i\right) | I_t^i, \theta_t^i\right) = E\left(\bar{W}_{t+1}^i | I_t^i, \theta_t^i\right) - \mu V\left(\bar{W}_{t+1}^i | I_t^i, \theta_t^i\right), \tag{22}$$

where μ is the measure of risk aversion. Then the agent's wealth at the next period $t+1$ is given by:

$$\begin{aligned} \bar{W}_{t+1}^i &= (1+\rho_{t+1})\bar{s}_{t+1} f_t^i + (W_t^i - s_t f_t^i)(1+r) \\ &= (1+r)W_t^i + f_t^i\left((1+\rho_t)\bar{s}_{t+1} - (1+r)s_t\right). \end{aligned} \tag{23}$$

So the investor has the following problem

$$\max_{f_t^i} E\left(U\left(\bar{W}_{t+1}^i\right) | I_t^i, \theta_t^i\right) = E\left(\bar{W}_{t+1}^i | I_t^i, \theta_t^i\right) - \mu V\left(\bar{W}_{t+1}^i | I_t^i, \theta_t^i\right). \tag{24}$$

Recall that the expectations of the investor depend on whether she is following the fundamentalist or chartist rule i.e. on $\bar{\theta}_t^i$ and, whichever rule she follows the expectations will be conditioned on the current exchange rate s_t. Hence using first order conditions we can write the demand for agent i as:

$$f_t^i = f_t^i(s_t, \theta_t^i) = \frac{(1+\rho)E(\bar{s}_{t+1} | I_t^i, \theta_t^i) - (1+r)s_t}{2\mu(1+\rho)^2 Var(\bar{s}_{t+1} | I_t^i, \theta_t^i)}. \tag{25}$$

Clearly the expectations in (24) and (25) are conditioned on which rule the agent is following and recalling the definition of the indicator function for the random variable given in (7) we can write:

$$f_t^i(s_t, \theta_t^i) = f_t^i(s_t, F)I_t(F) + f_t^i(s_t, C)I_t(C). \tag{26}$$

Finally we shall suppose that the demand for the individual i when she is using the rule F or C is given by replacing the expectation in (22) by the expectations for each rule which we have defined before, equations (11) and (12), and which we recall

$$E^f(s_{t+1} | I_t) = \bar{s}_t + \sum_{j=0}^{M_f} \nu_j(s_{t-j+1} - \bar{s}_{t-j}), \quad \sum_{j=0}^{M_f} \nu_j = 1,$$

for the fundamentalist and:

$$E^c(s_{t+1} | I_t) = \sum_{j=0}^{M_c} h_j s_{t-j},$$

for the chartist.

11 The Market Equilibrium

The next building block involves the choice of an equilibrium notion. Here we consider that the market clears at each period and hence that we are looking at a series of temporary equilibrium prices. This again is a compromise, agents could be less myopic and furthermore, some economists would insist that their expectations should be consistent with the stochastic price process. Doing this would change the dynamics completely and moreover, would remove the source of the fluctuations that characterise the dynamics of the prices in our model.

So, in order to establish our temporary equilibria, we have first to consider first the aggregate demand for foreign currency of the investors. This is given by:

$$\Phi_t = \sum_i f_t^i(s_t, \theta_t^i) = \sum_i f_t^i(F) I_t(F) + \sum_i f_t^i(C) I_t(C), \tag{27}$$

or if we define the number of investors who use the fundamental rule at time t as $N_t(F)$ and the total number of investors as N and $k_t = \frac{N_t(F)}{N}$ then we can rewrite aggregate demand as:

$$\Phi_t = N\left[f_t(F)k_t + f_t(C)(1-k_t)\right]. \tag{28}$$

The equilibrium exchange rate s_t^* is then given by:

$$\Phi_t - \underset{=}{X_t} 0, \tag{29}$$

where X_t is the random liquidity supply of foreign currency coming from underlying trade for example. Using (25) we obtain the following expression for the equilibrium exchange rate:

$$s_t^* = N\left[s_{t+1}^a(F)k_t + s_{t+1}^a(C)(1-k_t)\right] - \frac{2\mu(1+\rho)^2\sigma^2}{(1+r)} X_t, \tag{30}$$

where, $s_{t+1}^a(F)$ and $s_{t+1}^a(C)$ represent the expectations under each of the two rules, fundamentalist and chartist as to the next exchange rate that is:

$$s_{t+1}^*(F) = E(\bar{s}_{t+1} | I_t, F) \quad \text{and} s_{t+1}^*(C) = E(\bar{s}_{t+1} | I_t, C) \tag{31}$$

where X_t is the exogenous supply of foreign exchange, which we will take to be a random walk. Thus, we have the form for the exchange rate process originally given by Frankel and Froot.

12 Some Results and Some Properties of the Model

The model I have just described is close to that in Kirman et al.(2006) and is shown by them has the property that the temporary equilibrium is well defined at each point. Their model is more complicated in that there are two populations, foreign and domestic investors, each of whom considers the value of an investment in the other country to be stochastic. This has the advantage that it removes the necessity for introducing an exogenous "liquidity demand" for the foreign currency but considerably complicates the model from an analytical point of view.

Models of the sort described replicate many empirical stylised facts and a careful discussion of the values of the parameters for which these features appear are given in Teyssière (2003) and Kirman and Teyssière (2005). Depending on the particular rule used to determine the probabilities of choice, epidemiological, success of forecasting rules, either in terms of profit or forecast accuracy one can observe long memory, and there are almost always "bubbles" and periods of high volatility. The switching of regimes from ones dominated by chartists to ones dominated by fundamentalists is likely to provide a large part of the explanation for the existence of long memory and as suggested by Kirman and Teyssière (2002) this is really "spurious long memory". Teyssiere discusses the possibility of confusion between long-range dependence and change point GARCH processes. As he points out, much of the discussion has centered on the series with change points in the conditional mean process. In the sort of model just outlined, however, the interest lies in changes in the conditional variance. Periods of higher volatility typically correspond to the predominance of chartists. Teyssière (2003) indicates the tests for change points in the conditional variance proposed by Kokoszka and Leipus (2000), Horvàth, Kokoszka and Teyssière (2001) and Kokoszka and Teyssière (2002) as suitable benchmarks. Such tests can help to avoid the pitfall of incorrectly accepting long-range dependence in the face of change points. The early test proposed by Lo (1991) was vulnerable to this sort of mistake.

However, the purpose of this chapter is not to do any systematic econometric analysis, something which is dealt with by the econometricians and statisticians in this book. The purpose is to suggest a model structure capable of reproducing some of the features of financial asset price series, bubbles which always explode, kurtosis, and of course long memory whether real or spurious. Figure 4 shows just one simulation of the model but illustrates the fact that the bubbles correspond to periods where the chartists take over and hence where the price path becomes detached from the fundamentals. This is, of course, purely illustrative and does correspond to any careful testing of the simulated data.

13 Long Run Equilibrium

There remains one final question. Even if we accept the notion of a sequence of temporary equilibrium can we say anything about the time series that result? If we consider the previous model, for certain configurations of parameters

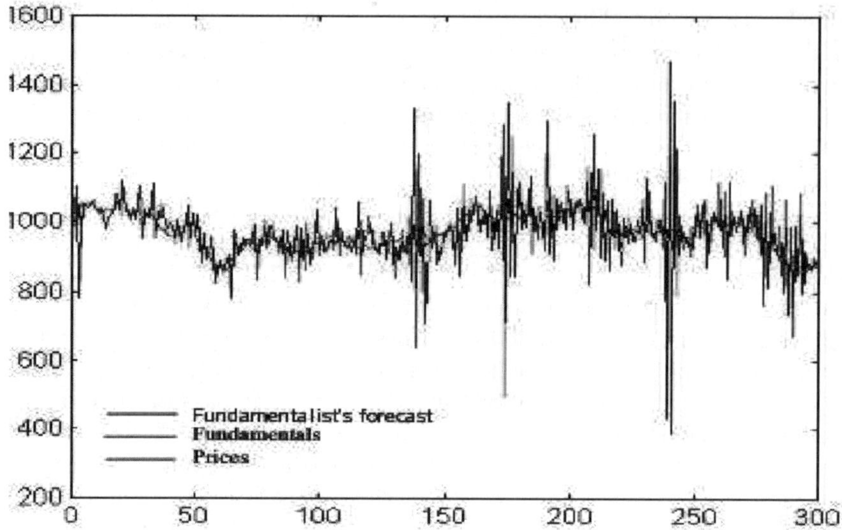

Fig. 4.

it could become explosive. There are two possible reactions to this. Since we will never observe more than a finite sample it could well be that the real DGP is actually explosive but this will not prevent us from trying to infer something about the data that we observe. Supposing however, that we are interested in being able, from a theoretical point of view, to characterise the long run behaviour of the system. Furthermore if we treat the process as being stochastic and do not make a deterministic approximation, then we have to decide what constitutes an appropriate long run equilibrium notion. With Hans Foellmer and Ulrich Horst, (Foellmer et al. (2005)), we have analysed formally the sort of price process discussed here and we have been able to produce some analytical results characterising the process. Furthermore, we can give a long run equilibrium notion which is not the convergence to a particular price vector. If prices change all the time how may we speak of "equilibrium"? The idea is to look at the evolving distribution of prices and to see if we can characterise its long-run behaviour. We have to examine the process governing the evolution of prices and profits and to see under what conditions we can show that it is ergodic i.e that the proportion of time that the price takes on each possible value converges over time and that the limit distribution is unique.. This means that, unlike the "anything" can happen often associated with deterministic chaos we can say that in the long run the price and profits process does have a well-defined structure.

If we think about the sort of model described up to this point it is clear that what destabilises the price process is the presence of chartists. Indeed, we know that if there were only chartists the process would explode. What we can show in the model that I will present here, is that the distribution of the time averages of prices converges, if the probability of any individual becoming a chartist is not too high. Thus although the presence of chartists prevents the price process from being bounded, if they do not dominate for too long, which is what is guaranteed by putting a bound on the probability of becoming a chartist, then the distribution of the time averages of prices is stable. In other words, although prices are always changing and cannot be bounded, structure emerges in the time series and this is probably the appropriate notion of equilibrium for such models.

14 The Model

The model presented here is taken from Foellmer et al. (2005) Consider a finite set A of agents who trade a single risky financial asset on a market. At each point in time agent $a \in A$ has a log-linear excess demand function given by:

$$e_t^a(p, \omega) = c_t^a \left(\hat{S}_t^a(\omega) - \log p \right) + \eta_t^a(\omega), \tag{32}$$

where $\hat{S}_t^a$ and η_t^a represent the agents current reference. This is already a departure from the previous models in that the excess demand is not derived from any underlying utility function but such demand or excess demand functions have been widely used in the mathematical finance literature, see Foellmer and Schweizer (1990) for example. The advantage of working in logs is evident since it avoids the problem of the underlying prices becoming negative. The excess demand function can be envisaged as having two components, an excess demand based on the difference between the current price and a reference level price and a liquidity demand which is random. The latter convention is often used, but can be dispensed with in two ways. Either one can introduce a second market, if for example one thinks of the foreign exchange market this would be the "rest of the world" or one introduces an aggregate excess demand for the asset or for foreign denominated liabilities, which arises not from the financial market itself but from simple trade requirements. At each point in time the temporary equilibrium price is that which makes the excess demand for the asset equal to zero.

$$S_t = \frac{1}{c_t} \sum_{a \in A} c_t^a \hat{S}_t^a(\omega) + \eta. \tag{33}$$

Thus equilibrium prices are a weighted average of individual reference prices and their liquidity demand. Where do the reference prices of the individuals come from. A simple idea is that they are provided by m experts or "gurus" among whom agents choose. An alternative interpretation is that these reference prices are forecasts obtained by using one of m forecasting rules. This can be represented as

$$\hat{S}_t^a \in \{R_t^1, \ldots, R_t^m\}. \tag{34}$$

The proportion of agents who choose rule or guru at time t is then given by rules. This can be represented as

$$\pi_t^i = \frac{1}{c_t} \sum_{a \in A} c_t^a \mathbf{1}_{\hat{S}_t^a = R_t^i}, \tag{35}$$

where the last term is the indicator function for the choice of guru, that is, it takes on value 1 if guru i is chosen and 0 otherwise. Given this, we can rewrite the expression for the log equilibrium prices as rules. This can be represented as

$$S_t = \sum_{i=1}^{m} \pi_t^i R_t^i + \eta_t. \tag{36}$$

Now the question arises as to how the reference prices are determined. The idea is that each guru $i \in \{1, \ldots, m\}$, has a fundamental value F^i but also takes account of the price trend. Thus the reference value for guru i is given by

$$R_t^i = S_{t-1} + \alpha_i \left[F_i - S_{t-1}\right] + \beta^i [S_{t-1} - S_{t-2}]. \tag{37}$$

Now we can look at the evolution of the price of the asset over time which is given by

$$S_t = F(S_{t-1}, S_{t-2}, \tau_t) = \left[1 - \alpha(\pi_t) + \beta(\pi_t)\right] S_{t-1} - \beta(\pi_t) S_{t-2} + \gamma(\pi_t, \eta_t), \tag{38}$$

and in these expressions there are two random elements one of which, the choice probabilities of the agents is endogenous and the other, the liquidity demand is exogenous. Thus the random environment in which the process evolves is given by,

$$\{\tau_t\} = \{(\pi_t, \eta_t)\}. \tag{39}$$

Now let me look a little closer at the characteristics of the gurus and see how they can be assimilated to the "chartists" and "fundamentalists" of the previous models. If we think of several "gurus" or fundamentalist rules, what identifies rule i is a fundamental value F_i to which the exchange rate or asset price will return. So the forecast will have the form is given by,

$$R_t^i = S_{t-1} + \alpha^i [F^i - S_{t-2}]_t, \quad \alpha^i \in (0, 1). \tag{40}$$

Suppose, for the moment that there were only fundamentalist rules or gurus then the evolution of prices would behave as

$$S_t = [1 - \alpha(\pi_t)] S_{t-1} + \gamma\left((\pi_t, eta_t)\right), \quad \sum_{i=1}^{m} \alpha^i \pi_t^i \tag{41}$$

Since $\alpha_i \in (0, 1)$ the sequence of temporary price equilibria will be mean-reverting. The process can be considered as an Ornstein-Uhlenbeck process in a random environment. In a sense fundamentalists stabilise the market dynamics. This is clear since the fundamental values which are finite in number bound the range of possible prices.

Now, consider the chartists who are simply extrapolating from previous prices. They may do so in any more or less sophisticated way, but what is important is that their forecast should only be based on previous prices. Consider the simplest case in which chartists forecast that the change in the future will be based on the last change. Forecasts will differ in the proportion of that change that is expected next time. Formally the recommendations take the form

$$R_t^i = S_{t-1} + \beta^i \left[S_{t-1} - S_{t-2}\right] + \eta_t, \quad \beta^i \in (0,1). \tag{42}$$

Now, we can ask what would happen if there were only chartists on the market. In this case with m chartist rules or gurus the process would be given by

$$S_t - S_{t-1} = \beta(\pi_t)[S_{t-1} - S_{t-2}] + \eta_t, \quad \beta(\pi_t) = \sum_{i=1}^{m} \beta_i \pi_t^i. \tag{43}$$

What is interesting here is that now, returns become mean reverting but prices will explode. Thus, in contrast to fundamentalists, chartists have a destabilising effect on prices.

Now put the two pieces together and the process becomes

$$S_t = n[1 - \alpha((\pi_t)) + \beta(\pi_t)]S_{t-1} - \beta(\pi_t)S_{t-2} + \gamma(\pi_t, \eta_t). \tag{44}$$

It should be clear, from the above, that when there are few chartists the process will be stable but that the price process will become explosive if the chartists dominate. There will be bubbles and crashes but these will be temporary if the probability of the chartists continuing to dominate is not "too high".

Before proceeding a further building block has to be specified as should be clear from the earlier part of the chapter. We now need to know how the probability that an agent i will follow guru j is determined. In this model we will assume that the probability of using a rule is dependent on the past experience the agent had, using that rule. To evaluate the agent's experience define the profits obtained at time t associated with the rule i as

$$P_t^i = \left(R_t^i - S_{t-1}\right)\left(e^{s_t} - e^{s_{t-1}}\right), \tag{45}$$

and then, by discounting past profits, we can define cumulated profits as

$$U_t^i = \alpha U_{t-1}^i + P_t^i = \sum_{j=0} \alpha^{t-j} P_j^i. \tag{46}$$

It will be these profits that determine the choice probabilities, i.e.

$$\pi_{t+1} \sim Q(U_t, : \cdot). \tag{47}$$

We make the assumption that the probabilities of choosing a guru depend on the experience with that guru in the past and will assume, in particular, the better the performance, the higher the probability of choosing that guru.

Two features then emerge from the model:

1. The more agents adopt a guru's recommendation, the more impact that guru will have on the evolution of prices.
2. The stronger a guru's impact on prices the better will be his performance. Thus, the simple fact that agents prefer successful gurus will lead to a self reinforcing move to the guru who is currently doing best.

Now that we have established all the building blocks we can see the overall structure

The asset price evolves according to the linear stochastic difference equation defined in (44) i.e.

$$S_t = [1 - \alpha(\pi_t) + \beta(\pi_t)]S_{t-1} + \gamma(\pi_t, \eta_t), \tag{48}$$

and this in a random environment$\{(\pi_t, \eta_t)\}$. The latter has an endogenous component which depends on past prices and performances and an exogenous component, the liquidity demand which is exogenously given. Note that the latter may be made endogenous by introducing a second population of agents, which has a natural interpretation in the exchange rate context. We assume that the aggregate liquidity demand is an i.i.d. process. The past prices have an influence on the environment because the choices made by the agents depend on profits, which, in turn are dependent on prices.

The process governing the evolution of prices and of profits is a Markov chain which we can write as

$$\xi_t = (S_t, S_{t-1}, U_t). \tag{49}$$

The dynamics of this stochastic process are given by

$$\xi_{t+1} = \begin{bmatrix} F(S_t, S_{t-1}, \tau_t) \\ S_t \\ \alpha U_t + P(S_t, S_{t-1}, \tau_t) \end{bmatrix}, \quad \tau_t \sim Z(U_t, : \cdot). \tag{50}$$

This process is far from trivial to analyse and, in particular, it should be noted that the map

$$(S_t, S_{t-1}) \to P\left((S_t, S_{t-1}, \tau_t)\right), \tag{51}$$

is non-linear, and that this means that some of the standard methods for analysing such problems cannot be applied.

Where is the fundamental problem with this model? It is quite simply that prices might explode and that, as a consequence, no long run properties of the system could be established. To show that this is not the case we need to have a property known as a "mean contraction" condition. The details of this are spelled out in Foellmer et al. (2005) from which this model is taken.

The condition can be translated into putting a bound on the probability that an individual becomes a chartist. For example, if only fundamentalists are present on the market the condition holds automatically. However, what is interesting is that it is enough to bound the probability that individuals switch to chartism, since one might intuitively think that at some point even though it is a low probability event almost all agents could become chartists. However, this situation will not persist if the condition holds since the probability that the system will stay in that configuration is very low. Thus, although a chartist bubble may start, it inevitably collapses. The beauty of this is that it allows for temporary swings to the dominance of chartism but guarantees that such periods will not last. Thus without having to bound prices or the proportion of chartists we can still get the results that we are looking for.

There are two main results. Firstly, the price and profit process will not explode. This is expressed by the following result:

Theorem 1. *(Foellmer et al. (2005)). Under the mean contraction condition the Markov chain is tight i.e.*

$$\lim_{c \to \infty} \sup_t P[|\xi_t| \geq c] = 0.$$

Now, the second result says that not only do prices not explode, they actually have a stable structure in the long run:

Theorem 2. *(Foellmer et al. (2005)). Under the mean contraction condition, the Markov chain has a unique stationary distribution μ and the time averages of prices and profits converge to their expected value under μ.*

The important thing to emphasise here is that we place no a priori bounds on prices nor on the capital gains that can be made with the rules, but the limitation on the probability that an agent will become a chartist is sufficient to guarantee that radical departures from fundamentals will be relatively short in duration. Although prices may, in principle "explode", they only exhibit this sort of transient behaviour for limited periods of time. Nevertheless, the influence of these episodes on long run average behaviour is sufficient to modify the distribution which would have been obtained had there been only fundamentalists and, what is more, it is this modification that produces the features such as long memory which are the subject of this book. It is also this sort of behaviour that makes the sort of statement that one hears from traders, such as, "only in the long run do fundamentals matter" meaningful.

We should not look for the system to settle to some steady state but we can see certain long run regularity in the behaviour of prices. However, in addition to showing that, with both fundamentalists and chartists present, we can guarantee the existence of a unique limit distribution it is also interesting to see how the characteristics of that distribution are affected by the presence of chartists. We can see, using simulations of a simple example, that chartists add to the noise in the system and that bubbles and crashes appear. The translation of this in terms of the limit distribution is increased kurtosis and variance. Thus, loosely speaking, chartists increase the volatility of the price process. The switching of agents' expectations between fundamentalist and chartist provides an answer to the "excess volatility" puzzle mentioned in the introduction. The following figures enable us to see this clearly

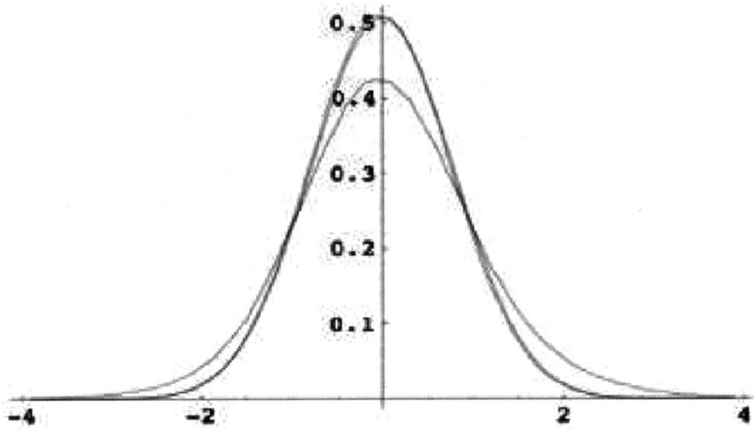

Fig. 5. The distribution of Stock Prices

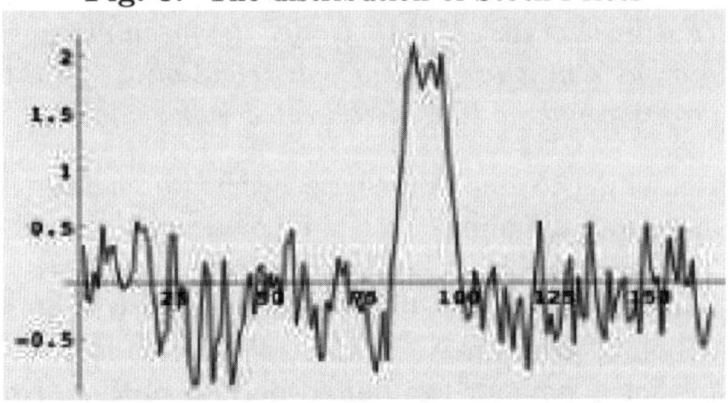

Fig. 6. Log-price of asset

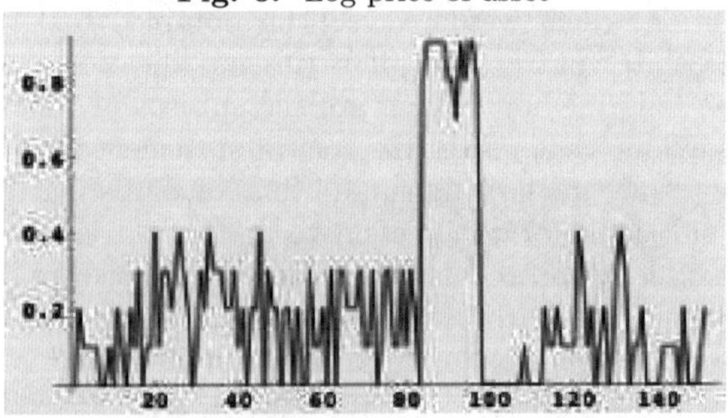

Fig. 7. Proportion of chartists

It is worth observing here that some features of the model differ from that described in the previous section. The demand function is taken as a building block rather than being derived from an underlying utility function. The function is of logarithmic form which rules out negative prices something which may occur in standard models. Secondly, the demand has a stochastic, "liquidity demand" component. This can also be assimilated to the random aggregate supply of foreign exchange that we had in the previous model. However, to obtain the analytic results, the stochastic process governing the supply of exchange cannot be a simple random walk.

Another thing is worth noting. The model, as did the previous one has different forecasting rules and a probabilistic way of choosing between them. Although the formulation is much more general, the logit or quantal response rule used earlier, can, in a modified form, be a candidate rule for this model. This in turn, determines the relative numbers of agents which follow each rule and there can be any finite number of the latter. The crucial feature here again, is the success, or performance of the rules or gurus. What we show, in Foellmer et al (2005) as we have seen, is that the joint price performance process is ergodic and this means that there is long-term structure in both the prices and the performances of the rules. The presence of bubbles, which burst fairly rapidly, is due to the restriction that we put on the probability of following the chartist rule. There is, as we have observed, a self-reinforcing trend towards chartism as soon as the price moves out of the limits defined by the fundamental values. However there is always a sufficiently high probability that sufficient agents will adopt fundamentalist rules to prevent the process from exploding without limit. As always, the expected time that it will take to return to the fundamental region can be determined, but does not depend on the current price nor on how long the process has been out of the fundamental region.

15 Conclusion

A major puzzle in the analysis of the price series of financial markets is the existence of bubbles, crashes, and excess volatility. These phenomena themselves and the associated statistical properties such as "long memory" are very hard to generate with standard models. Faced with this conundrum, there are two basic possibilities. Either to attribute the phenomena in question to irrational behaviour of the actors in the markets or to try to construct models of the behaviour of the traders which do generate the sort of phenomena that we observe empirically.

Many attempts have been made to do this. These vary from endowing individuals with substantial memories and complicated forecasting rules with sophisticated lag structures, to the introduction of a complicated process generating external shocks to the system. Both such approaches have been used in representative agent models with rational expectations. The failure of these models to capture some of the salient features of financial time series is well documented. An alternative approach has been to allow for heterogeneous agents. A classic example of this is given by the contributions of De Long et al (1989). They introduced so-called "noise traders" into their models. The objection to this sort of analysis is that it involves assuming that some agents are systematically wrong. Other approaches have been tried and some of the contributions appear in this book. A different avenue is that opened by Brock (1993), for example, who shows that the process generating prices can become unstable if agents adopt simple forecasting rules once the price converges to a stable path. The reason for their switch to the simple rule is that it is less costly than the more complicated rule that led the process to converge in the first place.

The approach I have outlined in this chapter shares some of the characteristics of Brock's analysis but is rather different. The driving feature of the price process is also the switching of agents from one forecasting rule to the other. However, this can be attributed to the relative success of the rules. The switching process has the characteristic that the great majority of agents will, at any point in time, herd on one rule. When this happens agents' forecasts are self-reinforcing. Thus agents are not systematically wrong and their behaviour cannot be described as irrational.

Furthermore, neither in the model spelled out nor in the version proposed in Foellmer et al. (2005) does the process depend on a particular form of the demand functions, nor on any specific process generating the "fundamentals". Lastly there is freedom in specifying the function, which maps past profits from the different forecasting rules into the probability of choosing those rules. Thus the specification of the models, which generate the sort of phenomena in which we are interested, is rather flexible.

Using this simple but general framework, we were able to capture certain stylised facts of empirical financial price time series and to show the existence of a long run equilibrium, which is different from the equilibrium notions commonly, used in the literature. It is worth repeating that an important feature of the model is the inclusion of chartists or trend chasers who may induce temporary bubbles and crashes. Whilst this introduces such characteristics as "excess volatility" and "long memory" in the price process, see Kirman and Teyssiere (2002), it presents a major difficulty from an analytical point of view. When chartists predominate they may generate bubbles but there is no a priori bound on the levels that prices may attain during a bubble. Thus it is not obvious that the price process will not explode, and the most appealing feature of the last model is that it prevents this without artificially bounding the process.

In this chapter I have also spelled out the basic building blocks that most models of financial markets share. It is then a question of deciding which of each of these should be changed from the features of the standard model in order to obtain more realistic time series. I have placed the emphasis here on the importance of interaction and the mutual influencing of economic agents.

There are many ways of modelling this and many different ways of characterising the interaction between different agents. I have emphasised the idea that agents modify their expectations over time and that there is a positive feedback as more agents converge on one forecasting rule. This generates self-reinforcing swings and if for some of the time the agents follow extrapolatory rules then many of the features that we observe in market data will emerge. One can think of many other ways in which interaction could take place and it would not be useful to detail all the different models that have been proposed here. It should be clear that within the class of models proposed here, by changing the rules available or by changing the horizon and utility functions of the actors in the markets one can generate rather a large range of possibilities.. However, the rather simple interaction in the models that I have described, between agents who believe that prices follow some "fundamentals" and those who extrapolate already captures many of the observed features of financial time series. This is surely not the whole story but may prove to be a useful start.

In concluding it might be worth recalling the purpose of the whole exercise. The basic goal from the economist's point of view is to produce market models, which are capable of generating the statistical properties of financial time series. The econometrician, on the other hand, is more concerned with capturing the essence of the underlying stochastic process and developing powerful tests to discriminate between various candidate types of process. These two activities are complementary. It is of limited interest to understand the mathematical properties of stochastic processes with realistic properties if we have no idea as to the nature of the economic activity that produces these characteristics. Nor, on the other hand, is it very interesting to produce economic models, which share, in some loose sense, the statistical properties of empirical time series. We need to test as rigorously as possible, the data generated by the sort of economic models that have been spelled out in this chapter and this has been the thrust of the work with Gilles Teyssiere, (see e.g.Kirman and Teyssiere, (2002)) More ambitiously, we would like to prove formally results which enable us to pin down the characteristics of the series. This is the goal pursued in Foellmer et al. (2005), for example. The interest of this approach is that one can develop a notion of equilibrium for the sort of time series of prices actually observed which is basically different from that usually envisaged in economics. Prices never settle to an equilibrium price, they change all the time, but the distribution of prices in the long run exists and is unique. Thus prices do have a well-characterised structure in the long run and we do not have to content ourselves with such statements as "anything can happen". Even if we have some notion of what the fundamental value of an asset is, we should not expect to see the distribution of prices concentrating itself more and more around this price. Prices will only stray far from underlying fundamentals for relatively short periods of time but they will never settle to those fundamental values. Being near to the latter is no guarantee of staying near and no exogenous shocks are required to move the prices. This seems to be a reasonable representation of economic reality and, what is more, seems to be consistent with the statistical properties of financial time series. Moreover, as can be seen from the contributions to this book, many of the models based on the building blocks explained in this chapter produce data which survive the tests recently developed by the econometricians, for detecting "long memory". This does not mean that the interaction models developed here are the sole explanation for the interesting properties of financial market time series but they do, at least, seem to explain facts which escape the standard model of financial markets.

References

1. Allen, F., Morris, S. and Shin, H-S. (2003). Beauty contests, bubbles and iterated expectations in asset markets capital adequacy regulation. In: *Search of a Rationale*, Center for Financial Institutions Working Papers **03-06**, Wharton School Center for Financial Institutions, University of Pennsylvania.
2. Anderson, S., De Palma, A., and Thisse, J. (1993). *Discrete Choice Theory of Product Differentiation.* MIT Press, Cambridge, MA
3. Avery C., and Zemsky, P. (1998). Multidimensionnal uncertainty and herd behaviour in financial markets, *American Economic Review*, **88,** 724–748.
4. Bancchetta, P. and van Wincoop, E. (2003). Can information heterogeneity explain the exchange rate determination puzzle, Preprint
5. Blanchard, O. and Watson, M.W. (1982). Bubbles, rational expectations and financial markets. In: Wachsel P. (ed) *Crisis in the Economic and Financial System.* Lexington Books, Lexington Mass.
6. Blume, L. (1993). The statistical mechanics of strategic interaction. *Games and Economic Behaviour*, **5,** 387–424.
7. Brock, W A. (1993). Pathways to randomness in the economy: emergent nonlinearity and chaos in economics and finance. *Estudios Economicos*, **8,** No 1, El Colegio de Mexico,
8. Brock, W A. and Hommes, C.H. (1997). A rational route to randomness, *Econometrica*, **65,** 1059–1095.
9. Brock, W.A , and Durlauf, S. (2001a). Interaction-based models. In: J. Heckman and E. Leamer (eds) *Handbook of Econometrics*, Vol V, North-Holland, Amsterdam.
10. Brock, W.A , and Durlauf, S. (2001b). Discrete choice with social interactions. *Review of Economic Studies*, **68,** 235–260.
11. Brooks, C, and Katsaris, A. (2003). Rational speculative bubbles: an empirical investigation of the London stock exchange. *Bulletin of Economic Research*, **55,** 319–346.
12. Chamley, C. (2002). *Rational Herds.* Cambridge, Cambridge University Press.
13. Cheung, Y-W., Chinn, M.D., and Pascual, A.G. (2002). Empirical exchange rate models in the nineties: Are they fit to Survive?, *NBER Working Paper*, **9393**.
14. Chionis, D. and MacDonald, R. (2002). Aggregate and disaggregate measures of the Foreign Exchange risk premium. *International Review of Economics and Finance*, **11,** 57–84.

15. De Long, J. Bradford, Schleifer, A., Summers, L. H. and Waldmann, R. J. (1989). The size and incidence of the losses from noise trading. *Journal of Finance*, **44,** 681–699.
16. De Long, J. Bradford, Schleifer, A., Summers, L. H. and Waldmann, R. J. (1990). Noise trader risk in financial markets. *Journal of Political Economy*, **98,** 703–738.
17. Domowitz, Ian. (1993). A taxonomy of automated trade execution systems. *Journal of International Money and Finance*, **12,** 607–631.
18. Donaldson, R.G. and Kamstra, M. (1996). A new dividend procedure that rejects bubbles in asset prices: the case of the 1920's stock crash. *The Review of Financial Studies*, **9,** 333–383 .
19. Elster, Jon. (1989). *Nuts and Bolts for the Social Sciences.* New York: Cambridge University Press.
20. Flood, R.P., Garber, P.M., (1980). Market fundamentals versus price level bubbles: the first tests. *Journal of Political Economy*, **88,** 74–97.
21. Flood, R.P. and Hodrick, R.J. (1990). On testing for speculative bubbles. *Journal of Economic Perspectives*, **4,** 85–102.
22. Foellmer, H. and Schweizer, M. (1993). A microeconomic approach to diffusion models for stock prices. *Mathematical Finance*, **3,** 1–23
23. Foellmer, H. Horst, U. and Kirman, A. (2005). Equilibrium in financial markets with heterogeneous agents: a new perspective. *Journal of Mathematical Economics*, **41,** 123–155.
24. Frankel, J. A., and Froot, K. (1986). The dollar as an irrational speculative bubble: a tale of fundamentalists and chartists. *The Marcus Wallenberg Papers on International Finance*, **1,** 27–55.
25. Frankel, J. A. and Rose, A. K. (1995). Empirical research on nominal exchange rates. In: G. M. Grossman and K. Rogoff (eds.) *Handbook of International Economics*, **III,** North Holland, Amsterdam.
26. Frederick, S., Loewenstein, G. and O'Donoghue, T. (2002). Time discounting and time preference: a critical review. *Journal of Economic Literature*, **40,** 351–401.
27. Garber P.M., (2000). *Famous First Bubbles.* M.I.T Press, Cambridge Mass.
28. Grandmont J.M. (1983). Money and value. *Econometric Society Monographs*, Cambridge University Press.
29. Horvàth, L. Kokoszka, P.S. and Teyssière, G. (2001). Empirical process of the squared residuals of an ARCH sequence. *The Annals of Statistics*, **29,** 445–469.
30. Kirman, A. P. (1991). Epidemics of opinion and speculative bubbles in financial markets. In: M. Taylor (ed.), *Money and Financial Markets*, 354–68, Macmillan, London.
31. Kirman, A. (1993). Ants, rationality and recruitment. *The Quarterly Journal of Economics*, **108,** 137–156.
32. Kirman, A., Ricciotti, R. and Topol, R. (2006). Bubbles and foreign exchange markets : it takes two to tango. forthcoming *Macroeconomic Dynamics.*
33. Kirman, A. and Teyssière, G. (2002). Microeconomic models for long–memory in the volatility of financial series. *Studies in Non-Linear Dynamics and Econometrics*, **5,** 281–303.
34. Kirman, A. and Teyssière, G. (2005). Testing for bubbles and change–points. *Journal of Economic dynamics and Control*, **29,** 765–799.
35. Kokoszka, P.S. and Leipus, R. (2000). Change-point estimation in ARCH models, *Bernoulli*, **6,** 513–539.

36. Kokoszka, P. S. and Teyssière, G. (2005). Change point detection in GARCH models: asymptotic and bootstrap tests, Revised preprint, *Université Paris 1.*
37. Lo, A. (1991). Long term memory in stock market prices. *Econometrica*, **59,** 1279–1313.
38. Lux, T. and Sornette, D. (2002). On rational speculative bubbles and fat tails. *Journal of Money, Credit, and Banking*, **34,** 589–610.
39. McFadden, D., (1974). Conditional logit analysis of qualitative choice behaviour. In: Zarembka, P. (Ed.), *Frontiers in Econometrics.* New York: Academic Press
40. Meese, R.A. (1986). Testing for bubbles in exchange markets. *Journal of Political Economy*, **94,** 345–373.
41. Meese, R.A. and Rogoff, K. (1983). Empirical exchange rate models; do they fit out-of-sample? *Journal of International Economics* **14,** 3–24.
42. Shiller, R.J. (1981). Do stock prices move by too much to be justified by subsequent changes in dividends? *American Economic Review*, **71,** 421–456.
43. Shiller, R. (2000). *Irrational Exuberance.* Princeton, Princeton University Press
44. Stiglitz, J.E. (1990). Symposium on bubbles. *Journal of Economic Perspectives*, **4,** 13–18.
45. Teyssière, G. (2003). Interaction models for common long range dependence in asset price volatility. In: Rangarajan G, Ding M, (eds), *Processes with Long Range Correlations:Theory and Applications* Lecture Notes in Physics, **621,** Springer Verlag, Berlin, pp. 231–269.
46. Tirole, J. (1985). Asset bubbles and overlapping generations. *Econometrica*, **53,** 1490–1528.
47. Townsend, R.M. (1983). Forecasting the forecast of others. *Journal of Political Economy*, **91,** 546–588.
48. Vriend, N.J. (2000). An illustration of the essential difference between individual and social learning, and its consequences for computational analyses. *Journal of Economic Dynamics and Control*, **24,** 1–19.
49. Weisbuch, G., Chenevez, O., Nadal, J-P. and Kirman, A.(1998). A formal approach to market organization: choice functions, mean field approximation and maximum entropy principle. In J. Lesourne and A. Orlean (eds.) *Advances in Self-Organization and Evolutionary Economics.* Economica pp.149–159.
50. Weisbuch, G., Kirman, A. and Herreiner, D. (2000). Market organisation and trading relationships. *Economic Journal*, **110,** 411–436.
51. West, K.D. (1988). Bubbles, fads and stock price volatility tests: a partial evaluation. *Journal of Finance*, **43,** 639–656.
52. Woo, W.T., (1987). Some evidence of speculative bubbles in the foreign exchange markets. *Journal of Money, Credit and Banking*, **19,** 499–514
53. Woodford, M. (1990). Learning to believe in sunspots. *Econometrica*, **58,** 277–307.
54. Young, P. and Foster, D. (1991). Cooperation in the short and in the long run. *Games and Economic Behavior*, **3,** 145–156.

A Minimal Noise Trader Model with Realistic Time Series Properties

Simone Alfarano[1] and Thomas Lux[2]

[1] Dept. of Economics, University of Kiel. alfarano@bwl.uni-kiel.de
[2] Dept. of Economics, University of Kiel. lux@bwl.uni-kiel.de

Summary. Simulations of agent-based models have shown that the stylized facts (unit-root, fat tails and volatility clustering) of financial markets have a possible explanation in the interactions among agents. However, the complexity, originating from the presence of non-linearity and interactions, often limits the analytical approach to the dynamics of these models. In this paper we show that even a very simple model of a financial market with heterogeneous interacting agents is capable of reproducing realistic statistical properties of returns, in close quantitative accordance with the empirical analysis. The simplicity of the system also permits some analytical insights using concepts from statistical mechanics and physics. In our model, the traders are divided into two groups: fundamentalists and chartists, and their interactions are based on a variant of the herding mechanism introduced by Kirman (1993). The statistical analysis of our simulated data shows long-term dependence in the auto-correlations of squared and absolute returns and hyperbolic decay in the tail of the distribution of the raw returns, both with estimated decay parameters in the same range like empirical data. Theoretical analysis, however, excludes the possibility of "true" scaling behavior because of the Markovian nature of the underlying process and the finite set of possible realized returns. The model, therefore, only *mimics* power law behavior. Similarly as with the phenomenological volatility models analyzed in LeBaron (2001), the usual statistical tests are not able to distinguish between true or pseudo-scaling laws in the dynamics of our artificial market.

1 Introduction

In the last couple of years, the study of behavioral models of dynamic interaction in financial markets has brought about a better understanding of some of the key stylized features of financial data, namely the *fat tails* of the distribution of returns and the *autoregressive dependence* in volatility. Although these statistical features have counted as almost universal findings for practically all financial time series for a long time and appear to be extremely uniform across assets and sampling horizons, economic explanations of their behavioral origins were nonexistent until very recently. However, the recent rush of interest in heterogeneous agents models with a diversity of interacting traders, the availability of fast computers for simulations of markets with a large number of agents, and the introduction of new analytical and computational tools (often adapted from statistical physics) in the analysis of multi-agent systems has led to a variety of new contributions in which the above stylized facts (one of them or both) are shown to be emergent properties of interacting agent dynamics. Some of these contributions show that besides the realistic distributional properties of their models, the overall dynamics is also undistinguishable from a unit-root process. Hence, despite having identifiable behavioral roots (in terms of the assumed speculative behavior of the agents), no immediately recognizable traces of predictability can be found in the presented time series, and the dynamics appears to be observationally equivalent to a martingale process.

Early papers in this area have often been the results of collaborations between economists and physicists, e.g. Takayasu *et al.* (1992), Palmer *et al.* (1994) and Bak *et al.* (1997). While they made important contributions to get this literature started, the similarity of the resulting time paths with empirical data was limited. Later studies have merged this multi-agent approach with the type of noise traders - fundamentalists' interaction introduced by Beja and Goldman (1980) and Day and Huang (1990). Papers along this line included the microscopic stock market models of Lux and Marchesi (1999, 2000), Chen *et al.* (2001, 2002), Iori (2002), Farmer and Joshi (2000), LeBaron (2000) as well as the adapting belief dynamics of Gaunersdorfer and Hommes (2000) and Gaunersdorfer, Hommes and Wagener (2000). A related variant can be found in the artificial foreign exchange markets of Arifovic and Gencay (2000) and Georges (2001) in which agents' behavior is modelled using genetic algorithmic (GA) for their selection of strategies.

Interestingly, some possible general explanations seem to emerge from this literature: first, volatility clustering and fat tails may emerge from indeterminacy in the equilibrium of the dynamics. In particular, with different strategies performing equally well in some kind of steady state, stochastic disturbances lead to continuously changing strategy configurations which at some point generate bursts of activity. This type of dynamics can be found already in Youssefmir and Huberman (1997) in the context of a resource exploitation model and can be identified in both the papers by Lux and Marchesi (1999, 2000) and the otherwise quite different GA models by Arifovic and Gencay (2000), Lux and Schornstein (2002) and Georges (2001).

Another more general approach can be attributed to Gaunersdorfer and Hommes (2000), who show that volatility clustering can also emerge from stochastic dynamics with multiple attractors. Small amounts of noise added to a deterministic dynamics with two or more attractive states can lead to recurrent switches between these attractors. As these different regimes often have different degrees of volatility, volatility clustering is a somehow natural result of these dynamics. Interestingly, both of these mechanisms are sometimes identified as examples of **intermittency** which might, therefore, be thought of as a general conceptual framework for the explanation of the particular characteristics of financial markets.

While the above results have - due to their origin from the behavioral finance literature - more or less complicated descriptions of agents' expectations and strategy choice, physicists have rather tried to reduce the dynamics to a few basic principles able to generate the required time series characteristics. Recent models with only a few ingredients for activation and frustration of agents leading to realistic simulated output include Eguiluz and Zimmermann (2000), Bornholdt (2001) and variants of the so called minority game (Challet *et al.*, 2001). Our aim in this paper is similar to these studies: we are interested in whether an extremely simplified model of interaction of noise traders and fundamentalists is already sufficient to reproduce the key stylized facts: unit roots, fat tails and volatility clustering. The model we investigate in this paper is a simple variant of the herding dynamics introduced by Kirman (1993) and Lux (1995). We distinguish between two groups and allow for mimetic contagion among agents by simply postulating that they will move from one group to the other, with a certain probability depending on group size. This leads to the natural emergence of majority opinion with all agents sharing one of two available opinions. However, the stochasticity of the dynamics also leads to recurrent switches between both opinions, so that we find a bistable system with a bimodal ergodic distribution of states. Adding a simple price adjustment rule, bi-modality carries over to prices as well. Simulations of this model show that it can *mimic* in surprising quantitative accuracy the above stylized facts. The simplicity of the model also allows some analytically insights into its dynamics. In particular, it is straightforward to show that the model does not exhibit 'true scaling', neither concerning the distribution of large returns, nor the temporal dependence structure. The apparent scaling, in fact, results from a kind of 'regime switching' between the two modes of its ergodic stationary distribution. This is a phenomenon similar to the difficulty of distinguishing between apparent and true scaling in certain stochastic processes; see Anderson *et al.* (1999), Granger and Teräsvirta (1999), Diebold and Inoue (2001) and LeBaron (2001). Our analysis thus demonstrates that 'apparent' scaling is not confined to a particular class of appropriately constructed stochastic models, but might also prevail in behavioral models of interacting agents.

2 A Simple Model of Contagion

Our market is populated by N agents, each of them being either in state A or in state B. The number of agents in both groups are denoted N_A and N_B. The state of the system can be described by an intensive variable:

$$x = \frac{N_A - N_B}{N}. \tag{1}$$

Every agent has a probability of switching per unit of time from one state to the other regulated by the transition rates:

$$\phi_{A\to B} = \nu \frac{N_B}{N} \phi_{B\to A} = \nu \frac{N_A}{N}. \tag{2}$$

where ν is a time-scaling parameter and

$$N_A = \frac{1+x}{2} N N_B = \frac{1-x}{2} N. \tag{3}$$

The number of agents that change state per unit of time is:

$$\omega_{A\to B} = \nu \frac{N_B}{N} N_A \omega_{B\to A} = \nu \frac{N_A}{N} N_B. \tag{4}$$

The previous transition rates with the finite realizations of x specify a reversible Markov chain, therefore the detailed balance condition holds, see Aoki (1996):

$$\omega(x \to x + \Delta x) P_e(x) = \omega(x + \Delta x \to x) P_e(x + \Delta x), \tag{5}$$

where the subscript e denotes the stationary probability distribution, that can be written as a Gibbs distribution (Aoki, 1996):

$$P_e(x) \propto exp^{U(x)}. \tag{6}$$

Using (5) and (6), we have:

$$exp^{[U(x+\Delta x)-U(x)]} = \frac{(1-x)(1+x)}{[1-(x+\Delta x)][1+(x+\Delta x)]}, \tag{7}$$

where

$$\Delta x = \frac{2}{N}. \tag{8}$$

For large N [3], we can rewrite (7) in the limit:

$$\Delta x \to 0. \tag{9}$$

At the end, the result is a simple differential equation for $U(x)$:

$$\frac{dU(x)}{dx} = -\frac{d}{dx} \ln[(1-x)(1+x)] U(x) = -\ln(1-x^2) + c. \tag{10}$$

From (6), it is straightforward to derive the equilibrium distribution:

$$P_e(x) = \frac{1}{L} \frac{1}{1-x^2}, \tag{11}$$

where the normalization constant L is given by:

$$L = \int_{-1+\varepsilon}^{1-\varepsilon} \frac{1}{1-x^2} dx = \ln \frac{2-\varepsilon}{\varepsilon} \tag{12}$$

ε is a positive number [4].

[3] In our simulations $N = 100$.

[4] In our simulation, we leave at least one agent in each group to avoid total extinction, i.e. we implemented reflecting boundaries at $x = 1$ and $x = -1$; so $\varepsilon = \frac{2}{N}$.

The system is bistable and generates transitions between the two 'equilibria' $x \approx +1$ and $x \approx -1$; we can derive the first passage time T_0 (Aoki, 1996):

$$T_0 = \sum_{x=-1+\varepsilon}^{1-2\varepsilon} \frac{1}{P_e(x)w(x \to x + \Delta x)} \sum_{y=x+\Delta x}^{1-\varepsilon} P_e(y). \tag{13}$$

In the limit of a large number of agents, the previous sum can be written as an integral; therefore we can calculate explicitly the first passage time:

$$T_0 = N \ln(N) + o(\varepsilon). \tag{14}$$

This model can be used to simulate interaction between economic agents (traders) based on **imitative behavior**.

The previous mechanism is inspired by Kirman's analysis of opinion formation (Kirman, 1993). The main difference with respect to Kirman's model is the absence of a constant term in the probability transitions (2), introduced by the author to prevent the absorbing states at $|x| = 1$; we replaced it by setting reflecting boundary conditions. Consequently, the only possible scenario is a distribution concentrated in the extreme values (U shaped distribution), while in the later case a flat distribution and a distribution with a pick around the mean are also possible, depending on the particular set-up of the parameters.

3 The Financial Market Model

We now use this two-state dynamics as the main ingredient in a financial market model with interacting heterogenous agents. Our market participants are divided into two groups:

N_F *fundamentalists* (**F**), who buy (sell) a fixed amount of stocks T_F when the price is below (above) its fundamental value p_F.

N_C *noise traders* (**C**), who are driven by herd instinct.

Depending on their expectation about future price movements, noise traders can be either *optimists* (buyers or O) or *pessimists* (sellers or P). T_C represents the fixed number of stocks that noise traders buy or sell. While the number of F and C is constant in time (i.e. there are no switches between them), switches from O to P and vice versa are allowed. The two-state model, detailed in sec.2, regulates the transition rates. The fundamental price is assumed to be constant in time.

Assuming sluggish price adjustment, the dynamics of the price is given by

$$\frac{dp}{pdt} = \beta[N_F T_F(p_F - p) + N_C T_C x]x = \frac{N_O - N_P}{N_C}, \tag{15}$$

where β is the reaction parameter of the market (speed of price adjustment).

As an approximation to the resulting non-equilibrium dynamics we assume instantaneous market clearing (which can be interpreted as an adiabatic approximation in physics terminology). We can solve (15) for the equilibrium price:

$$p = p_F + \frac{N_C T_C}{N_F T_F} x. \tag{16}$$

Without loss of generality, we focus attention on the following set of parameters values:

$$N_C = N_F = N T_C = T_F = 1 \beta = 1. \tag{17}$$

By (16), the average of the price is p_F because the mean of x is zero. We can observe, however, pessimistic phases in which the asset is undervalued (compared to the fundamental price), and switching to optimistic phases in which it is overvalued. In the first case the majority of the noise traders is in the 'pessimistic mood', while the second case most of them are in an 'optimistic mood'.

This model can be viewed as a simplified version of the more complex artificial stock market introduced by Lux and Marchesi (1999). However, their switching mechanism is influenced by two factors: the opinion of others, modelled via an opinion index similar to our x, and the local dynamics of the price, entering via the averaged trend. In addition to this, the complexity is increased by the possibility of switching among three different states and not only two as in our model.

As it turns out, the model is able to reproduce some of the salient characteristics of financial markets. Figures 1 and 2 on p. 351–352 illustrate the results of the model. Volatility clusters are visible in the time series of returns in correspondence to the change of majority, and the unconditional distribution of returns seems to be leptokurtic. Dependence of absolute and squared returns (as a measure of the volatility) is positive over an extended time horizon, while the raw returns show almost no correlation. All these features are in qualitative agreement with empirical findings.

4 Statistical Analysis of Simulated Data

In order to see how closely the statistical results from our simulated data match empirical observations, we performed a series of experiments with a long data set of $1,000,000$ integer time steps. Tables 1 and 2 give some elementary statistics from the whole sample. As can be seen, the resulting distribution is characterized by significant excess kurtosis and slight positive skewness. The Bera–Jarque test for normality leads to a strong rejection of its null hypothesis.

To investigate the auto-correlation structure, we applied the Box - Ljung test to the auto-correlations up to lags 8, 12 and 16 for the raw data as well the squares and absolute values of returns. In harmony with empirical

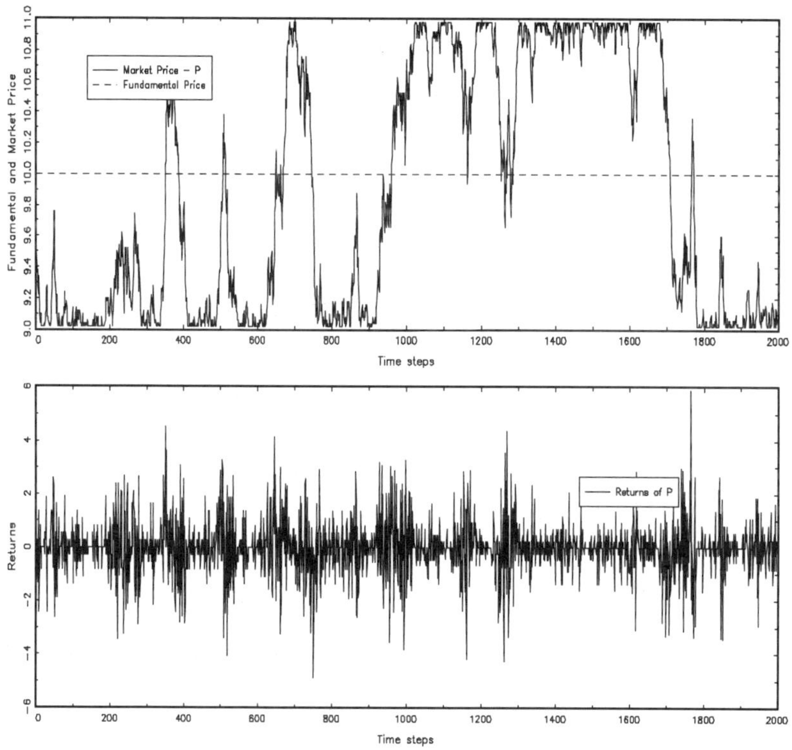

Fig. 1. The upper panel shows the behavior of the fundamental (simply assumed to be constant) and the market price from a typical simulation. The lower panel shows the returns of the market price, computed as log increments over unit time interval. Underlying parameters of this run: $N = 100$, $\nu = 1$, $p_F = 10$

Table 1. Sample statistics of returns

(Probability)	(0.000)
Mean	$4.14 \cdot 10^{-7}$
Variance	$7.52 \cdot 10^{-5}$
Kurtosis	2.67
Skewness	0.057
Bera-Jarque test	59, 571

records, there is only slight auto-correlation in the returns themselves, but highly significant auto-correlation in the squares and absolute values. Since with samples of that size, we are able to detect even very small degrees of auto-correlation with high reliability, we would not expect the results of the

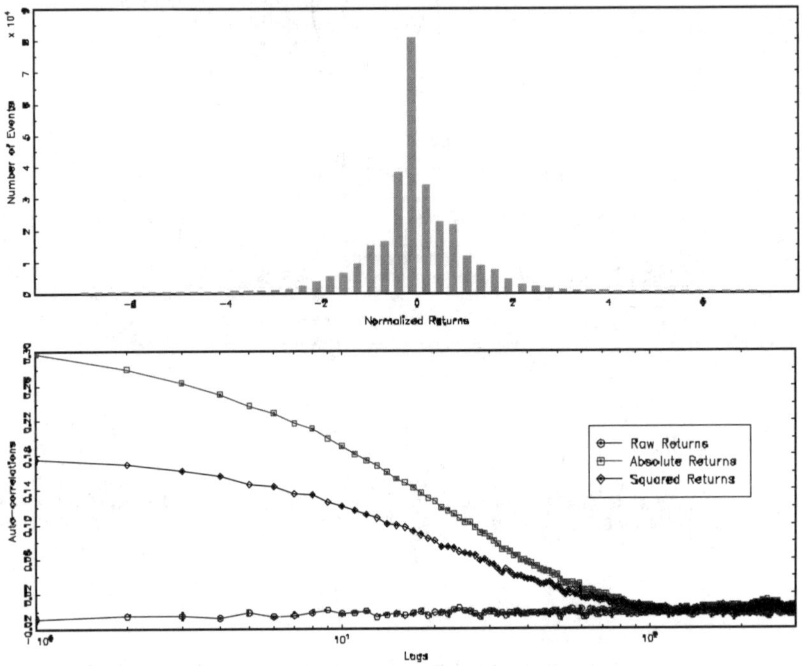

Fig. 2. The upper panel shows the distribution of normalized returns; notice the leptokurtic shape. The lower panel shows the auto-correlation function of raw, squared and absolute returns. Parameters: $N = 100$, $\nu = 1$, $p_F = 10$, number of events $3 \cdot 10^5$

Table 2. Results of Box-Ljung test

	8	12	16
R_t	58.59	96.17	116.74
R_t^2	40, 198	52, 270	61, 611
$\|R_t\|$	107, 132	138, 035	160, 551

Box-Ljung test to be insignificant (in fact, they allow rejection of the null of no auto-correlation even for the raw returns). However, what is interesting here is that the statistics are orders of magnitude larger for the squares and absolute values of returns.

The highly significant entries for the latter transformations lead to the questions of whether these time series are able to mimic the empirical observations of long-term dependence, defined as an hyperbolic decline of the auto-correlation function:

$$ACF(\tau) \approx \tau^{-\gamma}, \tag{18}$$

where γ is the decay constant. To this end we estimate the parameter of fractional differencing, denoted by **d**, from a regression in frequency space following the approach by Geweke and Porter - Hudak (1983) (GPH), and also the Hurst exponent **H** from Detrended Fluctuation Analysis (DFA), (Pcng *ct al.*, 1994), see tables 3 and 4.

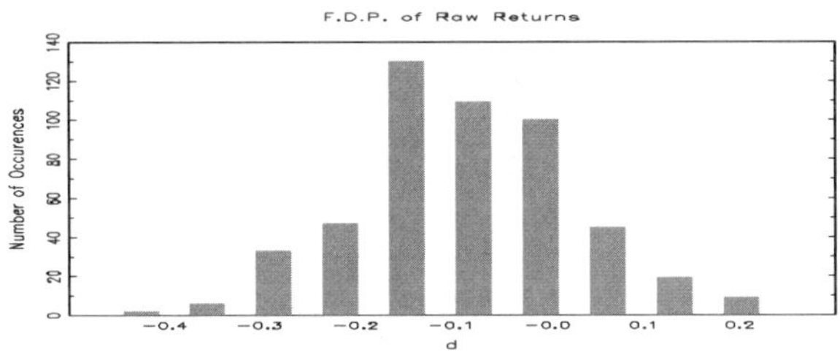

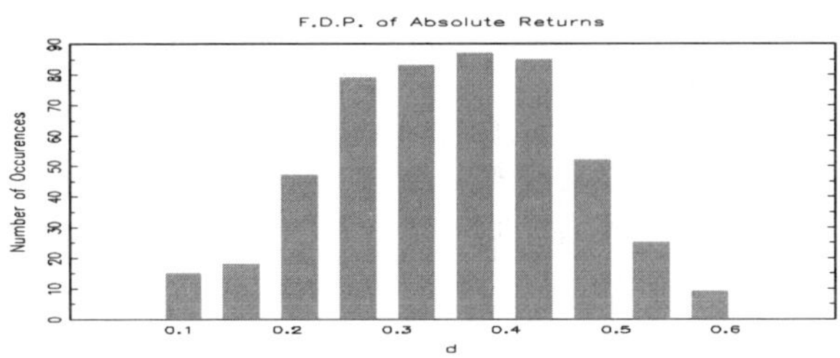

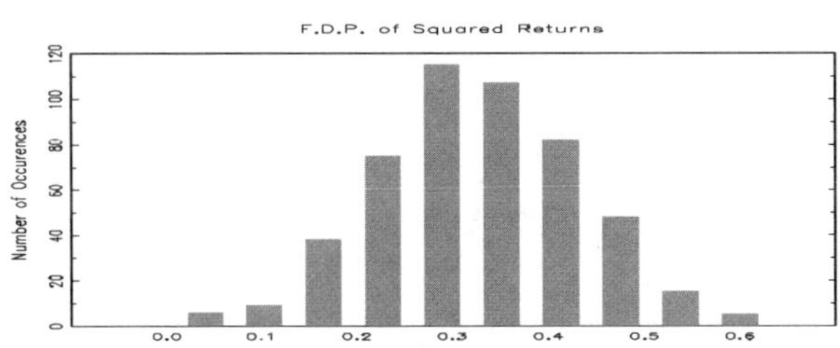

Table 3. Estimated parameters of fractional differencing for 500 sub-samples

	Mean	Minimum	Maximum
R_t	−0.09	−0.46	0.24
R_t^2	0.33	0.01	0.63
$\|R_t\|$	0.35	0.08	0.61

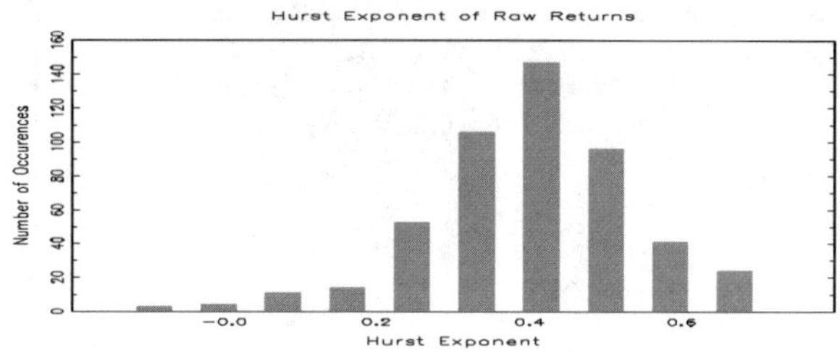

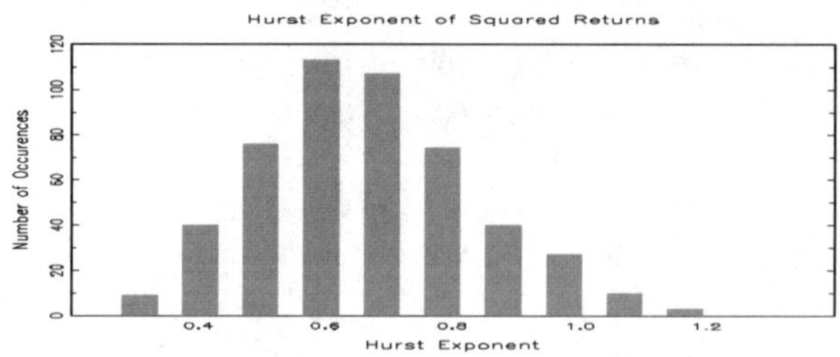

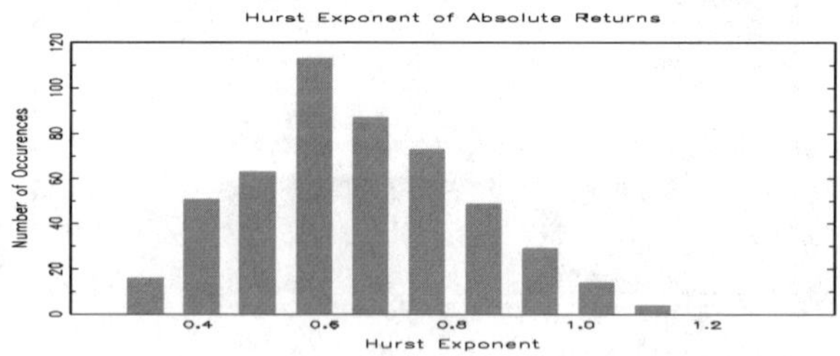

The GPH method is based on the linear regression of the log–periodogram on transformations of low frequencies of the Fourier spectrum. The estimated parameter d is related to the decay rate of the auto-correlation function by:

$$\gamma = 1 - 2d \tag{19}$$

A value of $d = 0$ would indicate absence of long memory, while d significantly above zero speaks in favor of long-term dependence. Table 3 gives summarizing results from 500 sub-samples of 2,000 observations each, and the histograms show the distribution of the 500 estimates. As it turns out, we get results in the vicinity of zero for the raw data, but on average much higher values for the squares and absolute returns. In fact, the latter are very close to typical empirical estimates obtained with returns of various financial markets (Lux and Ausloos, 2002).

Table 4. Hurst Exponent from DFA for 500 sub-samples

	Mean	Minimum	Maximum
R_t	0.40	−0.13	0.71
R_t^2	0.66	0.26	1.21
$\|R_t\|$	0.65	0.27	1.15

Estimates from the alternative DFA methods (shown in table 4 and relative histograms) confirm these results. The relationship between the two parameters is:

$$H = 2d + 0.5 \tag{20}$$

therefore for the two methods we have a satisfactory agreement for raw returns; for absolute and squared returns, results from both methods are qualitatively similar, albeit with some divergence in the numerical values. The later might be explained, however, by different small sample biases of both estimators.

The appearance of long term dependence is particularly interesting since simple inspection of the model, in fact, indicates that it does not exhibit this feature: any memory in the system is wiped out by stochastic fluctuations between the two modes of the distribution and the time needed to switch from one mode to the other is depended on the past behavior of the system. However, it is well known that Markov regime-switching models can indeed 'erroneously' give the impression of long - term dependence; see Lobato and Savin (1998), Anderson *et al.* (1999), Granger and Teräsvirta (1999), Diebold and Inoue (2001). The mechanism here is similar to Markov switching processes, which might explain the impression of long - term memory. A similar result is found in Kirman and Teyssière (2002), who study a more complicated foreign exchange market model in which Kirman's herding model is combined with a monetary model *á la* Frankel and Froot (1986).

Let's turn now to the unconditional distribution of the synthetic data. To complement the results for kurtosis, we compute the so-called tail index to get information about the heaviness of the tails of the simulated data. Empirical research indicates again a hyperbolic relationship for the decay of the probability in the outer part of the return distribution, following:

$$P(|R_t| > X) \approx X^{-\alpha}, \tag{21}$$

with α usually in the range of $[2.5, 5]$, (Lux and Ausloos, 2002). Here we applied the usual maximum likelihood estimator proposed by Hill (1975), using the same 500 sub-samples and tail sizes of 10, 5 and 2.5%. Both the range of the estimators and the tendency towards slightly increasing numbers are in good harmony with empirical results (see table 5 and the pertinent histograms for more details).

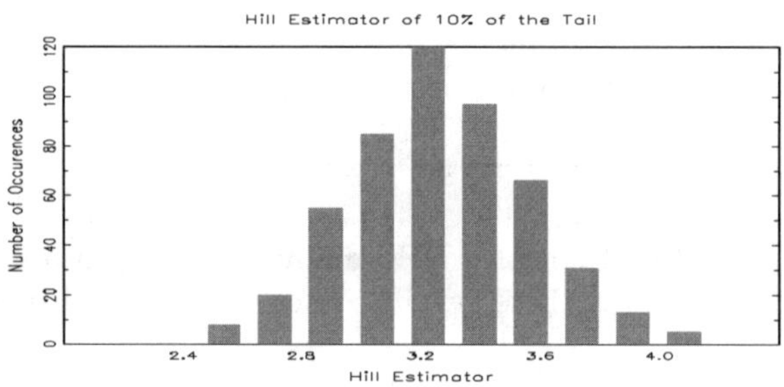

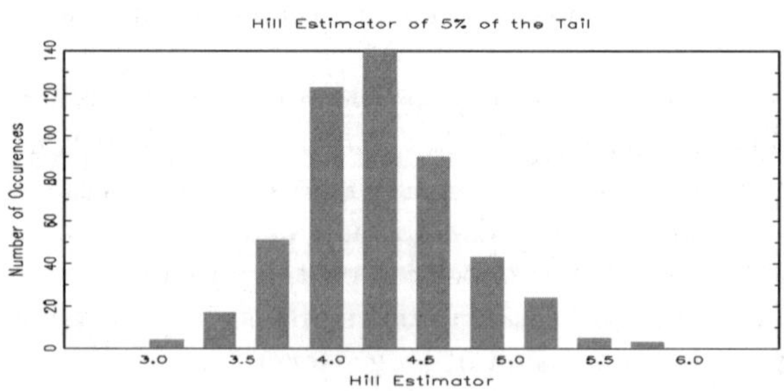

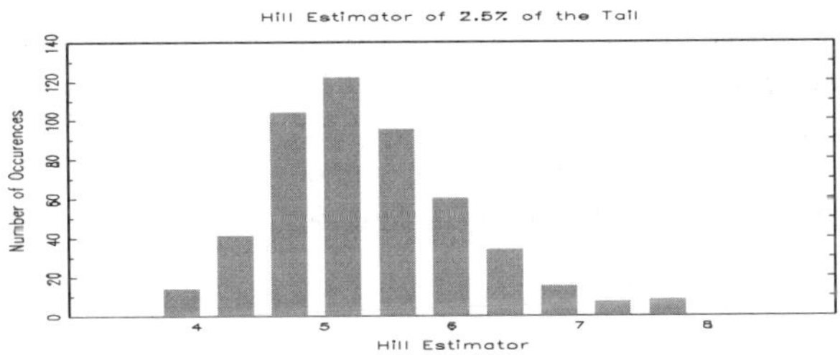

Table 5. Tail index estimators for 500 sub-samples

	Mean	Minimum	Maximum
10%	3.27	2.45	4.17
5%	4.27	2.92	5.92
2.5%	5.33	3.66	7.90

The unit-root hypothesis of the financial data is another well-established stylized fact of asset prices (de Vries, 1994), usually interpreted as a consequence of market efficiency. In other words, one is usually not able to reject the null hypothesis that the price of financial assets follows a random walk or martingale process. To test the unit root, we applied the famous Dickey-Fuller test to sub-samples of different lengths, (from 500 to 10000, see table 6), in order to check wether the simulated time series show the same pattern as the empirical data.

Table 6. Results of Unit-Root test. (a) Number of rejections at 95% level

Size of the sub-sample	Range of ρ	One-sided test [a]	Two-sided test [a]
500	0.99998279 – 1.00000171	0(2000)	615(2000)
2000	0.99999562 – 1.00000042	0(500)	114(500)
5000	0.99999824 – 1.00000016	0(200)	28(200)
10000	0.99999909 – 1.00000008	0(100)	0(100)

As can be seen from table 6, we cannot reject the null hypothesis of unit-root using a one-sided test for all the sub-samples considered. Conversely, applying a two-sided test, we observe several cases of rejections in favor of an explosive root of the dynamics. Inspection shows that these cases of rejection of a unit root in favor of explosive dynamics are driven by switching between the two modes of the distribution. The fast change in the majority of the noise traders creates the impression of an exponential increase of the price ($\rho > 1$) for particular choices of the size of the sub-samples, even tough the time series of the price is bounded. However, with longer sizes we observe fewer rejections also for a two-sided test, since the time series, then, runs over several transitions between the two "equilibria".

5 Discovering the Asymptotic Behavior

The incongruity between the theoretical properties of the model (absence of long memory) and the results of the statistical investigation, described in the previous paragraph, at the end should be a 'finite size' effect (even though one might recover the 'true' behavior only with immense amounts of data). To show the transition towards its true behavior in the case of apparent long-term dependence of volatility, it is necessary to study the asymptotic correlation properties of the time series.

To this end, figure 3 shows the Hurst exponent, estimated with DFA, as a function of different time windows (ranging from 10 to $5 \cdot 10^5$ time steps) for raw, squared and absolute returns.

Concerning the raw returns for a time window of few thousands data points, we observe a vanishing Hurst exponent, that approaches zero for longer time windows. This behavior can be explained by the boundedness of the time series of the price, which leads to a constant variance of returns. But if we restrict our time horizon to few hundreds time steps, the Hurst exponent is close to 0.5, the typical value for a random walk.

The estimation for the time series of squared and absolute returns shows different properties. In the first part it has a value greater than 0.5, within the characteristic interval for long memory processes, but for $\sim 10^4$ time steps, it declines to the typical value for the random walk; and at the end we observe a convergence to zero that means indication of a bounded time series.

The explanation of these results lies in the oscillatory pattern of the price. These oscillations create a characteristic time scale T_0 (see equation (14)), inside which the time series is a random walk, with a linear increase of the variance over time. But in the case of longer time series (the size of the sample T several times greater than T_0), the variance reaches a constant value since it, then, constitutes average over numerous oscillations.

In terms of absolute returns, these oscillations create a kind of regime switching between a calm period and a turbulent one, giving the impression of a long memory process, at least for time windows not too large compared to T_0. This effect vanishes as soon as the size of the sample is long enough.

The time scale T_0, therefore, regulates the necessary amount of data for recovering the true behavior of the model. Samples of smaller size, on the other hand, give rise to different 'spurious' characteristics, which are, in fact,

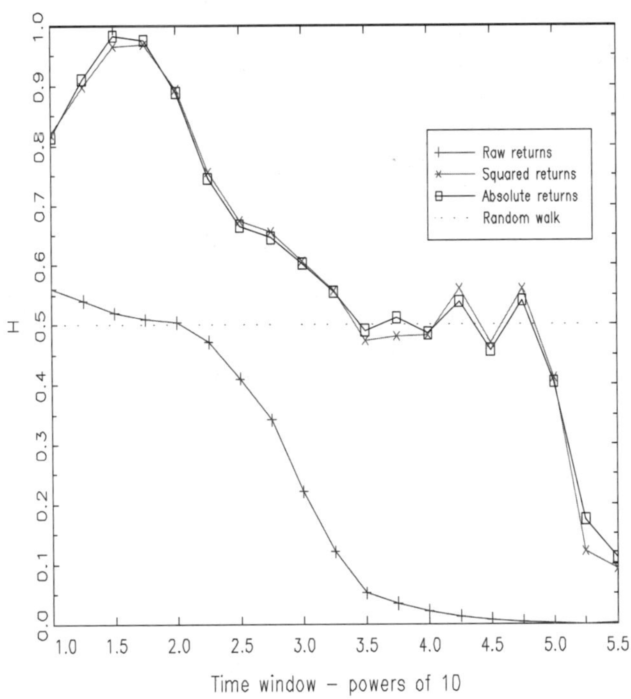

Fig. 3. Estimated Hurst exponent for raw, squared and absolute returns as a function of different time windows calculated via detrented fluctuation analysis. The dashed line is the typical value for a random walk. $T_0 = 461$ for $N = 100$, which corresponds to $10^{2.65}$

in good agreement with the empirical data; from (14) we can even calculate the scaling of the necessary sample size with respect to the parameters of the model.

6 Conclusions

This paper has presented an extremely simple variant of a noise trader/infection model. In contrast to many other contributions in the literature on artificial financial markets, it belongs to a class of models whose dynamical behavior is well understood. In particular, we know that as a bounded Markovian process with a bistable limiting distribution, the model should lack any 'true' scaling properties. Nevertheless, applying the usual statistical tests to simulated data, we find 'apparent' scaling with quite close agreement with empirically observed exponents. This shows that the difficulty to distinguish between true and spurious scaling is not confined to particular stochastic processes, but may also emerge in the area of multi-agent behavioral models. We argue that such apparent scaling might also occur in other models presented in the literature.

References

1. Anderson, M.K., Eklund, B. and Lyhagen, J. (1999). A simple linear time series model with misleading nonlinear properties. *Economic Letters*, **65,** 281–284.
2. Aoki, M. (1996). *New Approaches to Macroeconomic Modeling: Evolutionary Stochastic Dynamics, Multiple Equilibria, and Externalities as Field Effects.* Cambridge University Press.
3. Arifovic, J. and Gencay, R. (2000). Statistical properties of genetic learning in a model of exchange rate. *Journal of Economic Dynamics and Control*, **24,** 981–1006.
4. Bak, P., Paczuski, M. and Shubik, M. (1997). Price variations in a stock market with many agents. *Physica A*, **246,** 430–453.
5. Beja, A. and Godman, M.B. (1980). On the dynalic behavior of prices in disequilibrium. *Journal of Finance*, **35,** 235–248.
6. Bornholdt, S. (2001). Expectation bubbles in a spin model of markets: Intermittency from frustration across scales. *International Journal of Modern Physics C*, **12,** 667–674.
7. Challet, D., Chessa, A., Marsili, M. and Zhang, Y.-C. (2001). From minority game to the real markets. *Quantitative Finance*, **1,** 168–176.
8. Chen, S.H., Lux, T. and Marchesi, M. (2001). Testing from nonlinear structure in an "artificial" financial market. *Journal of Economic Behavior and Organisation*, **46,** 327–342.
9. Chen, S.H. and Yeh, C.H. (2002). On the emergent properties of artificial stock markets: The Efficient Market Hypothesis and the Rational Expectation Hypothesis. *Journal of Economic Behavior and Organisation*, **49,** 217–239.
10. Day, R.H. and Huang, W. (1990). Bull, bears, and market sheep. *Journal of Economic Behavior and Organisation*, **14,** 299–329.
11. Diebold, F.X. and Inoue, A. (2001). Long memory and regime switching. *Journal of Econometrics*, **105,** 131–159.
12. Eguiliz, V.M. and Zimmermann, M.G. (2000). Transmission of information and herd behavior: an application to financial markets. *Physical Review Letters*, **85,** 5659–5662.
13. Farmer, J.D. and Joshi, J. (2000). The price dynamics of common trading strategies. *Journal of Economic Behavior and Organisation*, forthcoming.
14. Frankel, J. and Froot, K. (1986). The dollar as an irrational speculative bubble: A tale of fundamentalists and chartists. *Marcus Wallenberg Papers in International Finance*, **1,** 27–55.
15. Gaunersdorfer, A. and Hommes, C.H. (2000). A nonlinear structural model for volatility clustering. *Preprint.*

16. Gaunersdorfer, A., Hommes, C.H. and Wagener, F.O.O. (2000). Bifurcation routes to volatility clustering. *Preprint.*
17. Georges, C. (2001). Learning dynamics in an artificial currency market. *Preprint.*
18. Geweke, J. and Porter–Hudak, S. (1983). The estimation and application of long memory time series models. *Journal of Time Series Analysis*, **4,** 221–238.
19. Granger, C.W.J. and Teräsvirta, T. (1999). A simple nonlinear time series model with misleading properties. *Economic Letters*, **62,** 161–165.
20. Hill, B.M. (1975). A simple general approach to inference about the tail of a distribution. *The Annals of Statistics*, **3,** 1163–1173.
21. Iori, G. (2002). A micro-simulation traders' activity in the stock market: the role of heterogeneity, agents' interaction and trade friction. *Journal of Economic Behavior and Organisation*, **49,** 269–285.
22. Kirman, A. (1993). Ants, rationality and recruitment. *Quarterly Journal of Economics*, **108,** 137–156.
23. Kirman, A. and Teyssière, G. (2002). Microeconomic models for long–memory in the volatility of financial time series. *Studies in Nonlinear Dynamics and Econometrics,* **5,** 281–302.
24. LeBaron, B. (2000). Agent based computational finance. Suggested readings and early research. *Journal of Economic Dynamics and Control*, **24,** 679–702.
25. LeBaron, B. (2001). Stochastiv volatility as a simple generator of apparent financial power laws and long memory. *Quantitative Finance*, **1,** 621–631.
26. Lobato, I.B. and Savin, N.E. (1998). Real and Spurious long–memory properties of stock market data. *Journal of Business and Economic Statistics*, **16,** 261–283.
27. Lux, T. (1995). Herd behaviour, bubbles and crashes. *Economic Journal*, **105,** 881–896.
28. Lux, T. and Marchesi, M. (1999). Scaling and criticality in a stochastic multi–agent model of a financial market. *Nature*, **397,** 498–500.
29. Lux, T. and Marchesi, M. (2000). Volatility clustering in financial markets: A microsimulation of interacting agents. *International Journal of Theoretical and Applied Finance*, **3,** 675–702.
30. Lux, T. and Ausloos, M. (2002). Market fluctuations I: Scaling, multiscaling and their possible origins. In: *Theories of Disasters- Scaling Law Governing Weather, Body, and Stock Market Dynamics*, Bunde, A., Kropp, J. and Schellnhuber H.J. editors, pp. 373–409. Springer Verlag, Heidelberg.
31. Lux, T. and Schornstein, S. (2002). Genetic learning as an explanation of stylized facts of foreign exchange markets. *Preprint.*
32. Palmer, R.G., Arthur, W.B., Holland, J.H., LeBaron, B. and Tayler, P. (1994). Artificial economic life: A simple model of stock market. *Physica D*, **75,** 264–274.
33. Peng, C.K., Buldyrev, S.V., Havlin, S., Simons, M., Stanley, H.E., and Goldberger, A.L. (1994). Mosaic organisation of DNA nucleotides. *Physica Review E*, **49,** 1685–1689.
34. Takayasu, H., Miura, H., Hirabayashi, T. and Hamada, K. (1992). Statistical properties of deterministic threshold elements - the case of market price. *Physica A*, **184,** 127–134.
35. de Vries, C.G. (1994). Stylized facts of nominal exchange rate returns. In: *The Handbook of International Macroeconomics*, van der Ploeg, F. editor, pp. 348–389. Blackwell, Oxford.
36. Youssefmir, M. and Huberman, A. (1997). Clustering volatility in multi-agent dynamics. *Journal of Economic Behavior and Organisation*, **32,** 101–118.

Long Memory and Hysteresis

Christian de Peretti *

Department of Economics, Finance, and International Business
London Metropolitan University.
c.deperetti@londonmet.ac.uk

Summary. The aim of this chapter is to determine whether the hysteretic series can be confused with long memory series, since the hysteretic effect is a persistence in the series like the long memory effect.

Nevertheless, the long term behavior of the hysteretic series is very different from the long term behavior of the long memory series: the hysteretic series are not mean reverting whereas the long memory series are (if correctly differencied). Since the mean reverting property is crucial for many economic models for checking the stability of equilibria, distinguishing between hysteresis and long memory is very important. This difference is due to the fact that hysteresis models have in fact a short memory, since dominant shocks erase the memory of the series, and the persistence is due to permanent and nonreverting state changes at a microstructure level. For checking whether hysteretic series can display long memory property, a model possessing the hysteresis property is used for simulating hysteretic data. Statistical tests for short memory against long memory alternatives are applied to these simulated data.

1 Introduction

The past 25 years have witnessed an explosion of interest in non-linear dynamical systems, in mathematics as well as in applied sciences. This chapter deals with a particular non-linear system modelling the hysteretic effect, emphasising its relevance to economics and finance. According to the mathematical definition of hysteresis (see Krasnosel'Skii and Pokrovskii, 1989, Mayergoyz,

* I thank very much Alan P. Kirman (University of Aix Marseille III - EHESS) for giving me the opportunity to submit this paper for publication in this book. I thank very much Gilles Teyssière (SAMOS) for his useful comments. I thank very much Carole Siani (INSERM U379 - University of Aix-Marseille II) for encouraging me to continue my research in this way and for her careful reading. I thank very much Dany Lang (CEDERS - University of Aix-Marseille II) for having shown me the existence of the hysteresis effect.

1991), if a process has a memory of past shocks, to be hysteretic, this memory possesses two key properties: remanence and a selective, erasable memory (this definition of the term is the one used in *all* other sciences, like physics or biology). Remanence occurs when the application of a shock of the same extent as an initial shock, but in the opposite direction, does not bring the system back to its initial position. The selective, erasable memory is that only the non-dominated extremum values of the shocks remain in the bank of memory of the system. In economics, this "strong" hysteresis definition has been used for modelling the behaviour of exchange rates changes (see Amable *et al.*, 1991, Göcke, 2002), and the unemployment (see Cross *et al.*, 1998, Lang and de Peretti, 2003). Conversely, van Dijk *et al.* (2002) argued for non-linear long memory in US unemployment. Similarly, Coakley *et al.* (2002) studied in detail the persistence and structuralism of unemployment for several countries. In addition, long memory property of volatility of exchange rates and financial indexes is well known. Thus the analysis of the dynamics of these series is of interest

The aim of this chapter is to determine whether the hysteretic property can be confused with the long memory property. This idea originally comes because hysteretic effect is a persistence in the series as long memory effect. Nevertheless, the long term behaviour of hysteretic series is very different from the long term behaviour of long memory series: the hysteretic series are not mean reverting whereas the long memory series are (if correctly differentiated). Since the mean reverting property is crucial for many economic models for checking the stability of equilibria, distinguishing between hysteresis and long memory is very important. This difference is due to the fact that hysteresis models have in fact a short memory, since dominant shocks erase the memory of the series (see section 2), and the persistence is due to permanent and non-reverting state changes at a microstructure level. For checking whether hysteretic series can display long memory property, a model possessing the strong hysteresis property is constructed following Lang and de Peretti (2003), inspired by the work of Cross *et al.* (2000) for building their own algorithm. Then, the model is tested econometrically by applying empirical tests for long memory based on de Peretti's (2003) tests, and modified for accounting for heteroskedasticity, on simulated hysteretic series.

The remainder of this paper proceeds as follows. In section 2, the analytical framework for hysteresis is presented, where a hysteretic model of agent-switching activity is constructed. This section also presents the algorithm for constructing hysteretic variables. In section 3, the various long memory tests are presented as well as the results of the Monte Carlo experiments. Asymptotic and bootstrap tests for long memory in the conditional mean and in the conditional volatility are applied on various simulated hysteretic series. Section 4 draws the concluding remarks.

2 Hysteresis Model

In this part, a hysteresis model possessing the property of selective memory is presented. For modelling a macroeconomic variable, denoted the *output variable*, with hysteretic effect, we first consider the "activity function" of individual heterogeneous agents (e.g. firms). Each agent has two different threshold values: one to become active, the other one to become inactive. For example, in the case of unemployment study, the agent is a firm, and the firm is active if it has, or hires, an employee. In the case of financial markets, an agent is active if s/he buys or sells. Agents are heterogeneous in the sense that these threshold values are different from one agent to another due to structural reasons. The macroeconomic behaviour of the whole economy in response to another macroeconomic economic variable, denoted the *input variable*, (for example the aggregate growth fluctuations in the case of unemployment) is then analysed for discovering its properties.

2.1 The Hysteretic Features

The term hysteresis has been used in economics to cover several distinct phenomena:

- A first usage is to describe *persistence* in deviation from equilibria. Consequently, for example, if shocks cause unemployment to deviate from natural rate equilibria, actual unemployment remains in disequilibrium for some time, though the natural rate remains an attractor point in some long-run sense (Layard *et al.*, 1991 for example).
- A second usage is to describe the presence of *unit or zero roots* in linear difference or differential equations. In this case, the application of a shock changes the equilibrium path of the system, neither the initial nor the shocked position of the system being forgotten. This property arises from the Cauchy-Lipschitz theorem on the existence of solutions of systems of linear differential equations, but it also implies that the application of equal but opposite shocks will leave the equilibrium unchanged (see Amable *et al.*, 1995).

Neither of these usages corresponds to the use of hysteresis in physics and mathematics, where the term was first coined, and so can be considered *inappropriate* usages. shocks continue to affect current economic equilibria can be attributed to Marshall (1890). In hysteresis models, only the non-dominated extremum values of the shocks remain in the memory bank. The term "hysteresis" was originally applied to systems that display *remanence*.

- In contrast to *persistence* in deviations from equilibria, *remanence* implies that the application and removal of a shock changes the equilibrium of a system.

- In contrast to *zero/unit root* usage, *remanence* implies that the application of equal but opposite shocks generates effects that do not cancel out, so that the equilibrium changes.

2.2 Analytical Framework

In order to display *remanence* and the other characteristics of systems with hysteresis, two basic hypotheses are required:

H1 the outputs of the system must respond on a *non-linear* way to input shocks,
H2 the elements that make up the system must be *heterogeneous*: at least some of them respond differently to a common shock (Krasnosel'Skii and Pokrovskii, 1989).

The presence of these two hysteresis characteristics implies that the system under consideration has a selective, erasable memory in which only the *non-dominated extremum values* of the past shocks affect the current output of the system (see Cross, 1993, for explanation of this property in an economic context).

The Microeconomic Assumptions

In the economic field, the hysteresis definition has been applied to effects of exchange rates changes (Amable *et al.*, 1991, Göcke, 2002), and to unemployment (Cross *et al.*, 1998, Lang and de Peretti, 2003). The various microeconomic assumptions of the hysteretic model deal with the nature of the agents (firms, traders, or others type of agent) and their relevant decision variables. The assumptions under consideration are the following:

A1 When making their decisions as to whether or nor to be active, or to maintain their previous activity level (and therefore produce output), agents take into account the input variable. Since the individual activity variable is binary, the agents adjust their activity level discontinuously, jumping from 0 to 1, and reversibly (see Figure 1). The existence of this discontinuity has been analysed by, among other authors, Dixit (1989), Hammermesh (1989), and Zoega *et al.* (2002).
A2 This discontinuity is modelled here as the presence of the two different threshold values for entering into activity, and for out.
A3 The agents are heterogeneous, since each of them has different threshold values. Heterogeneity is one of the sufficient conditions for a system to be "strongly" hysteretic (see Cross, 1993).

A model satisfying these assumptions can capture the 'selective memory' feature of hysteretic behaviour, that is, the influence only of certain past events (typically, non-dominated sequences of previous peaks and troughs). These assumptions are the original characteristics that permit a model to display a non-linear dynamic with hysteretic properties.

Individual Behaviour of the Agents

Under these assumptions, the behaviour of the agents can be modelled in a simple way. Let a_i be the threshold in terms of the *input variable*, g, required for the individual agent to become active, and b_i the threshold for g below which it becomes inactive [2]. By construction, for every agent i, $b_i < a_i$. Let e_i be an "activity dummy", which indicates whether the agent is active or not. The variation of the activity dummies in response to the fluctuations of the *input variable*, g, is therefore determined as follows:

- if $g \geq a_i$, the agent becomes active ($e_{i,t} = 1$),
- if $g \leq b_i$, the agent becomes inactive ($e_{i,t} = 0$),
- if $b_i < g < a_i$, the agent maintains its previous activity ("inaction zone").

The activity that prevails at the level of the agent then depends, inter alia, on the previous economic climate. The variation of the activity (dummy) function is represented in figure 1.

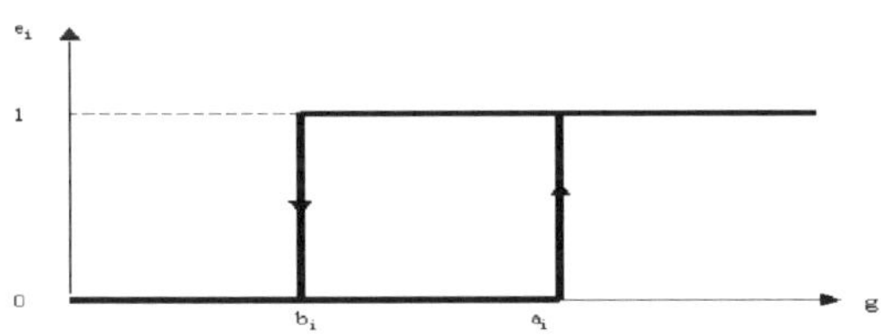

Fig. 1. Illustrates this microeconomic behaviour of the agent: the activity dummy function of an individual agent

Therefore, the activity (dummy) function can be formalised as following:

$$e_{i,t} = \begin{cases} 1 \text{ if } e_{i,t-1} = 0 \text{ and } g \geq a_i \\ 1 \text{ if } e_{i,t-1} = 1 \text{ and } g > b_i \\ 0 \text{ if } e_{i,t-1} = 0 \text{ and } g < a_i \\ 0 \text{ if } e_{i,t-1} = 1 \text{ and } g \leq b_i \end{cases} \tag{1}$$

which can be summarised as:

$$e_{i,t} = e_{b_i a_i}[g(t), e_{i,t-1}] \equiv e_{b_i a_i}[g(t), g(t-1), \ldots, g(1), e_{i,0}], \tag{2}$$

where $e_{b_i a_i}[\,.\,]$ is the hysteresis operator, or "hysteron". It should be noted that the past history of the system matters.

An important feature is that there is a *remanence* effect, illustrated on figure 2. Let us suppose that the *input variable* rises from g_0 to g_1, and then

[2] In this model, the thresholds are assumed to be fixed. An extension of the model can allow the threshold to vary. See Cross *et al.* (2001) for a time-dependent Preisach model.

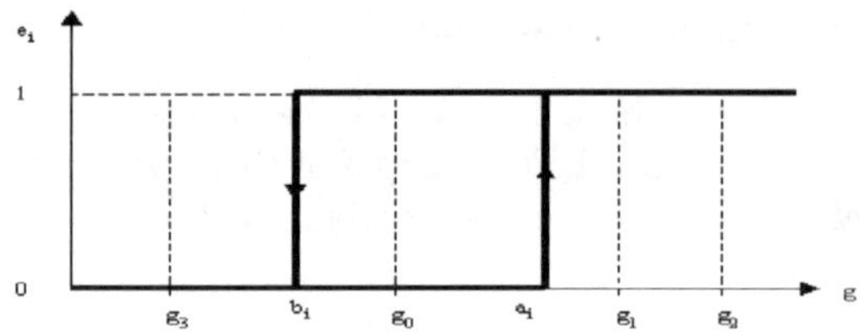

Fig. 2. An illustrative example of the *remanence* effect

goes back to its initial value. If, initially, the agent was inactive, it will become active, and remains so if the *input variable* returns to its initial value. Nevertheless, if the *input variable* varies from g_3 to g_0, or from g_3 to g_1, then in the opposite direction, the agent will be inactive in the end. The *remanence* does not necessarily depend on the amplitude of the variations in the *input variable* rates: the most crucial thing is whether the threshold values have been crossed, or not. So, if the input variable rises from g_0 to g_2, then it goes back to the initial level, this will have the same effect as a variation from g_0 to g_1, and a return to the initial value.

Aggregate Behaviour

At the aggregate level, there are many agents with significant variations in switching values [3]. Each agent i can be represented by its threshold values, (b_i, a_i): see Mayergoyz's diagram (Mayergoyz, 1991) presented in Figure 3.

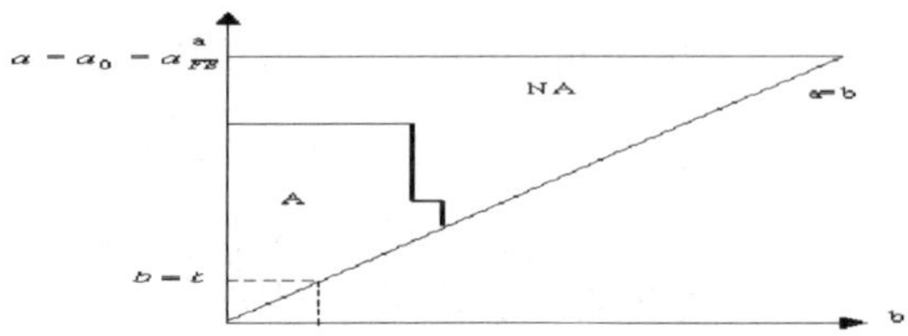

Fig. 3. The border between the "A" and "NA" areas

The active agents belong to the domain marked "A", the others corresponding to the "NA" area (inactive, or "non active" agents).

The set of micro-elements (described by hysterons) all behaving hysteretically. For a continuous set of hysterons, the model can be written as follows:

[3] No *a priori* hypothesis is made here about the distribution of the threshold values at the aggregate level.

$$h\left[g(t)|I_{t-1}\right] = \int\int_{a\geq b} e_{ba}\left[g(t)|I_{t-1}\right] f(a,b)\mathrm{d}a\mathrm{d}b, \tag{3}$$

where $f(a,b)$ is the weighting function, that is the probability density function of the threshold values, and I_{t-1} is the information set on the state of the system at time $t-1$. $h\left[g(t)|I_{t-1}\right]$ is in fact the proportion of active agents in the economy. For example, in the context of unemployment, g is the growth rate, f is the distribution function of the thresholds that make a firm hiring or firing, and $1-h$ the unemployment rate.

The model is hysteretic at a macro-level: the economy will retain in its memory the sequence of past shocks of the *input variable*, in a selective way. The memory is erasable, in that only non-dominated shocks remain in the memory of the system. For example, if a negative shock occurs in the growth rate, the unemployment rate will jump to a higher level, and will persist at this level even if the growth rate turn back to its original level.

2.3 Computing the Hysteretic Variable

As to test the model, the hysteretic transformed variable, denoted the "h" variable, has to be computed. Cross *et al.* (2000) suggest a four steps procedure to construct such a variable. However, the procedure is developed analytically only in the case where $f(a,b)$ is uniform, and deriving the procedure analytically for other distributions is in general difficult. Therefore, Lang and de Peretti (2003) propose a general procedure based on simulation methods.

First Step: Choosing the Weight Function

First, the bivariate probability density function $f(a,b)$ for the pair of thresholds is chosen. This function is a weight function specifying how much each of the agent contributes to the h variable. Since cross-sectional information on the distribution function of switching points among the agents is not available empirically, and would be rather difficult to collect, we have to specify the weight function ourselves. In this chapter, two density functions are considered. The first one is such that b_i is the minimum of two i.i.d. uniform random variables, and a_i is the maximum of both the same variables. The second density function is such that b_i is the minimum of two i.i.d. Gaussian variables, and a_i is the maximum of both the same variables.

Both the uniform and Gaussian univariate distributions have to be calibrated. We chose the following calibration: the uniform distribution takes values over $[\min g(t), \max g(t)]$ (following Cross *et al.*, 2000), where g is the input variable; the Gaussian distribution has $\bar{g} = \frac{1}{T}\sum_{t=1}^{T} g(t)$ mean and $\frac{1}{T}\sum_{t=1}^{T}\left(g(t)-\bar{g}\right)$ standard deviation. The justification of this calibration is that the agents are adapted to the economy (if the thresholds are all smaller than $\min g(t)$ or larger than $\max g(t)$, the agents cannot react to a change into the input variable).

Second Step: Simulating the Agents

The number of simulated agents, denoted by N, is then chosen. It should be as large as possible, so that the empirical distribution function of the threshold pair is close to the theoretical probability density function. Consequently, N simulated heterogeneous agents are generated in accordance to our microeconomic model (see assumption A2): they are entirely determined by the draw of the b_i, a_i pairs, for $i = 1, \ldots, N$.

Third Step: Defining the "Initial Conditions"

For computing the activity of the agents over time, the initial activity has to be determined arbitrarily, since the economy is not known before time $t = 1$. The activity of each agent, $e_{i,0}$, is chosen randomly with probability 0.5.

Consequently, the first value of the input variable, $g(1)$, is used for determining the first value for the activity dummies in the following way:

$$e_{i,1} = e_{b_i a_i}[g(1), e_{i,0}].$$

The activity, as well as the input variable, are then considered from $t = 1$ to $t = T$.

Fourth Step: Computing the Hysteretic Transformation

For each period of time, the activity dummy for each of the agents is computed by the program using equation 2.

The hysteretic variable, that is the proportion of active agents, is given by the mean of the activity over the periods of time:

$$h\left[g(t)|I_{t-1}\right] = \frac{1}{N}\sum_{i=1}^{N} e_{i,t}. \tag{4}$$

In the algorithm developed in Lang and de Peretti (2003), it is not necessary to select, for each period of time, the non-dominated extrema of the "input" variable, since it is done implicitly in the recursive equation 2.

The hysteretic transformation defined in Equation 4 displays *remanence.*

2.4 Two Examples of Hysteresis Application

Application to Unemployment

A *strong* hysteresis version of the link between the fluctuations of unemployment and growth is built and tested empirically for some countries in Lang and de Peretti (2003). The non-hysteretic version of this model is the *growth rate* version of Okun's Law. The underlying assumption that unemployment responds to growth shocks in a *linear* way can be regarded as open to question.

Arguably, fluctuations of unemployment when growth varies do not need to be the same during booms as during recessions, and should depend on the intensity of the economic fluctuations, and possibly on the *past history* of the economic system. For example, at the microeconomic level, if the economic circumstances deteriorate, employers may want to make sure this deterioration is not just temporary, before laying off employees. This behaviour is rational. As argued in Oi (1962), labour is a quasi-fixed factor: there are costs associated with redundancies, and with hiring, and training new employees; the firm may want to make sure it will not lose key personnel to other firms during a temporary decline in the demand for its output. Therefore, firms may wait some periods before firing, and, for symmetrical reasons, before hiring persons (see Zoega *et al.*, 2002). Another reason for this non-linearity is the well-known phenomenon of "discouraged workers": during recessions, some unemployed workers stop looking for work, therefore dropping out of the working population, where they come back during booms. This implies that the reactions of the unemployment to growth shocks can be *asymmetric*: if employers fire a certain number of workers after a negative growth shock, they may not hire exactly the same number of workers after a positive shock of the same amount as the original one. Therefore, for all these reasons, the link between the growth and the fluctuations of unemployment may be *hysteretic*.

In the theoretical model, a large number of heterogeneous firms adjusts discontinuously their activity level in reaction to the fluctuations of the aggregate demand level. The firms are heterogeneous since each of them has a couple of *entry-exit* switching points, different from the other firms. According to Dixit and Pindyck (1994), this heterogeneity arises in part from differences between firms, in the sunk costs associated with their investments, and from differences in the value of waiting to gain more information about market opportunities. Last but not least, "intractable uncertainty" can be the source of this heterogeneity: firms assess the uncertain future in different ways, a problem arising because "they simply do not know" what the future will be (see Keynes, 1936) [4].

When making their decisions as to whether or nor to be active —and therefore produce output and employ labour—, firms take into account the "growth shocks" variable. We concentrate on the decision to become active, to become inactive, or to maintain the previous activity level that depends on the aggregate growth g. The macroeconomic consequence of this behaviour is the presence of hysteresis in the "growth-employment fluctuation" relationship. The aggregate growth rate, which is relatively easy to obtain, obviously gives an indication about the future state of demand and general "health" of the economy. In fact, a firm reacts to its own demand in its market. Anyway, this

[4] See Göcke (2002) for a multi-period, uncertainty and option value effect microeconomic modelling of hysteresis.

demand is not independent of the growth rate of the economy: when there is some growth in one sector, there will be a demand for products in other sectors, and, on the contrary, when the economy is depressed, the demand for goods coming from the other sectors will diminish. In a dynamic economy, a market where the demand is independent from the general health as measured by the economic growth rate may exist, but that configuration is rather an exception than the general rule.

Furthermore, the total amount of demand in a particular market may be difficult or impossible to know for a firm, or, if it is possible to know, this knowledge may be expensive: the firm may have to employ an economist, to find the relevant statistics, etc. On the contrary, the aggregate growth rate can be known readily and freely, and may give precious information concerning the amount of future demand. Following Keynes' General Theory (Keynes, 1936), the general growth rate may also influence the expectations of the employers concerning the future state of demand (Z curve in the "principle of effective demand"). As we said before, information about the state of demand in some markets are difficult to find. Therefore, collecting all data for all sectors in a particular economy would be an impossible task. That would prevent us from being able to run the empirical tests. Seen in this light, the growth rate of the economy can be considered as a proxy of the demand addressed to the firm. For all these reasons, we have chosen to make the implicit assumption that the demand in all particular markets is linked to the general growth rate, an assumption which is underlying not only in the original law coined by Okun, but in many macroeconomic models. And, as for all the hypotheses, this one has, of course, its limitations.

Application to Financial Markets

One of the most important 'stylized facts' in financial series is that the asset returns exhibit *volatility clustering* and exhibit the *long memory property.* The microstructures that should cause these empirical properties are discussed below: the *heterogeneity* and the *non-linearity.* It should be noted that these two properties can generate hysteresis property according to the explanations given in this Section.

Heterogeneity in Financial Markets

In a perfect rational *efficient market hypothesis* world, [5] all traders are rational, and it is *common knowledge* that all traders are rational. In real financial

[5] "An 'efficient' market is defined as a market where there are large numbers of rational, profit-maximizers actively competing, with each trying to predict future market values of individual securities, and where important current information is almost freely available to all participants. In an efficient market, competition among the many intelligent participants leads to a situation where, at any point in time, actual prices of individual securities already reflect the effects of information based both on events that have already occurred and on events which, as of now,

markets, however, traders are different, especially with respect to their expectations about future prices and dividends. A quick glance at the financial pages of newspapers is sufficient to observe that *difference of opinions* among financial analysts is the rule rather than the exception. In the last decade, a rapidly increasing number of structural heterogeneous agent models have been introduced in the finance literature, see, Zeeman (1974), Frankel and Froot (1988), De Long *et al.* (1990), Kirman (1891), Chiarella (1992), De Grauwe *et al.* (1993), Brock (1993, 1997), Wang (1994), Dacorogna *et al.* (1995), Lux (1995), Brock and LeBaron (1996), Arthur *et al.* (1997), Kurz (1997), Brock and Hommes (1997b,1998), Farmer (1998), Le Baron *et al.* (1999), Lux and Marchesi (1999a,1999b), Chiarella and He (2000), Farmer and Joshi (2000), Gaunersdorfer (2000), Gaunersdorfer and Hommes (2005), Kirman and Teyssière (2005) and LeBaron (2000). Some authors even talk about a *Heterogeneous Market Hypothesis* as a new alternative to the *Efficient Market Hypothesis*. In all these heterogeneous agent models, different groups of traders, having different beliefs or expectations, co-exist.

Non-Linearity in Financial Markets

The *Heterogeneity* of expectations among traders introduces an important *non-linearity* into the market. In an *Adaptative Belief System* [6] (ABS), there are also two important sources of *noise*: model approximation error and intrinsic uncertainty about economic fundamentals. Asset price fluctuations in an ABS are characterised by an irregular switching between phases of close-to-the-fundamental-price fluctuations, phases of optimism where most agents follow an upward price trend, and phases of pessimism with small or large market crashes follow an upward price. Temporary speculative bubbles can occur, *triggered* by noise and *amplified* by evolutionary forces.

3 Long Memory Property of Hysteretic Series

3.1 Tests for Detecting Long Memory

Many financial and macroeconomic time series exhibit the *long memory* property, that is the autocorrelations of the series (or of a power of the series) are

---

the market expects to take place in the future. In other words, in an efficient market at any point in time the actual price of a security will be a good estimate of its intrinsic value." (Fama, 1965).

[6] In a series of papers, Brock and Hommes (1997a, 1997b, 1998, 1999) propose to model economic an financial markets as *Adaptative Belief System* (ABS). An ABS is an evolutionary competition between trading strategies. Different groups of traders have different expectations about future prices and future dividends. Traders choose their trading strategy according to an evolutionary 'fitness of measure', such as accumulated past profits.

significantly positive, even at high order lags, and decay slowly, possibly following a scaling law. de Peretti(2002) shows that the tests for long memory based on the asymptotic distributions display very large size distortion and have to be corrected. A way for correcting the distortions is to use bootstrap methods based on simulation techniques. The foundation of the bootstrap is presented in Efron (1979), developments are presented in Davidson and MacKinnon (1993), and further analyses in Davidson and MacKinnon (1996a, 1996b). Bootstrap tests for long memory in volatility robust to ARCH effect are proposed in Andersson and Gredenhoff (1998). de Peretti (2003a) propose bilateral bootstrap tests for long memory in the conditional mean robust to ARMA effect. This bilateral bootstrap P value greatly improves the size-power properties of the tests and can be extended straight to tests for long memory in the conditional mean simultaneously robust to ARMA and ARCH effects, and to tests for long memory in volatility robust to ARCH effect. This test is presented in this Section.[7]

Long Memory Model

We restrict attention to univariate, linear fractionally integrated models of the ARFIMA[8] form:

$$\phi(L)(1-L)^d x_t = \theta(L)\varepsilon_t \qquad t \in \{1,\ldots,T\}, \tag{5}$$

where

- $\{\varepsilon_t\}$ can follow a stationary GARCH process (non necessarily Gaussian),
- ϕ and θ are polynomials that have all roots outside the unit circle,
- $\sigma_\varepsilon^2 < \infty$,
- L is the lag operator,
- d is the differentiating parameter and takes a real value,

In some circumstances, a long-memory process may be approximated by a fractionally integrated model; hence testing for long-memory can be done by a test on d. Such tests are applied to stationary and invertible series (requiring that $|d| < 1/2$), and H_0: $d = 0$ is thus a natural null hypothesis.

Bootstrap Test for Long Memory in Heteroskedastic Returns

For financial data, that often display heteroskedastic effect, the classical bootstrap tests have to be modified for taking account the heteroskedasticity. For the return series, the null hypothesis is an ARMA$(p';q')$ against the alternative of an ARFIMA$(p;d;q)$ process, where p' and p are the auto-regressive

[7] This kind of test is proposed and briefly presented in de Peretti (2003b) for different values for the long memory parameter.

[8] The ARFIMA model is presented in greater detail in Granger and Joyeux (1980) and Hosking (1981).

parameters, d is the long memory parameter, and q' and q are the moving average parameters. However, the heteroskedasticity effect has to be corrected for making robust inference. Consequently, the *Data Generating Process* (DGP) under the null used in the bootstrap tests has to be changed from an ARMA$(p';q')$ process to a ARMA$(p';q')$-GARCH$(p'';q'')$ (mean-variance) process. The modified procedure is the following.

1. Estimate the long memory parameter, denoted by $\hat{d}$, and then compute the test statistic, which will be denoted by $\hat{\tau}$. Robinson's statistic test is chosen here, however, another can be used, e.g. re-scaled adjusted range statistic first introduced by Hurst (1951), modify re-scaled range statistic suggested by Lo (1991), the modified Higuchi's method based on the measure of the fractal dimension Higuchi (1988), de Peretti (2002) the wavelet method (Jensen, 1994). For choosing of the bandwidth, the heuristic rule proposed in de Peretti (2002) is used [9].
2. Estimate the model 5 by maximum likelihood under the null H_0: $d = 0$ to obtain the estimated parameters and the residuals. The model is therefore reduced to an ARMA$(p';q')$-GARCH$(p'';q'')$ model.
3. Draw B sets of bootstrap error terms, ε^b, following the estimated GARCH $(p'';q'')$. The error terms of the GARCH$(p'';q'')$ process are drawn by re-sampling with replacement from the vector of residuals of the GARCH $(p'';q'')$ estimation of the series.
   Use ε^b to generate B bootstrap samples x^b. The elements of x^b are generated recursively from the ARMA$(p';q')$ model using the estimated parameters and ε^b.
4. For each bootstrap sample, estimate the long memory parameter, denoted by $\hat{d^b}$, and then compute the test statistic, denoted by τ^b with x^b instead of x.
5. Then compute the bilateral bootstrap P value, denoted by $\hat{p}$, as follows:

$$\hat{p}(\hat{\tau}) = 2\min\{\hat{p}_{uni}(\hat{\tau}), 1 - \hat{p}_{uni}(\hat{\tau})\}, \tag{6}$$

   where

$$\hat{p}_{uni}(\hat{\tau}) = \frac{1}{B}\sum_{b=1}^{B} I(\tau^b > \hat{\tau}),$$

   and $I(\ .\ )$ is is an indicator variable.

Bootstrap Test for Long Memory in the Volatility

Consider the volatility measured, for example, by the absolute value of the returns:

[9] The effectiveness of this rule is sufficient for the purpose of Monte Carlo experiments. This rule permits to avoid complicating the procedure in the context of large number of Monte Carlo replications. For the choice of an optimal bandwidth, see Henry and Robinson (1996).

$$v_{1,t} = |r_t|. \tag{7}$$

For testing the presence of long memory in the volatility, the null hypothesis is that the absolute value of the returns follows the absolute value of a GARCH$(p'; q')$ process against the long memory alternative $d > 0$ [10]. Similarly to the case of long memory test in the conditional mean, the bootstrap DGP under the null has to be changed to the absolute value of a GARCH$(p'; q')$ process rather than an ARMA$(p; q)$ process. The same reasoning holds if the volatility is measured by the logarithm of the absolute returns or the squared returns for example:

$$v_{2,t} = r_t^2,$$
$$v_{3,t} = \ln(|r_t|).$$

The procedure is similar to the previous one:

1. Compute the test statistic $(\hat{\tau})$ with $|x^b|$, $\left(x^b\right)^2$, or $\ln(|x^b|)$ instead of $|x|$, x^2, or $\ln(|x|)$ as in Subsection 3.1.
2. Estimate the series using the GARCH$(p'; q')$ model by maximum likelihood.
3. Generate B bootstrap samples x^b. The elements of x^b are generated recursively from the GARCH$(p'; q')$ model using the estimated parameters.
4. For each bootstrap sample, compute the test statistic (τ^b), with $|x^b|$, $\left(x^b\right)^2$, or $\ln(|x^b|)$ instead of $|x|$, x^2, or $\ln(|x|)$.
5. Then compute the bilateral bootstrap P value defined as previously in Equation 6.

3.2 Monte Carlo Experiments: Conditional Mean Analysis

In these Monte Carlo experiments, (non-hysteretic) short memory series are studied. Long memory tests are run for verifying that the short memory effect does not induce a detection of long memory effect. Simultaneously, the same non-hysteretic short memory series are transformed using the hysteretic transform defined in Equation 4. Long memory tests are then run on these transformed series to see whether the hysteretic series can be confused with long memory processes. The null hypothesis is short memory, that is $d = 0$, against long memory, that is $d \neq 0$.

In these first experiments, the hysteretic transform is applied directly to the series, and the tests are applied to seek long memory in the conditional mean of the series.

[10] We assume that the mean of the series is memoryless as it is often the case in finance. Consequently, there is not the ARMA component. Otherwise, the test can also be applied on the residuals of a regression.

Monte Carlo Methodology

Non-Hysteretic Short Memory DGPs

To be more realistic, DGPs containing autocorrelation with leptokurtic error terms are chosen. Zero mean AR(p) processes with Gaussian and then $t(5)$ error terms [11] having the characteristics presented in Table 1 are studied.

Table 1. Monte Carlo DGPs

DGP 1: AR(1) process

Characteristic	Value
Autoregressive parameter	0.3
Standard deviation	1
Sample size	512
Error terms distribution	$N(0, \sigma^2)$ or $t(5)$

Hysteretic DGPs

The previous short memory and non-hysteretic series are transformed using hysteretic transform. Two distributions for the threshold pair, (a_i, b_i), of the "activity function" are used (see Section 2.2). The first distribution is the uniform distribution, the second one is Gaussian distribution: see subsection 2.3, first step.

Bootstrap Procedure

In these Monte Carlo experiments, two tests for long memory are run: the asymptotic test with a bootstrap estimation of the standard deviation of the long memory estimator, and the bootstrap test (again with the bootstrap estimation of the standard deviation).

The first bootstrap procedure deals with the estimation of the standard deviation of the long memory estimator in the asymptotic test for long memory. $B_2 = 123$ bootstrap replications are used. The bootstrap DGP parameters of this process are estimated for each series. The procedure is asymptotic in that the distribution of the test statistic is the asymptotic one, even if the statistic contains a bootstrap estimate.

The second bootstrap procedure deals with the bootstrap test for long memory that is a test using the bootstrap distribution of the statistic. The number B of bootstrap replications in the bootstrap loop for computing the

[11] The $t(5)$-AR(p) processes are fourth order stationary and satisfy the assumption S by Giraitis *et al.* (2003), that is necessary for making inference. The Gaussian AR(p) processes satisfy obviously this assumption

P value of the test has to be specified. $B = 245$ is chosen [12]. For these Monte Carlo experiments, the bootstrap DGP under the null is fixed to be an AR(2)-GARCH(1; 1) process for all the simulated series. The ARMA and GARCH orders are not estimated for each series for saving computation time. Nevertheless, the parameters of this process are estimated by maximum likelihood for each simulated series.

Monte Carlo Results

The number of Monte Carlo replications is $S = 1000$.

Density Functions

In a first step, the density functions of the long memory estimator are plotted on Figure 4 according to the different DGPs defined in Table 1 and its two hysteretic transforms in the case of Gaussian error terms and then in the case of $t(5)$ error terms. Figure 4 shows that in the case of non-hysteretic short memory series, the realisations of the long memory estimator are around the true value of the long memory parameter under the null hypothesis, that is $d = 0$. In the cases of hysteretic series, Figure 4 clearly shows a tendency for the long memory estimator to obtain values for the long memory parameter larger than 0. This feature appears for both Gaussian and Uniform hysteretic transformations. When the error terms are $t(5)$ distributed, the feature is emphasised. Consequently, the hysteresis property can be confused with the long memory property, particularly when the distribution of the error terms are leptokurtic, as for financial series. It should be noted that since there is no over-estimation of the long memory parameter due to the short memory component of the process, the long memory detection for the hysteretic series is totally due to the hysteretic characteristic.

P Values of the Asymptotic Test

For getting a rigorous presentation of the results, the P values of the asymptotic long memory test are presented. The true probability of rejecting the null hypothesis of short memory against long memory is plotted against the significance level of the tests using the graphical techniques proposed in Davidson and MacKinnon (1998a) (see Figure 5).

The non-hysteretic short memory series under consideration display clearly short memory property: if the true Monte Carlo DGP is really an short memory process, the plot corresponds to the true size of the test for this DGP and the size curve has to be close to the 45° line. Figure 5 shows that there is

[12] The numbers B_2 of bootstrap replications for the estimate of the standard deviation of the long memory estimator used in the test statistic has also to be specified. $B_2 = 123$ is chosen. This second bootstrap loop is fitted into the second one.

Case of Gaussian error terms

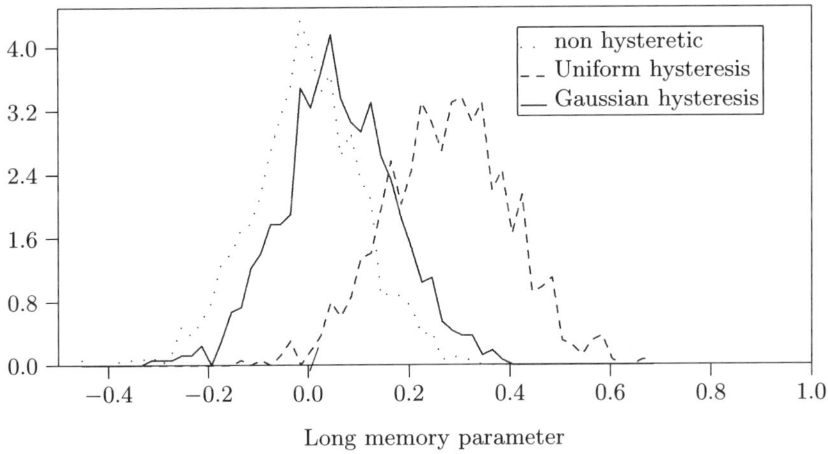

This picture presents the density functions for the long memory parameter estimator in the case of three DGPs: the first is a Gaussian AR(1) process defined in Table 1 as "DGP 1", the second is its uniform hysteretic transform, and the third is its Gaussian hysteretic transform.

Case of $t(5)$ error terms

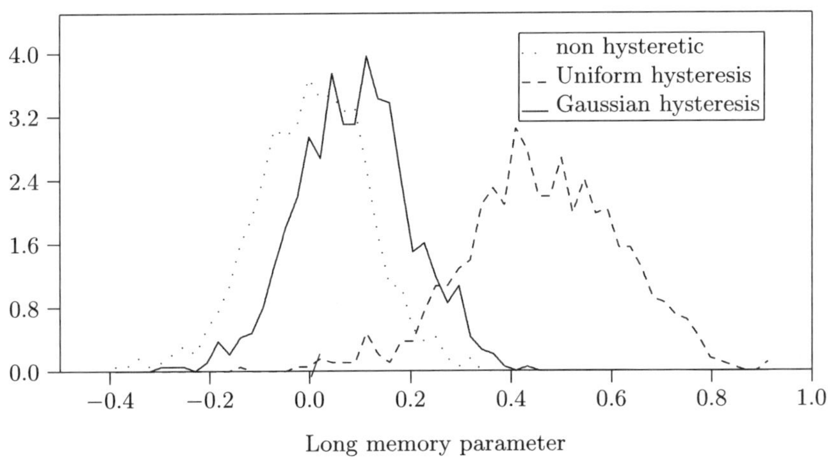

This picture presents the density functions for the long memory parameter estimator in the case of three DGPs: the first is a $t(5)$-AR(1) process defined in Table 1 as "DGP 1", the second is its uniform hysteretic transform, and the third is its Gaussian hysteretic transform.

Fig. 4. Density functions of the long memory estimator for DGP 1

Case of Gaussian error terms

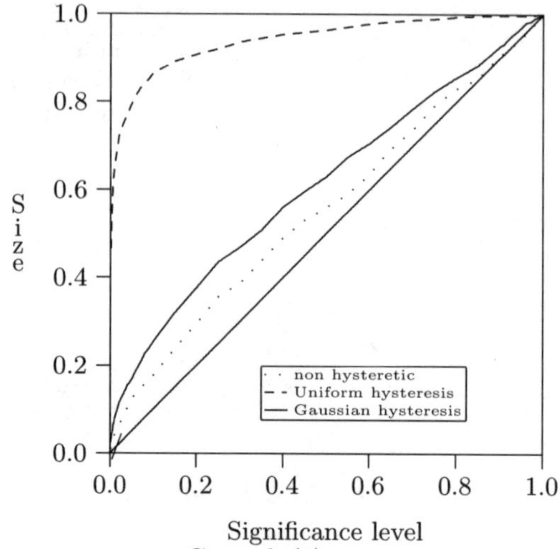

Case of $t(5)$ error terms

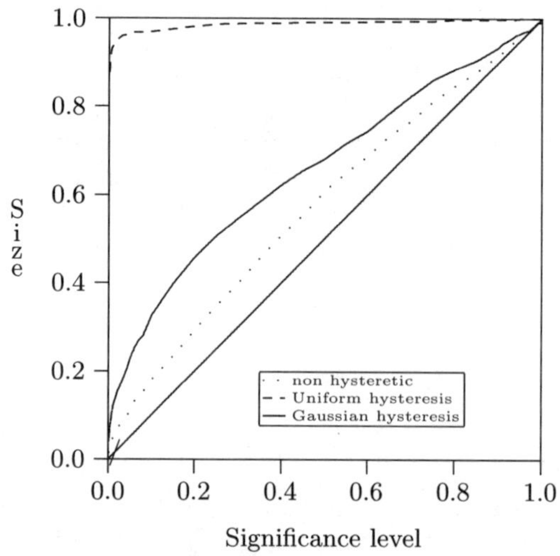

These pictures present the true probability of rejecting the null hypothesis of short memory against long memory plotted against the significance level. The used test is the asymptotic Robinson's (1995) test with a non-parametric bootstrap estimation of the standard deviation. It is applied in the cases of DGP 1 defined in Table 1 with Gaussian and $t(5)$ error terms and its hysteretic transforms (uniform and Gaussian distributions for the hysterons).

Fig. 5. True probability of rejecting the null hypothesis of short memory almost no excess of acceptation of the long memory hypothesis due to short memory effect (in our experiments) since the curves corresponding to the probability of rejection for the non-hysteretic short memory DGPs is close to the 45° line.

In the cases of hysteretic series, Figure 5 confirms the features found for the density functions in Figure 4: the hysteresis property can be confused with the long memory property, particularly when the distribution of the error terms are leptokurtic. If the true Monte Carlo DGP is a long memory process (that is not the case in our experiments), the plot corresponds the power of the test and should converge to 1 with the sample size. In our case, if the DGP is a hysteretic process, the plot indicates whether a long memory test has an excess of acceptation of the alternative hypothesis of "long memory", or, in other words, whether the hysteretic property is confused with long memory property.

Results for the Bootstrap Test

The results for the bootstrap long memory test are not presented here because they suggest that no there is no long memory in the series. However, these results have to be interpreted correctly in the case of hysteretic series, because when the bootstrap procedure estimates the DGP under the null having the form of an ARMA$(p'; q')$ process, a root close to unity is found. A root close to unity can be interpreted as more than long memory since it is "perfect" memory. In fact, the bootstrap procedure suffers from its best advantage: its adaptability. When the bootstrap test is applied, the test hypotheses are interpreted as "short memory" against "long memory". However, technically, the hypotheses are "ARMA$(p'; q')$" against "ARFIMA$(p; d; q)$" (with or without heteroskedasticity correction). Since a ARMA$(p'; q')$ process can contain a unit root, the hypotheses are in fact an "integer integration" model (with possible integration of order 0) against a "fractionally integration" model. Consequently, the bootstrap retains the hypothesis of a unit root in an ARMA model ("perfect" memory) rather than a fractionally integration (long memory).

It should be noted that a unit root is not an acceptable hypothesis for a hysteretic series, because an estimation of the long memory parameter should lead to a value close to 1, where as the estimates are around to 0.4 – 0.5 for the uniform hysteretic transformation for example. In fact, the bootstrap test encounters the same problem when the long memory parameter is close to 0.5 (see de Peretti (2003). Consequently, it can be conclude that the hysteretic series behave as long memory series with a long memory parameter close to 0.5, as found by the estimator on hysteretic series (see Figure 4).

Obviously, in practice, the bootstrap null DGP can be improved by taking into account sophisticated characteristics such as structural breaks or other features, but it should take us away from the purpose of this chapter: estimating hysteresis with long memory tools.

3.3 Monte Carlo Experiments: Conditional Volatility Analysis

In these second experiments, long memory tests are then run on the volatility of these series to see whether the hysteretic in the volatility can be confused with long memory volatility. The volatility is measured by $|r_t|$ here.

Hysteretic Volatility

In the context of financial series, no memory of a very short memory is generally found in the conditional mean of the returns. A "strong" persistence is generally found in the conditional volatility of the returns. The model presented in Subsection 3.2 is not able to capture this feature since it deals with the conditional mean. Consequently, building a model with hysteretic volatility is useful. The following hysteretic transform of the volatility of a series r_t is proposed:

$$h^{\star}(r_t) = \text{sign}(r_t)h(|r_t|), \tag{8}$$

$$h^{\star\star}(r_t) = \text{sign}(r_t)\sqrt{h(r_t^2)}, \tag{9}$$

where

$$\begin{aligned} \text{sign}(r) &= 1 \text{ if } r > 0, \\ &= 0 \text{ if } r < 0, \end{aligned}$$

and $h(\ .\ )$ is the hysteretic transform defined as in Section 2. For additional comments on the combination of sign effects and (weighted) amplitude effects see Heyde (2002).

Monte Carlo Methodology

Non-Hysteretic Short Memory DGPs for the Volatility

GARCH($p;q$) processes with Gaussian and then $t(5)$ error terms are studied, with the characteristics presented in Table 2.

Table 2. Monte Carlo DGPs for the volatility

DGP 2: GARCH(1; 0) process

Characteristic	Notation	Value
Constant parameter	δ_0	0.7
Autoregressive parameter	δ_1	0.3
Sample size	T	512
Error terms distribution	$\mathrm{N}(0,\sigma^2)$ or $t(5)$	

Hysteretic DGPs

The previous non-hysteretic short memory series are transformed using hysteretic transforms defined in subsection 3.3. The uniform distribution for the threshold pair, (a_i, b_i), of the "activity function" is used.

Bootstrap Procedure

The first bootstrap procedure, dealing with the estimation of the standard deviation of the long memory estimator in the asymptotic test for long memory. $B_2 = 123$ bootstrap replications are chosen, with a GARCH$(2;1)$ DGP that is estimated for each series.

The second bootstrap procedure, dealing with the bootstrap test for long memory. $B = 245$ bootstrap replications are chosen, with a GARCH$(2;1)$ DGP under the null that is estimated for each series (again $B_2 = 123$ bootstrap replications is used for the standard error estimate).

Monte Carlo Results

The number of Monte Carlo replications is $S = 1000$.

Density Functions

The density functions of the long memory estimator for the different DGPs are plotted. Figure 6 presents the density functions for the DGP defined in Table 2 and its hysteretic transforms in the case of Gaussian error terms and then in the case of $t(5)$ error terms. As for the conditional mean case, the long memory estimation of the hysteretic volatility series has a tendency to find values larger than 0. This long memory finding is not due to short memory effect, since the long memory estimation of the (non-hysteretic) short memory series seems to be not biased. For being rigorous, the results for the long memory tests have to be considered.

P values of Long Memory Test

The result for the asymptotic test is presented in Figure 7. The test for long memory in non-hysteretic short memory series is absolutely not biased since the corresponding curves are almost confused with the 45° line, even for the $t(5)$ error terms case. Consequently, the short memory component cannot be the cause of the long memory detection in hysteretic series. For the hysteretic series, the probability of retaining the long memory hypothesis is larger than the significance level of the tests, suggesting that the hysteretic property can be confused with the long memory property. The same feature is obtained for both the hysteretic volatility transforms defined by equations 8 and 9, and for Gaussian and $t(5)$ distributed error terms. However, the probability of retaining the long memory hypothesis is less for the hysteretic transform for the volatility than for the hysteretic transform of the returns.

Case of Gaussian error terms

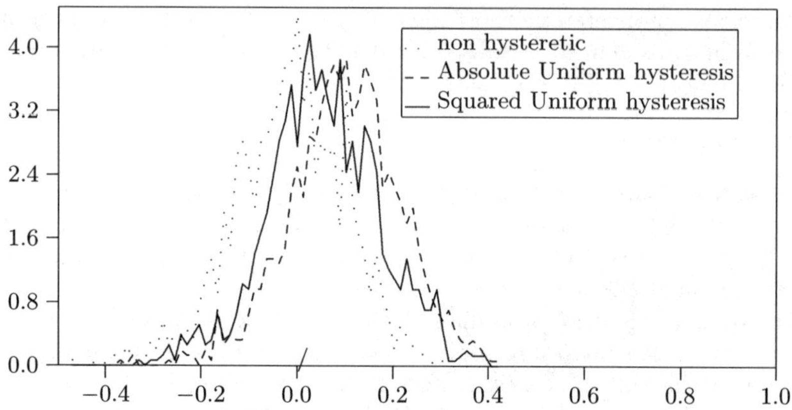

Long memory parameter

This picture presents the density functions for the long memory parameter estimator in the case of three DGPs: the first is a Gaussian GARCH(1; 0) process defined in Table 2 as "DGP 2", the second is the uniform hysteretic transform of its absolute value, and the third is the uniform hysteretic transform of its squared value.

Case of $t(5)$ error terms

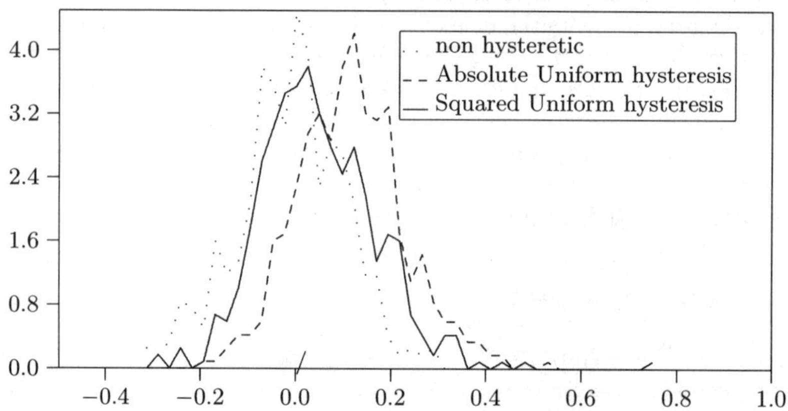

Long memory parameter

This picture presents the density functions for the long memory parameter estimator in the case of three DGPs: the first is a $t(5)$-GARCH(1; 0) process defined in Table 2 as "DGP 2", the second is the uniform hysteretic transform of its absolute value, and the third is the uniform hysteretic transform of its squared value.

Fig. 6. Density functions of the long memory estimator for DGP 2

Case of Gaussian error terms

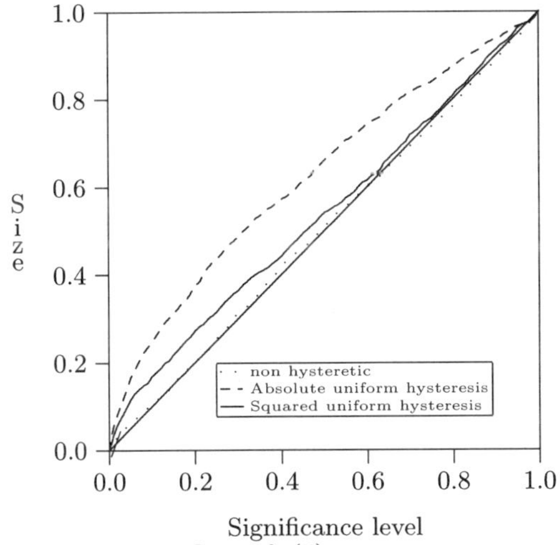

Case of $t(5)$ error terms

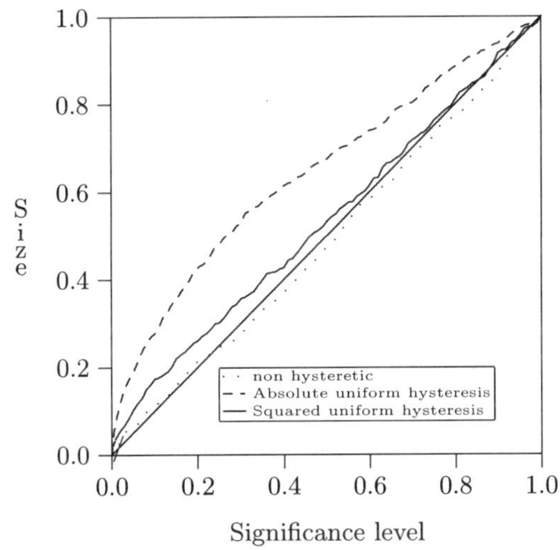

These pictures present the true probability of rejecting the null hypothesis of short memory against long memory plotted against the significance level. The used test is the asymptotic Robinson's (1995) test with a non-parametric bootstrap estimation of the standard deviation. It is applied in the cases of DGP 2 defined in Table 2 with Gaussian and $t(5)$ error terms and their hysteretic transforms (uniform distributions applied on the absolute and squared values of the series).

Fig. 7. True probability of rejecting the null hypothesis of short memory

4 Conclusion

In economics, the hysteresis definition has been applied to effects of exchange rates changes (Amable *et al.*, 1991, Göcke, 2002), and to unemployment (Cross *et al.* (1998), Lang and de Peretti (2003). On another hand, Coakley *et al.* (2002) studied in detail the persistence and structuralism of unemployment for several countries. Similarly, van Dijk (2002) argued for non-linear long memory in US unemployment. Consequently, the question of what is the dynamics of such series, is of interest.

The aim of this paper is to determine whether the hysteretic property can be confused with the long memory property or more generally, whether the hysteretic microstructure can be an economic justification of the long memory statistical stylised fact found in many macroeconomic and financial series. The long memory property of hysteretic series is studied via Monte Carlo experiments. Hysteretic conditional mean processes is analysed with long memory tests as well as hysteretic conditional volatility processes. It can be concluded that hysteretic series can very easily be confused with long memory series.

For determining whether a series is hysteretic or has long memory, in practice, the variables that explain the series of interest have to be found first. For example, the unemployment level should be explained by a transformation of the growth. We have then to test whether the series of interest is a hysteretic transform of the explanatory variables, or whether the series is a linear transform of long memory variables (long memory regression). However, discriminating between both these hypotheses has not necessarily a sense if the microstructure model for hysteresis explains in fact the observed long memory in the aggregate series.

An opening way for the research is to examine the components of the hysteretic transform that control the observed long memory feature. Intuitively, more a hysteretic transform is nonlinear, and thus persistent, more the estimated long memory parameter should be large. More rigorously, a hysteretic transform with uniformly distributed thresholds and a spread between the thresholds close to zero is close to a linear transformation, and thus does not cause any long memory effect. Consequently, a growing spread and distributions far from the uniform one, in a way that has to be studied, will generate long memory effect.

References

1. Amable, B., Henry, J., Lordon, F. and Topol, R. (1991). Strong Hysteresis: an Application to Foreign Trade. *OFCE Working Paper*, **91(3)**.
2. Amable, B., Henry, J., Lordon, F. and Topol, R. (1995). Hysteresis revisited: A methodological approach. In R. Cross (ed.), *The Natural Rate of Unemployment: Reflections on 25 Years of the Hypothesis*. Cambridge University Press, Cambridge.

3. Andersson, M. K. and Gredenhoff, M. P. (1998). Robust testing for fractional integration using the bootstrap. *Working paper series in economics and finance*, **218**.
4. Arthur, W. B., Holland, J. H., LeBaron, B., Palmer, R. and Taylor, P. (1997). Asset pricing under endogenous expectations in an artificial stock market. In: Durlauf, S. N., Arthur, W. B., and Lane, D. A. (Eds.), *The Economy as an Evolving Complex System II*, 15–44. Reading, MA: Addison-Wesley.
5. Brock, W. A. (1993). Pathways to randomness in the economy: emergent nonlinearity and chaos in economics and finance. *Estudios Económicos*, **8,** 3–55.
6. Brock, W. A. (1997). Asset price behavior in complex environments. In: Durlauf, S. N., Arthur, W. B., and Lane, D. A. (eds.), *The Economy as an Evolving Complex System*, Vol **II,** 385–423. Reading MA: Addison-Wesley.
7. Brock, W. A. and Hommes, C. H. (1997a). A rational route to randomness. *Econometrica*, **65,** 1059–1095.
8. Brock, W. A. and Hommes, C. H. (1997b). Models of complexity in economics and finance. In: C Hey *et al* (Eds.), *System dynamics in economic and financial models*, 3–41. New York: Wiley.
9. Brock, W. A. and Hommes, C. H. (1998). Heterogeneous beliefs and routes to chaos in a simple asset pricing model. *Journal of Economic Dynamics Control*, **22,** 1235–1274.
10. Brock, W. A. and Hommes, C. H. (1999). Rational animal spirits. In: Van der Laan, G., Herings, P. J. J., and Talman, A. J. J. (eds.) *The Theory of Markets*, 109–137. Amsterdam: North-Holland.
11. Brock, W. A. and LeBaron, B. (1996). A structural model for stock return volatility and trading volume. *Review of Economic Statistics*, **78,** 94–110.
12. Chiarella, C. (1992). The dynamics of speculative behaviour. *Ann. Operations Research*, **37,** 101–123.
13. Chiarella, C. and He, T. (2002). Heterogeneous beliefs, risk and learning in a simple asset pricing model. *Computational Economics*, **19,** 95-132.
14. Coakley, J., Fuertes, A.M. and Zoega, G. (2002). Evaluating the persistence and structuralist theories of unemployment. *Studies in Nonlinear Dynamics and Econometrics*, **5,** 1–22.
15. Cross, R. (1993). On the foundations of hysteresis in economic systems. *Economics and Philosophy*, **9,** April, 53–74.
16. Cross, R., Darby, J., Ireland, J. and Piscitelli, L. (1998). Hysteresis and unemployment: a preliminary investigation. *CEPR/ESRC Unemployment Dynamics Workshop Paper*, **February**.
17. Cross, R., Krasnosel'Skii, A. M. and Pokrovskii, A. V. (2001). A time-dependent Preisach model. *Physica B*, **306,** 206–210.
18. Cross, R., Piscitelli, L., Grinfeld, M. and Lamba, H. (2000). A test for strong hysteresis. *Computational Economics*, **15,** 59–78.
19. Dacorogna , M. M., Müller, U. A., Jost, C., Pictet, O. V., Olsen, R. B. and Ward, J. R. (1995). Heterogeneous real-time trading strategies in the foreign exchange market. *European Journal of Finance*, **1,** 383–403.
20. Davidson, R. and MacKinnon, J. G. (1993). *Estimation and inference in economics.* Oxford University Press, New York.
21. Davidson, R. and MacKinnon, J. G. (1996). The power of bootstrap tests. *Queen's University Institute for Economic Research, Discussion Paper*, **937**.
22. Davidson, R. and MacKinnon, J. G. (1998). Graphical methods for investigating the size and the power of hypothesis tests. *The Manchester School*, **66,** 1–22.

23. Davidson, R. and MacKinnon, J. G. (1999). The size distortion of bootstrap tests. *Econometric Theory*, **15,** 361–376.
24. DeGrauwe, P., DeWachter, H. and Embrechts, M. (1993). *Exchange Rate Theory, Chaotic Models of Foreign Exchange Markets.* Oxford: Blackwell.
25. De Long, J. B., Shleifer, A., Summers, L. H. and Waldmann, R. J. (1990). Noise trader risk in financial markets. *Journal of Political Economy*, **98,** 703–738.
26. van Dijk, D. and Franses, P. H. and Paap, R. (2002). A nonlinear memory model for US unemployment. *Journal of Econometrics*, **110,** 135–165.
27. Dixit, A. (1989). Entry and exit decisions under uncertainty. *Journal of Political Economy*, **93,**.
28. Dixit, A. and Pindyck, R. (1994). *Investment Under Uncertainty.* Princeton university press, Princeton.
29. Efron, B. (1979). Bootstrap methods: another look at the Jacknife. *Annals of Statistics*, **7,** 1–26.
30. Fama, E. F. (1965). Random walks in stock market prices. *Financial Analysts Journal*, **September/October** (reprinted January-February 1995).
31. Farmer, J. D. (1998). Market force, ecology, and evolution. *Santa Fe Institute Working Paper*, **98-12-117**.
32. Farmer, J. D. and Joshi, S. (2002). The price dynamics of common trading strategies. *Journal of Economic Behaviour Organization*, **49,**.
33. Frankel, J. A. and Froot, K. A. (1988). Chartists, fundamentalists and the demand for dollars. *Greek Economic Review*, **10,** 49–102.
34. Gaunersdorfer, A. (2000). Endogenous fluctuations in a simple asset pricing model with heterogeneous beliefs. *Journal of Economic Dynamics and Control*, **24**, 799–831.
35. Gaunersdorfer, A. and Hommes, C. H. (2005). A nonlinear structural model for volatility clustering. In: Teyssière, G. and Kirman, A. (Eds.), *Long-Memory in Economics.* Springer Verlag, Berlin. Appears in this volume.
36. Giraitis, L., Kokoszka, P., Leipus, R., and Teyssière, G. (2003). Rescaled variance and related tests for long memory in volatility and levels. *Journal of Econometrics*, **112,** 265–294.
37. Göcke, M. (2002). Various concept of hysteresis applied in economics. *Journal of Economic Surveys*, **16,** 167–188.
38. Granger, C. W. J. and Joyeux, R. (1980). An introduction to long-memory time series models and fractional integration. *Journal of Time Series Analysis*, **1,** 15–29.
39. Hammermesh, D. (1989). Labour demand and the structure of qdjustment costs. *American Economic Review*, **79,** 674–689.
40. Henry, M. and Robinson, P. M. (1996). Bandwidth choice in Gaussian semiparametric estimation of long range dependence. *Athens Conference on Applied Probability and Time Series Analysis*, **II**, *Lecture Notes in Statistics*, **115,** Springer: New York.
41. Heyde, C. C. (2002). On modes of long range dependence. *Journal of Applied Probability*, **39,** 882–888.
42. Higuchi, T. (1988). Approach to an irregular time series on the basis of the fractal theory. *Physica*, **D 31,** 277–283.
43. Hosking, J. R. M. (1981). Fractional differencing. *Biometrika*, **68,** 165–176.
44. Hurst, H. E. (1951). Long-term storage capacity of reservoirs. *Transac. Am. Soc. Civil Eng.*, **116,** 770–808.

45. Jensen, M. J. (1994). Wavelet analysis of fractionally integrated processes. *Dep. of Economics, Washington University, St. Louis,* **MO 63130**.
46. Keynes, J. M. (1936). *The General Theory of Employment, Interest and Money.* Macmillan Press.
47. Kirman, A. P. (1991). Epidemics of opinion and speculative bubbles in financial markets. In Taylor, M. (ed), *Money and Financial Markets*, London: Macmillan.
48. Kirman, A. P. and Teyssière, G. (2000) Microeconomic models for long-memory in the volatility of financial time series. *Studies in Nonlinear Dynamics and Econometrics*, **5**, 281–302.
49. Krasnosel'Skii, M. A. and Pokrovskii, A. W. (1989). *Systems wich Hysteresis.* Springer Verlag, Berlin.
50. Kurz, M. (1997). *Endogenous Economic Fluctuations.* Berlin: Springer.
51. Layard, R., Nickell, S. and Jackman, R. (1991). *Unemployment. Macroeconomic performance and the labour market.* Oxford University Press.
52. Lang, D. and de Peretti, C. (2003). A strong hysteretic model for Okun's law: Theory and preliminary investigation. *Preprint.*
53. LeBaron, B. (2000). Agent based computational finance: suggested readings and early research. *Journal of Economic Dynamics and Control*, **24,** 679–702.
54. LeBaron, B., Arthur, W. B. and Palmer, R. (1999). Time series properties of an artificial stock market. *Journal of Economic Dynamics and Control*, **23,** 1487–1516.
55. Lo, A. W. (1991). Long-term memory in stock market price. *Econometrica*, **59,** 1279–1313.
56. Lux, T. (1995). Herd behavior, bubbles and crashes. *The Economic Journal*, **105,** 881–896.
57. Lux, T. and Marchesi, M., (1999). Scaling and criticality in a stochatic multi-agent model of a financial market. *Nature*, **397,** 498–500.
58. Lux, T. and Marchesi, M. (2000). Volatility clustering in financial markets: a micro-simulation of interactive agents. *International Journal of Theoretical Applied Finance*, **3,** 675–702.
59. Marshall, A. (1890). *Principles of Economics*, 1st edition. Macmillan, London.
60. Mayergoyz, (1991). *Mathematical Models of Hysteresis.* Springer Verlag, Berlin.
61. Oi, W. (1962). Labour as a quasi-fixed factor. *Journal of Political Economy*, **70,** 538–555.
62. de Peretti, C. (2003a). Bilateral bootstrap tests for long memory: an application to the silver market. *Computational Economics*, **22,** 187–212.
63. de Peretti, C. (2003b). Graphical methods for investigating the finite-sample properties of confidence regions: Application to the long memory parameter.*Preprint.*
64. de Peretti, C. and Marimoutoo, V. (2002). Are the long memory tests really effective? *Preprint.*
65. Robinson, P. M. (1995). Gaussian semiparametric estimation of long range dependence. *The Annals of Statistics*, **23**, 1048–1072.
66. Wang, J. (1994). A model of competitive stock trading volume. *Journal of political Economy*, **102,** 127–168.
67. Zeeman, E. C. (1974). The unstable behavior of stock exchange. *Journal of Mathematical Economics*, **1** 39–49.
68. Zoega, G., Booth, A. L., and Chen, Y.-F (2002). Hiring and firing: a tale of two thresholds. *Journal of Labour Economics*, **20,** 217–248.